Vascular Biology Protocols

METHODS IN MOLECULAR MEDICINE™

John M. Walker, SERIES EDITOR

METHODS IN MOLECULAR MEDICINE™

Vascular Biology Protocols

Edited by

Nair Sreejayan
Jun Ren

University of Wyoming, School of Pharmacy,
Laramie, Wyoming, USA

HUMANA PRESS ✳ TOTOWA, NEW JERSEY

Production Editor: Tracy Catanese
Cover design by

Cover illustration:

Printed in the United States of America. 10 9 8 7 6 5 4 3 2 1
eISBN 978-1-59745-571-8

Library of Congress Control Number: 2007933145

Preface

Cardiovascular disease is the greatest scourge affecting the industrialized nations and is increasing rapidly in developing countries, responsible for high morbidity and mortality. Over the past decades, the pathogenesis, diagnosis, treatment, and prevention of cardiovascular diseases have been benefited significantly from intensive research activities. More molecules and novel techniques have been made available to better understand the progression and management of cardiovascular diseases. In order to provide a comprehensive "manual" in a field that has become as broad and deep as cardiovascular medicine, this volume of "Methods in Molecular Medicine" covers a wide spectrum of in vivo and in vitro techniques encompassing biochemical, pharmacological, and molecular biology disciplines which are currently used to assess vascular disease progression. Each chapter included in this volume focuses on a specific vascular biology technique and describes various applications as well as caveats of these techniques. The protocols included here are described in detail, allowing beginners with little experience in the field of vascular biology to embark on new research projects.

The first few chapters are focused on protocols for animal models commonly used in vascular biology research. These models include rat carotid artery balloon injury, plaque-rapture, atherosclerosis, ventricular hypertrophy, hypertension, and insulin resistance. Besides these in vivo protocols, several ex vivo and in vitro protocols often employed to assess vascular functions may be also found in this volume such as pseudo-working heart, analysis of glucose metabolism in perfused transgenic mice, evaluation of cardiovascular renin angiotensin aldosterone activation, detection and quantification of apoptosis in vasculature.

The next series of chapters include protocols for isolation, characterization, and assays to assess the function/dysfunction of various vascular cell types. We have then lined up a few chapters on detection of reactive oxygen species and post-translational modification of proteins including protein glycoxidation, assessment of kinase activity, and 2D gel electrophoresis. In order to keep pace with the cutting-edge research, we have also inserted, as the final chapter, a protocol that describes stem cell therapy in the heart and vasculature.

We hope that this protocol handbook, among all series of "Methods in Molecular Medicine" will prove useful information to those who wish to broaden their knowledge for cardiovascular research tools. To achieve this goal, credit must be given to all talented and dedicated scientists involved in this book series. Our deepest appreciation goes to Professor John Walker, the Series Editor for Methods in Molecular Medicine, for his encouragement and guidance throughout the preparation process. We are indebted to all contributing authors for their professional expertise, knowledge and devoted scholarship, which are at the very "heart" of this book.

Nair Sreejayan
Jun Ren

Contents

vii

Contributors

ANDREW D. BLANN, • *Haemostasis Thrombosis & Vascular Biology Unit, University Department of Medicine, City Hospital, England, UK*

WENDY A. BOIVIN, • *Pathology and Laboratory Medicine, The James Hogg iCAPTURE Centre for Cardiovascular and Pulmonary Research, Vancouver, Canada*

CHRISTOPHER J. BOOS, • *Haemostasis Thrombosis & Vascular Biology Unit, University Department of Medicine, City Hospital, England, UK*

KEITH R. BRUNT, • *Department of Physiology, Queen's University, Ontario, Canada*

HUA CAI, • *Division of Molecular Medicine, Departments of Anesthesiology & Medicine Cardiovascular Research Laboratories, David Geffen School of Medicine, University of California, Los Angeles, USA*

WANGDE DAI, • *The Heart Institute, Good Samaritan Hospital, Los Angeles, California, USA*

ALAN DAUGHERTY, • *University of Kentucky, Lexington, Kentucky, USA*

AMY J. DAVIDOFF, • *Department of Pharmacology, University of New England, College of Osteopathic Medicine, Maine, USA*

SERGEY DIKALOV, • *Division of Cardiology, Department of Medicine, Emory, University School of Medicine, Atlanta, Georgia, USA*

RAJAKUMAR V. DONTHI, • *Department of Pediatrics, Diabetes Research, University of Louisville, School of Medicine, Louisville, Kentucky, USA*

JOAN S. DOW. • *The Heart Institute, Good Samaritan Hospital, Los Angeles, California, USA*

PAUL N. EPSTEIN, • *Department of Pediatrics, Diabetes Research, University of Louisville, School of Medicine, Louisville, Kentucky, USA*

ALBERTO U. FERRARI, • *Dipartimento di Medicina Clinica e Prevenzione, Centro Interuniversitario di Fisiologia Clinica e Ipertensione, Università di Milano-Bicocca, Milano, Italy*

ULRICH FÖRSTERMANN, • *Department of Pharmacology, Johannes Gutenberg University, Mainz, Germany*

ÁNGEL GARCÍA • *Rede de Infraestruturas de Apoio á Investigación e ó Desenvolvemento Tecnolóxico (RIAIDT), Universidade de Santiago de Compostela, Spain and Oxford Glycobiology Institute, Department of Biochemistry, University of Oxford, Oxford, UK*

SANJOY GHOSH, • *Division of Pharmacology and Toxicology, The University of British Columbia, Vancouver, British Columbia, Canada*

DAVID J. GRANVILLE, • *Pathology and Laboratory Medicine, The James Hogg iCAPTURE Centre for Cardiovascular and Pulmonary Research, Vancouver, Canada*

KATHY K. GRIENDLING, • *Division of Cardiology, Department of Medicine, Emory, University School of Medicine, Atlanta, Georgia, USA*

GARY J. GROVER, • *Department of Pharmacology, Eurofins Product Safety Laboratories, Dayton, New Jersey, USA*

JAVAD HABIBI, • *University of Missouri, Harry S. Truman VA Hospital, Columbia, Missouri, USA*

SEAN R.R. HALL, • *Department of Physiology, Queen's University, Ontario, Canada*

DAVID G. HARRISON, • *Division of Cardiology, Department of Medicine, Emory, University School of Medicine, Atlanta, Georgia, USA*

STEVE HSU, • *Abbott Vascular, Santa Clara, California, USA*

MOHIT JAIN, • *Cardiovascular Division, Department of Medicine, Brigham and Women's Hospital and Harvard Medical School, Boston, Massachusetts, USA*

ROBERT A. KLONER, • *The Heart Institute, Good Samaritan Hospital, Los Angeles, California, USA*

MASAFUMI KUZUYA, • *Department of Geriatrics, Nagoya University Graduate School of Medicine, Nagoya, Japan*

GUIDO LASTRA, • *University of Missouri, Harry S. Truman VA Hospital, Columbia, Missouri, USA*

SHI-YAN LI, • *University of Wyoming, Center for Cardiovascular Research and Alternative Medicine, Division of Pharmaceutical Sciences, School of Pharmacy, Laramie, Wyoming, USA*

SONG LI, • *Department of Bioengineering, University of California, Berkeley, California, USA*

RONGLIH LIAO, • *Cardiovascular Division, Department of Medicine, Brigham and Women's Hospital and Harvard Medical School, Boston, Massachusetts, USA*

GREGORY Y. H. LIP, • *Haemostasis Thrombosis & Vascular Biology Unit, University Department of Medicine, City Hospital, England, UK*

HONG LU, • *University of Kentucky, Lexington, Kentucky, USA*

CAMILA MANRIQUE, • *University of Missouri, Harry S. Truman VA Hospital, Columbia, Missouri, USA*

GIUSEPPE MARANO, • *Dipartimento del Farmaco, Istituto Superiore di Sanità Rome, Italy*

TAKASHI MATSUI, • *Cardiovascular Research, Cardiovascular Division, Beth Israel Deaconess Medical Center, Harvard Medical School, Boston, Massachusetts, USA*

LUIS G. MELO, • *Department of Physiology, Queen's University, Ontario, Canada*

BRETT M. MITCHELL, • *Department of Internal Medicine Texas A&M Health Science Center-College of Medicine Texas, USA*

E. MATTHEW MORRIS, • *University of Missouri, Harry S. Truman VA Hospital, Columbia, Missouri, USA*

KAE NAKAMURA, • *Department of Geriatrics, Nagoya University Graduate School of Medicine, Nagoya, Japan*

KELLY R. PITTS, • *Molecular Pharmacology, Gilead Colorado, Inc. Westminster, CO, USA*

DEBRA L. RATERI, • *University of Kentucky, Lexington, Kentucky, USA*

JUN REN, • *University of Wyoming, Center for Cardiovascular Research and Alternative Medicine, Division of Pharmaceutical Sciences, School of Pharmacy, Laramie, Wyoming, USA*

BRIAN RODRIGUES, • *Division of Pharmacology and Toxicology, The University of British Columbia, Vancouver, British Columbia, Canada*

TAKESHI SASAKI, • *Department of Geriatrics and Department of Anatomy and Neurosciences, Nagoya University Graduate School of Medicine, Nagoya, Japan*

ROSARIO SCALIA, • Thomas Jefferson University, Jefferson Medical College, Department of Physiology, Philadelphia, Pennsylvania, USA

RAJNI SINGH, • *Department of Pharmacology, Eurofins Product Safety Laboratories, Dayton, New Jersey, USA*

JAMES R. SOWERS, • *University of Missouri-Columbia School of Medicine, Columbia, Missouri, USA*

NAIR SREEJAYAN, • *University of Wyoming, School of Pharmacy, Laramie, Wyoming, USA*

CRAIG S. STUMP, • *University of Arizona, Diabetes Research Center, Tucson, Arizona, USA*

RAHUL THAKAR, • *Department of Bioengineering, University of California, Berkeley, California, USA*

CHRISTOPHER F. TOOMBS, *Preclinical Pharmacology, Ikaria, Inc., Seattle, WA*

DAVID A. TULIS, • *Cardiovascular Disease Research Program, J.L. Chambers Biomedical/Biotechnology Research Institute, North Carolina Central University Durham, North Carolina, USA*

THOMAS WALLERATH, • *Department of Pharmacology, Johannes Gutenberg University, Mainz, Germany*

CHRISTOPHER A. WARD, • *Department of Physiology, Queen's University, Ontario, Canada*

YONGZHONG WEI, • *University of Missouri, Harry S. Truman VA Hospital, Columbia, Missouri, USA*

LOREN E. WOLD, • *Centre for Cardiovascular Medicine, Columbus Children's Research Institute and Department of Pediatrics, The Ohio State University College of Medicine, Columbus, OH of Osteopathic Medicine, Maine, USA*

XIAOPING YANG, • *University of Wyoming, School of Pharmacy, Laramie, Wyoming, USA*

1

Rat Carotid Artery Balloon Injury Model

David A. Tulis

Summary

Numerous and diverse experimental animal models have been used over the years to examine reactions to various forms of blood vessel disease and/or injury across species and in multiple vascular beds in a cumulative effort to relate these findings to the human condition. In this context, the rat carotid artery balloon injury model is highly characterized and commonly used for investigating gross morphological, cellular, biochemical, and molecular components of the response to experimentally induced arterial injury. The mechanical damage caused by the balloon catheter completely removes the intimal endothelial lining and creates a distending mural injury in the operated vessel. This elicits a reproducible remodeling response characterized by vascular smooth muscle cell (SMC) mitogenesis and migration (through phenotypic switching), SMC apoptosis, partial vascular endothelial cell regeneration, enhanced matrix synthesis, and establishment of an invasive neointima in time-dependent fashion. This multi-factorial process allows for investigation of these many important pathophysiological processes and can serve as a valuable "proof-of-concept" tool to verify and substantiate in vitro results; however, inherent anatomical and adaptive constraints of this in vivo model ration comparison to the diseased human system (*see* **Note 1**). In this chapter, brief overview of the materials needed and the methodologies commonly employed for successful routine performance of this important experimental animal model is provided. Individual sub-sections will cover animal care and handling, pre-operative and post-operative procedures, and the surgery proper. Protocols for histopathology and morphometry and procedures for data management and interpretation pertinent to the rat carotid artery balloon injury model are discussed in Chapter 2.

Key Words: Adventitia; Balloon Injury; Common Carotid Artery; Vascular Endothelial Cell; Extracellular Matrix; Media; Neointima; Rat; Remodeling; Vascular Smooth Muscle Cell.

From: *Methods in Molecular Medicine, Vol. 139: Vascular Biology Protocols*
Edited by: N. Sreejayan and J. Ren © Humana Press Inc., Totowa, NJ

1. Introduction

Investigation into the response of blood vessels to injury is of pivotal importance in understanding the pathophysiology of vascular disorders. Injury-based experiments are generally designed for examination of gross, cellular, biochemical, and/or molecular mechanisms that contribute to the vascular injury response, a phenomena particularly important in both basic science and clinical medicine. Injury-based approaches consist of mechanical or other artificial intervention(s) to elicit a primary or secondary injury. Responses to the injury, then, provide adaptive measures for study. This should not be confused with approaches used to study the pathophysiology of vascular disease which often employ genetic, dietary, and/or environmental induction.

The rat carotid artery balloon injury method was originally described in several seminal articles by Clowes and colleagues *(1–4)* and has since been employed and thoroughly characterized in a plethora of basic and clinical science research endeavors. Briefly and from a methodological perspective, this approach involves isolating a segment of carotid artery vasculature in an anesthetized laboratory rat, creating an arteriotomy incision in the external carotid branch through which the balloon catheter is inserted, advancement of the catheter through the common carotid artery, repeated inflation and withdrawal of the catheter to induce endothelial cell loss and mural distension, and removal of the catheter with closure of the arteriotomy and resumption of blood flow through the common carotid artery and internal carotid artery branch. As detailed in Chapter 1, histological protocols germane to this model can then be employed that allow acute measures of cellular and molecular changes and longer term qualification of neointima development and vessel wall restructuring along with morphometric analyses and quantification.

In this chapter, comprehensive methods are provided along with special considerations for practical use of this technique as suggested by the author; however, individual experiences and preferences may dictate alternate more suitable practices on a case-by-case basis. This chapter is not intended for use as a scientific review of the model nor of the mechanisms contributing to the vascular injury response. Excellent summaries of neointima formation and vascular remodeling following injury including the rat carotid artery balloon injury model are recommended for interested readers *(5–8)*.

2. Materials

2.1. Animals

A variety of rat strains has been utilized for this method; however, perhaps the most highly used and characterized strain is Harlan Sprague-Dawley (HSD, Harlan, Indianapolis, IN, USA). Male rats are preferred because of the potential impact of hormone levels on various cellular function(s) that has been identified in females. A wide range of animal weights and ages has been used in this approach; yet, most citations report body weights between 350 and 500 g. Nonetheless, it is important to use fully grown animals as vessel caliber will directly impact the severity of the injury from use of a standard-sized (2 French) inflated balloon catheter. Retired breeders may be used as long as they are of appropriate age/weight. Unless otherwise desired, animals can be kept on standard rodent chow and water ad libitum peri-operatively.

2.2. Pre-Operative Procedures

2.2.1. Solutions

1. Betadine solution (The Purdue Frederick Company, Stamford, CT, USA) or other topical anti-septic/ bactericide agent.
2. Absolute alcohol (to cleanse surgical area on animal skin).
3. Seventy percent alcohol (to cleanse surgery platform).
4. Ophthalmic ointment/lubricant (Phoenix Pharmaceutical, Inc., St. Joseph, MO, USA; *see* **Note 2**).
5. Anesthetic of choice (*see* **Note 3**).
6. Pre-warmed phosphate buffered saline (PBS), Lactated Ringers (LR) solution, normal saline solution, or alternate choice for supplemental fluids (*see* **Note 4**).

2.2.2. Supplies

1. Glass bead sterilizer (Germinator 500, Roboz Surgical Instrument Company, Inc., Gaithersburg, MD, USA) or other suitable instrument sterilizer.
2. Animal hair clippers (A-5 Clipper, size 40 blade, Oster, Sunbeam Products, Inc., McMinnville, TN, USA).
3. Scissors, medium (for removing fine hairs).
4. Needles (26 gauge for anesthetic if parenteral administration; 18–20 gauge for supplemental fluids).
5. Syringes (1 ml for anesthetic if parenteral administration; 3–5 ml for supplemental fluids).
6. Gauze.
7. Cotton-tipped applicators (for use in topically applying anti-septic/bactericide/virucide agent and ophthalmic ointment/lubricant).
8. Tape [cut to appropriate lengths (~3 in.) and placed nearby for easy access].
9. Rodent operating table or other surgery platform.
10. Surgical blanket, sterile.
11. Rodent limb tie-downs or restrainers (*see* **Note 5**).
12. Animal weighing scale (capable of weights up to ~600 g).
13. Sterile water (used to fill balloon catheter ahead of time).
14. Drinking water (to moisten tongue and mouth) (*see* **Note 2**).

2.3. Surgery Proper

2.3.1. Solutions

1. Lidocaine hydrochloride (*see* **Note 6**).
2. Ophthalmic ointment/lubricant (*see* **Note 2**).
3. Supplemental fluids (*see* **Note 4**).
4. Anesthetic, supplemental (*see* **Note 3**).

2.3.2. Supplies

1. Needles and syringes (*see* **Subheading 2.2.2., steps 4** and **5**).
2. Gauze, sterile.
3. Cotton-tipped applicators, sterile.
4. Tape.
5. Fogarty balloon embolectomy catheters, 2 French (Edwards Lifesciences Corp., Irvine, CA, USA).
6. Trocar guiding needle (18 gauge, thin-walled (TW), 1 1/2 in., Becton Dickinson and Company, Franklin Lakes, NJ, USA) or catheter sheath or suitable alternative (*see* **Note 7**).
7. Two-way stopcocks with luer-lock end (used for attachment of syringe to balloon catheter).
8. Inflation device for balloon catheter (Encore 26 Advantage, Boston Scientific, Natick, MA; *see* **Note 8**).
9. Suture: 4-0 black braided silk (Roboz Surgical Instrument Company, Inc., Gaithersburg, MD, USA), 6-0 Prolene blue monofilament (Ethicon, Inc., Somerville, NJ, USA; *see* **Note 9**).
10. Ample lighting (*see* **Note 10**).
11. Heating source (heat lamps and heating blanket, *see* **Note 11**).
12. Rodent operating table or other surgery platform.
13. Rodent limb tie-downs or restrainers (*see* **Note 5**).
14. Surgical blanket, sterile.
15. Micro-caliper (Roboz) or small metric ruler (to measure vessel lengths).

2.3.3. Surgical Tools (see **Note 12**)

1. Scissors: large, medium, small-micro (Roboz).
2. Small curved forceps (Micro-dissecting tweezers, Pattern 7S, Roboz).
3. Small arterial clamps (Roboz).
4. Retractors (large for skin, small for tissues, *see* **Note 13**).
5. Skin staples (7–9 mm) or skin suture.
6. Skin stapler.
7. Skin staple remover (*see* **Note 14**).

2.4. Post-Operative Procedures, Animal Recovery

2.4.1. Solutions

1. Topical anti-septic/bactericide/virucide agent.
2. Supplemental fluids (*see* **Note 4**).
3. Ophthalmic ointment/lubricant (*see* **Note 2**).
4. Water for drinking (*see* **Note 2**).
5. Surgical instrument cleaner, detergent.
6. Seventy percent alcohol.
7. Analgesic of choice (*see* **Note 15**).

2.4.2. Supplies

1. Gauze.
2. Sterile blanket.
3. Surgical drapes, towels (*see* **Note 16**).
4. Animal cage with cover and water supply.
5. Rodent chow.

3. Methods

3.1. Animals

All animal care and experimental procedures must adhere strictly to the recommendations of the Guide for Care and Use of Laboratory Animals [DHEW (NIH) 85-23, revised 1985] and the Public Health Service Policy on Humane Care and Use of Laboratory Animals (revised 1986) as well to the guidelines of the local institutional animal care and use committee. Continual monitoring of the state of anesthesia and well-being of the animal by the investigator(s) is imperative throughout this surgery, and guidelines for early euthanasia should be strictly followed (*see* **Subheading 3.4.**).

3.2. Pre-Operative Procedures

3.2.1. Anesthesia

Choice of appropriate anesthetic and route(s) of delivery are decisions of the investigator; however, the major consideration is that a surgical plane of anesthesia be maintained for the duration of the surgery. This can range from approximately 30 min for an experienced surgeon to over 90 min for a novice or in the case of unexpected problems. The level of anesthesia should be continually monitored through toe or tail pinch, inspection of breathing rate and pattern, and inspection of heart rate throughout the surgery to ensure that the rat is adequately sedated. If the animal appears to be coming out of anesthesia during the surgery, immediately provide supplemental anesthesia (following institutional guidelines or ~10% original dose) and pay particular attention to the level of sedation for the remainder of the surgery. Supplemental oxygen is not normally needed for this surgery; however, if the investigator deems it necessary, oxygen or supplemental air can be provided to the rat via nasal cannula without interference to the surgeon.

3.2.2. Setup

It is imperative that the investigator has everything prepared ahead of time and that the surgical area correctly be setup in order to avoid complications or emergencies during the surgery proper. Be aware that the more expediently the surgery is completed (in scientifically and medically sound fashion of course), the better chance for rapid recovery by the animal and the higher chance for success. Preparation and sterilization of the surgical area, preparation of all solutions and reagents, cleaning and sterilization of all necessary surgical tools, and correct setup of the surgical area with convenient placement of all materials and supplies that will be needed throughout the surgery should take place before sedating the animal. Also, try to have available any item that might be needed in case of emergency (*see* **Note 17**).

A major task to complete before surgery is preparation of the balloon catheter and trocar (if one is to be used). This can take 10 or 15 min to complete, so plan accordingly. Remove the new balloon embolectomy catheter from its package and remove packaging components and the luer-lock cap and discard. Fill a 1-ml syringe (without needle) completely with sterile water and remove air bubbles. Carefully fill a two-way stopcock with sterile water to create a water-filled system and similarly remove air bubbles. Attach the syringe to the stopcock without introducing air bubbles (*see* **Note 18**). In similar fashion, fill the luer-lock portion of the balloon catheter with sterile water (ejected from the syringe through the stopcock, again removing air bubbles) and remove any trapped air in the catheter opening. Maintaining a closed system of water, firmly attach the stopcock to the luer-lock end of the balloon catheter. Depress the syringe slightly to check for leaks and to ensure that balloon inflation occurs. A suggested capacity of the inflated balloon for these surgeries is 0.02 ml; however, be sure to note that the maximum inflation capacity (as stated in "Indications for Use" in the product insert) is 0.2 ml. For these studies, be sure to make a note regarding the degree of balloon inflation using 0.02 ml water and the appropriate markings on the syringe. Upon inspection, do not be surprised if a small bubble appears in the balloon itself. This is impossible to remove from the catheter and will soon dissipate. If using a manual barometer and inflation device, fill the device with water (through the luer-lock tubing) while removing air bubbles according to manufacturer's directions. Similarly, remove air from the luer-lock end of the balloon catheter and attach the catheter end to the luer lock of the inflation device. Manually inflate the device to an appropriate pressure (∼2.0 atm) and watch the gauge closely. Be careful because a little pressure from this device can inflate the balloon a large degree. Also, be cautious that there can be a delay in the pressure induced by the device (by rotating the handle) and the distension of the balloon. This can lead to over-pressurizing the balloon to higher than desired levels. A photo of an arterial balloon catheter attached to a 1.0-ml syringe via a stopcock along with an 18 gauge thin-walled guiding trocar and an automated balloon inflation device is shown in **Fig. 1**.

A surgical mask, surgical gloves, and/or protective goggles are recommended to be worn by the surgeon during the surgery proper. This will minimize the influence of human exposure on the animal as well as protect the animal from potential infection. This will also serve to protect the surgeon from animal dander and exposure. A glass bead sterilizer or other sterilizing apparatus (mini-autoclave) is recommended and should be pre-heated, and all surgical tools should be sterilized and placed conveniently on the surgery platform (*see* **Note 19**). All other reagents and treatment(s), solutions, gauze, cotton-tipped

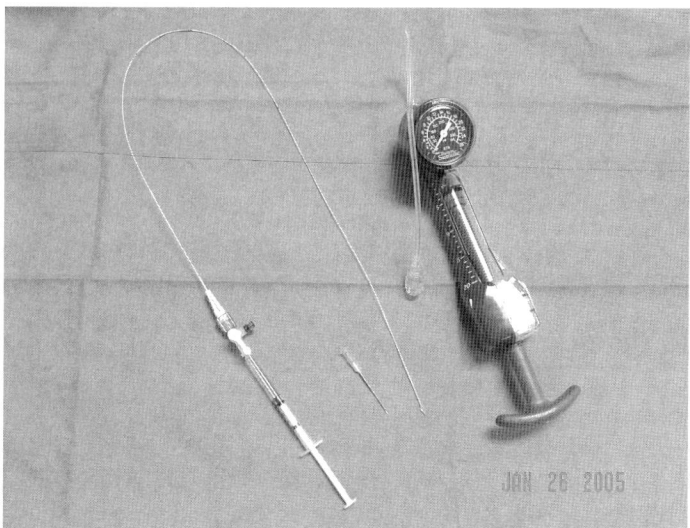

Fig. 1. Photograph of an arterial balloon embolectomy catheter attached to a water-filled 1.0-ml syringe via a stopcock. Also shown is an 18 gauge thin-walled needle that can be used as a trocar and a standardized inflation device for the balloon catheter.

applicators, tape, suture, and others should also be setup ahead of time. Cut appropriate lengths of suture and tape (*see* **Note 9**) and make sure gauze and ample cotton swabs are within easy reach. Cleanse and sterilize the operating surface on surgery area/platform/table with 70% alcohol. Lay out all items to be used for surgery as well as anything else that you might potentially need in case of emergency or other unexpected occurrence (*see* **Note 17**). A photo of a typical pre-operative surgical setup for a left-handed surgeon performing a rat carotid artery balloon injury study is shown in **Fig. 2**.

Once the animal is sedated to a surgical level of anesthesia (verified by toe or tail pinch and breathing and heart rate patterns), hold the animal supine in one hand (making sure to stabilize the head) and shave the surgery site with animal hair clippers. This area to be shaved is the ventral neck region from the chin down to just above the sternum. Be careful not to press too deeply while shaving, as any pressure here will compress the thorax and impede the animal's ability to breathe. Use both side-to-side and up-and-down motions

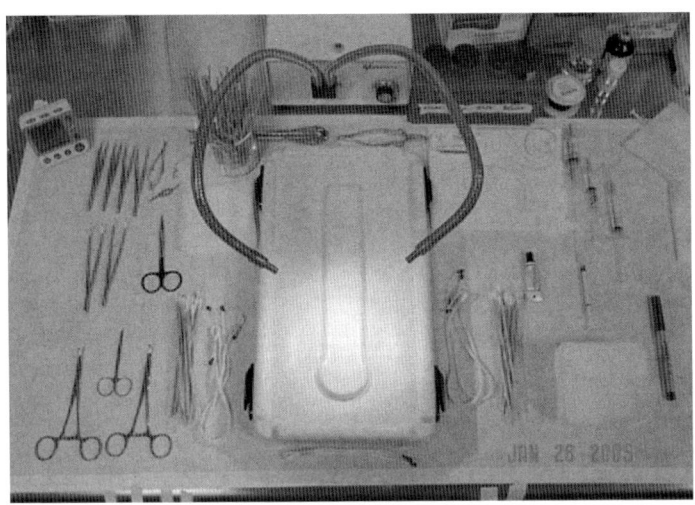

Fig. 2. Typical pre-operative surgical setup for rat carotid artery balloon injury studies. Suggested sites for convenient placement of all items needed as well as those potentially needed for successful completion of the surgery are included and depicted for a left-handed surgeon. It is imperative that comprehensive pre-operative procedures be performed in this method including preparation and calibration of the balloon catheter, cutting appropriate lengths of suture and tape, sterilizing all necessary surgical tools, preparation of all solutions and reagents, and proper well-situated location of these items for the surgery proper.

with the clippers and try to remove all hair in this region (*see* **Note 20**). Following use of the clippers, remove lingering fine hair with small scissors. Again, do not press too deeply. Empty all hair into the trash and place the animal supine on the surgery table in a heated environment (*see* **Note 11**) with head toward the surgeon. Harness the arms and legs and gently retract the head (*see* **Note 5**). Place a piece of gauze on the lower groin (covering the penis) to soak up urine flow that will occur. Swab the cervical area with a topical anti-septic/bactericide/virucide agent followed by 70% alcohol. An ophthalmic solution should be placed on the open eyes with a cotton swab in order to prevent drying. The tongue and mouth of the animal can also be moistened with drinking water to avoid drying. At this time, provide 3–6 ml supplemental fluids (*see* **Notes 2, 3, 4**).

3.3. Surgery Proper

Given the important details of this protocol and the many procedures involved, this section is presented in steps to simplify instruction and to make it easy to follow. At this point, the animal is sedated and lying supine on the operating platform with head toward the surgeon and appendages retracted, with neck area shaved and cleansed and all surgical items conveniently located nearby within easy access.

1. Using sharp/blunt serrated-edge scissors (*see* **Note 21**) and starting immediately below the chin of the animal, make a straight incision in a direction toward the tail all the way to the top of the sternum just above the rib cage. Make the incision as straight as possible and not too deep (this is for purposes of cutting the skin only). Keep the scissor tips up! *See* **Note 22**.
2. Using medium hemostats and/or dull forceps, blunt dissect underlying glandular tissue from skin. Keep the tips of your instruments parallel to the tissues during this process so as not to puncture the skin or the underlying tissue. With blunt dissection, go in with tips closed, gently separate the tips and remove the instrument with the tips wide open. In this fashion, gently separate the skin from underlying tissue circumferentially around the entire incision wound. This will aid in suturing the underlying tissues independently from the skin and in layers following surgery. Do not be afraid if small vessels break and if there is some minor bleeding. Swab these areas with cotton-tipped applicator to stop bleeding. Keep area moist with warm sterile PBS or other fluid. Following this step, the skin should be completely separated from underlying tissues all the way around the incision. Use of a medium-to-large skin retractor at this stage to keep the skin out of the way is recommended (*see* **Note 13**).
3. Using medium scissors, cut through the fascia overlying the glandular tissue (again, keep tips up) to expose underlying glands. Again, make this incision as straight

as possible, as this will be sutured following surgery. Gently separate glands to expose underlying muscular layer via blunt dissection.

4. Carefully separate muscular tissues with dissection using sharp 7S forceps (be careful not to dig too deep, as these forceps are very sharp and can easily penetrate tissues). For purposes of this protocol, procedures are indicated for surgery to be performed on the left common carotid artery (*see* **Note 23**). Blunt dissect along the longitudinal left aspect of the central and adjacent muscular tissues (sternocleidomastoid, omohyoid, thyrohyoid, and sternohyoid) and remember to avoid pressure on these muscles as below them lay the thorax and the animal's ability to breath! Gently separate the central muscle from parallel neck muscles and the diagonal thin muscular band (omohyoid) lying directly over the carotid vasculature (*see* **Note 24**). At this point, retraction of skin and muscular tissues is highly recommended and virtually essential for visualization of the underlying carotid artery vasculature. During all of these procedures, keep the tips of the instruments up and keep all tissues moist. Following thorough dissection and retraction of tissues, at this stage, the surgeon will be able to view the left common carotid artery, the vagus nerve (a thin white sheath lying adjacent to the carotid artery), and adjacent nerves and vessels (*see* **Note 25**).

5. Continue blunt dissection alongside the left carotid artery distally toward the head to expose the carotid artery bifurcation into the internal and external branches. Use of lidocaine hydrochloride at this stage on the exposed carotid vasculature is recommended to keep it moist (*see* **Note 6**). As the surgeon looks at it, upon exposure, the artery lying on top (ventral aspect) distal to the bifurcation will be the external carotid artery branch, whereas the internal carotid artery branch digs dorsally and is not usually apparent unless the carotid vasculature is moved aside. To ensure identification of the internal carotid artery and location of the bifurcation, gently move the distal portion of the common carotid artery to the right and the internal branch should appear directly below it. Make a note of where the common carotid bifurcation exists for each animal (*see* **Note 26**). At this point, the surgeon might also be able to visualize the hyoid cartilage directly under the chin. Avoid damage to the hyoid cartilage and to all nerves that are present, especially the vagus nerve and nerve plexus. These nerves control many aspects of physiologic function, and perturbation can result in severe cardiovascular depression. As before, use retraction to keep the tissues separated. Keep in mind to moisten all exposed tissues throughout the surgery and to use liberal amounts of lidocaine on exposed vasculature.

6. Expose the left common carotid artery the entire length of the skin incision down to the sternum. In an adult rat, this should be ∼20 mm in length proximally from the bifurcation (some variation exists between animals). The entire length of the adult common carotid artery is estimated to be 35–40 mm. The surgeon will be able to visualize vasodilation with each arterial bolus (heartbeat) if the vasculature is tented up with forceps (be careful not to over-manipulate the vessel). Add lidocaine on top of the exposed common carotid artery and let it incubate for

several minutes (*see* **Note 6**). At this stage, lidocaine will serve to dilate the blood vessel and its branches thus allowing easier vascular access. Swab away excess lidocaine before the next step.

7. Before placement of sutures around specific sections of the carotid vasculature can take place (*see* **Fig. 3**) *(9)*, the surgeon must make sure that these vessel sections are completely separated from all adjacent tissues. If this is not the case, then during the retraction and/or clamping procedures complete cessation of blood flow will not occur because of the presence of additional tissue in the clamp or retractor. This will result in unexpected bleeding, severely hamper progress of the surgery, and cause undue stress on the animal. At each site for suture placement (*see* **Fig. 3**), carefully blunt dissect away all adjacent tissues from the vessel so that at least 2–3 mm of the vessel is free from extraneous tissue.

8. This step involves placement of sutures around certain sections of the carotid artery vasculature to be used for retraction and/or clamping and hemostasis (*see* **Note 27**). For exact anatomical location of these sutures, *see* **Figs 3** and **4**. When wrapping sutures around the vessels, use the small forceps (7S micro-dissecting tweezers). At the most proximal site on the left common carotid artery (as close to

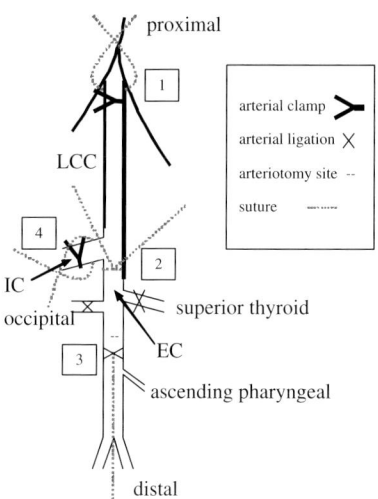

Fig. 3. Simplified diagram of the rat carotid artery vascular anatomy along with sites for placement of arterial clamp(s), suture, and ligations *(9)*. Proximal and distal anatomical locations are indicated, as well as steps (indicated by numbers) identified in the protocol. The arteriotomy site for insertion of the balloon catheter is on the external carotid artery branch between the bifurcation of the common carotid artery and the site of distal ligation and retraction. EC, external carotid artery; IC, internal carotid artery; LCC, left common carotid artery.

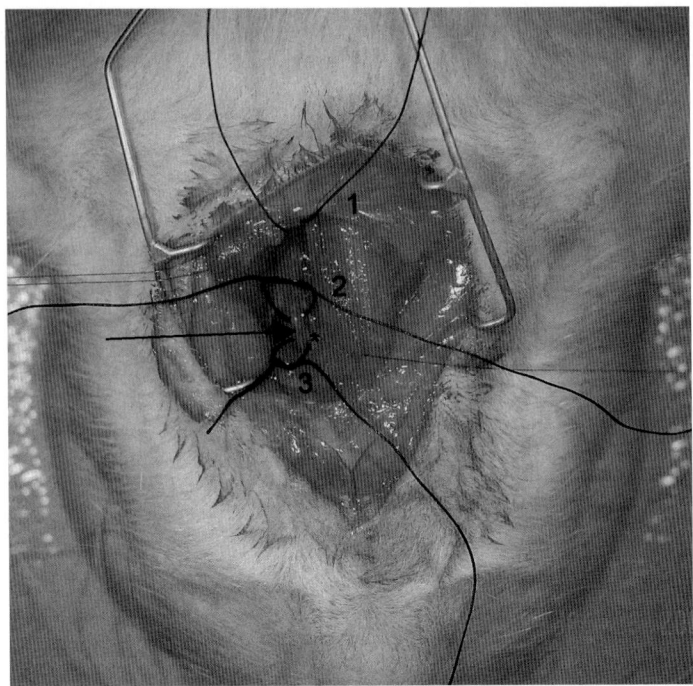

Fig. 4. Photograph of the surgical site on the ventral aspect of the neck of a rat undergoing carotid artery balloon injury. At this point in the procedure, skin and underlying muscles have been retracted and sutures have been carefully placed around the most accessible proximal site on the left common carotid artery (1) and looped but not tied around the external carotid artery branch immediately distal to the bifurcation (2) and as distally as possible (3). The suture at (2) will be tied immediately upon removal of the balloon catheter following surgery, while the suture at (3) will be tied and used to retract the carotid vasculature before performing the arteriotomy. These numbers correspond to those included in **Fig. 3**. The internal carotid artery branch was not accessed in this animal and therefore is not indicated. Also shown are sites for the arteriotomy on the external carotid branch (arrowhead) and the unligated superior thyroid artery (*) which will be tied off before the surgery.

the sternum as possible) and having separated the vascular fascia and vagus nerve from the artery, loop a single approximately 3 in. suture (size 4-0, black braided silk) around the artery [*see* **Note 9; Figs 3** (denoted as 1) and **4**]. Next, immediately distal to the bifurcation of the common carotid artery, loop and loosely tie one approximately 3 in. 4-0 silk suture around the external carotid artery branch [the branch that lies on the surface in the prone animal; **Figs 3** (denoted as 2) and

4]. Do not tie this off but keep it loosely looped for later (this will be the suture that will be pulled tight immediately after removal of the balloon catheter from the arteriotomy incision). Next, loop and loosely tie one 5–6 in. 4-0 black silk suture on the external artery branch as distally and far away from the bifurcation as possible [*see* **Figs 3** (denoted as 3) and **4**]. When loosely tieing this suture, keep one end of the wrapped suture as long as possible as this will be used to retract the vessel in a direction toward the head. In this effort, try to obtain as long a section on the external carotid artery branch between the two sutures as possible (*see* **Note 28**). A longer isolated section of the external carotid artery will significantly improve chances for successful balloon intervention. At this point, three individual pieces of 4-0 silk suture have been loosely placed around the carotid vasculature: one as proximal on the common carotid as possible, one on the external carotid branch immediately distal to the bifurcation and loosely looped, and one as distal as possible on the external branch and loosely looped with one long end. No sutures are tied yet and blood flow is still patent throughout the entire vasculature at this step.

9. As the surgeon is dissecting around the bifurcation and along the external carotid artery branch, several small arteries appear (in most cases) and, if not ligated, could be sources for retrograde blood loss during balloon intervention. If arterial branching off the external carotid artery is apparent, the ascending pharyngeal, occipital, and/or superior thyroid arteries should be completely tied off using short lengths of 6-0 Prolene blue monofilament suture (*see* **Note 29**; *see* **Figs 3** and **4**).

10. The surgeon now must decide whether to access the internal carotid artery branch which, as the surgeon is looking at the animal, lies immediately below the carotid bifurcation and the external carotid artery branch and quickly moves into deep tissue in a dorsal direction [*see* **Fig. 3** (denoted as 4)]. The reason to access the internal artery is that it can be a significant source for retrograde blood loss if its flow is not adequately controlled. Access to this branch can be difficult as there is not much space to work, and the available length of the internal branch is usually minimal. The surgeon must determine if it is in the best interest for the animal and for the research study itself to try to access this vessel to provide another degree of localized hemostasis. The author does not routinely access the internal carotid artery during the surgical technique; moreover, the small amount of bleeding that occurs from internal carotid artery retrograde blood flow is minimal in his hands. This choice for the surgeon must be taken seriously, especially for a novice who might take more time to successfully introduce the catheter into the arteriotomy (hence, allow more time for blood loss via internal carotid artery retrograde flow). It is recommended that the surgeon attempt this surgery both with and without accessing the internal carotid artery and choose whichever approach works best in his/her hands. If the surgeon chooses to isolate the internal carotid artery branch and control its flow, then the exact site of the common carotid artery

bifurcation needs to be determined. At this site, the middle suture (or the more proximal suture on the external carotid branch) should be easily accessible, and the surgeon can use this to gently retract the overlying vessels to the right in order to visualize the underlying internal carotid artery. Using the small forceps (7S), gently blunt dissect adjacent tissues from the internal carotid branch immediately adjacent to the bifurcation, exposing as much of the internal carotid as possible. Loop a section of 4-0 silk suture around the artery and keep it untied. Control of the internal carotid artery blood flow can then be achieved with either arterial micro-clamps or retraction of the suture (*see* **Note 13**; *see* **Fig. 3**). If using an arterial clamp, do not place it too close to the common carotid bifurcation, as this will physically impede advancement of the balloon catheter and the clamp will have to be removed and placed at a more distal site on the internal carotid artery. Similarly, if using suture to retract the internal carotid branch, be careful not to alter the geometry of the carotid vasculature and create an angle that might obstruct insertion and advancement of the catheter through the vessel lumen.

11. Now, the internal carotid artery may or may not be retracted and/or clamped [*see* **Fig. 3** (denoted as 4)], and it is time to tie the other sutures already in place (one on the common carotid and two on the external carotid) and to clamp the common carotid artery in advance of the arteriotomy incision and balloon catheterization. First, using a double-knot tie the most distal suture on the external branch, retract it toward the head, and adhere it to the operating surface with tape [*see* **Fig. 3** (denoted as 3)]. Be careful here not to pull too tightly and constantly watch the carotid artery during retraction to ensure that the vasculature is not being stressed too much. Avoid undue pressure on the vasculature and try to maintain normal vessel geometry (avoid angles). Gently retracting the carotid artery during this step lifts up the external carotid artery segment for easier access.

12. Gently retract the proximal suture on the common carotid artery and place an arterial clamp on the vessel in order to stop common carotid artery blood flow [*see* **Notes 30** and **31**; *see* **Fig. 3** (denoted as 1)]. Be careful here and try not to traumatize the vessel. Try to place the clamp exactly perpendicular to the vessel and avoid catching adjacent tissues in the clamp. Lidocaine can be applied to the vasculature and incubated for several minutes. At this point, the external carotid artery has been retracted distally, the common carotid artery blood flow has been stopped via clamping, most or all (if internal carotid artery is involved) retrograde blood flow is controlled via ligation of miscellaneous small arteries, and suture is loosely looped around the external branch just distal to the bifurcation of the common carotid artery. The common and external carotid arteries should now be lying straight with easy access to the portion on the external carotid artery between the two sutures. This is the critical site for performing an arteriotomy incision and introducing the balloon catheter.

13. The balloon catheter should have been calibrated well ahead of time (*see* **Subheading 3.2.**) and should be located within easy reach of the surgeon. With a cotton swab in one hand and small micro-scissors in the dominant hand, *very delicately* snip a portion of the external carotid branch perpendicular to its axis and as distally as possible (toward suture nearest the head, *see* **Figs 3** and **4**). This will allow sufficient length of external carotid artery branch if there presents a problem with this initial arteriotomy incision (*see* **Note 28**). The cut should be straight with a length approximately one-fourth to one-third of the circumference of the vessel. Do not make the cut too deep or cut completely through the vessel, as this will make things more difficult for introduction of the catheter (*see* **Notes 32** and **33**). Immediately following the arteriotomy gently swab it to stop bleeding (there should be minimal bleeding if adequate hemostasis was achieved). It may help to gently retract the distal external carotid artery ligature to make the arterial section taut before making the incision.

14. Several investigators recommend flushing the lumen of the common carotid artery (after adequate hemostasis is ensured) with 1 ml heparin solution (50 IU/ml) or with PBS via the arteriotomy incision before inserting the balloon catheter as well as immediately following the injury and removal of the catheter *(10)*. This practice is not performed nor recommended by the author as heparin has been shown to inhibit vascular smooth muscle cell (SMC) proliferation, an essential component of the neointimal response to injury *(3,11)*. Additionally, other investigators recommend removing the adventitia from the carotid artery before performing balloon injury *(12)*. This is not encouraged as the adventitia is a valuable source for sensory nerves, immune elements including macrophages and mast cells, and fibroblasts that have been implicated in the response to balloon injury in rats *(13,14)*.

15. Following a successful arteriotomy incision on the external carotid artery branch and holding the balloon catheter alone or inserted through a trocar guiding needle (*see* **Note 7**), gently insert the uninflated balloon into the arteriotomy hole and advance it all the way to the arterial clamp on the common carotid artery using 7S forceps held sideways to guide the catheter. Be careful not to damage the balloon with the forcep tips and hold the catheter at the same plane as the blood vessel during this process. Remove the clamp on the common carotid to allow passage of the uninflated balloon catheter all the way down the common carotid to the aortic arch ($\sim$ 35–40 mm total length). Try to feel the resistance when the balloon tip reaches the back wall of the aortic arch. Make sure to watch the markings on the catheter sheath when this happens, which generally occurs 10–15 mm between the first black band and the arteriotomy incision (*see* **Fig. 5**). Also, watch for excess bleeding out of the arteriotomy incision around the catheter, which can occur with an uninflated balloon.

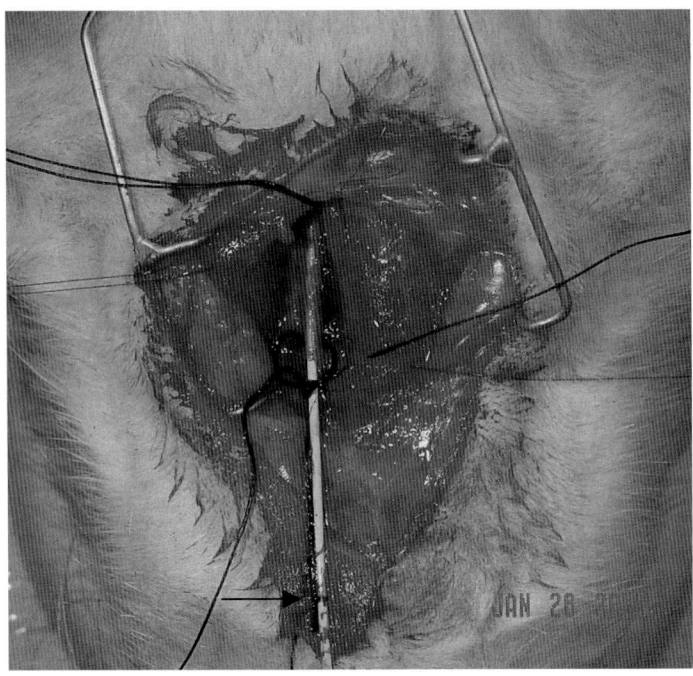

Fig. 5. Photograph of the rat left carotid artery vasculature during balloon injury. Illustrated is the distal end of a balloon catheter inserted into an arteriotomy incision on the external carotid artery, with a section of suture wrapped around the catheter and an arrowhead indicating the first black band from the balloon tip on the catheter. The suture will be tied following injury and removal of the balloon catheter.

16. Now, slowly inflate the catheter to a pre-determined volume (suggested 0.02 ml) or pressure (~ 2.0 atm) manually (via syringe) or with the use of an inflation device and barometer. If performing this step manually, remember to lock the stopcock between the syringe and the catheter to maintain appropriate pressure in the balloon. Keep an eye on the catheter and its markings here because if the catheter is inserted too far into the aortic arch, aortic flow will take the inflated balloon and pull it down the thoracic aorta! With the inflated balloon in the common carotid artery, get a feel for the degree of resistance as it is slowly withdrawn. Gently remove the catheter using rotation (*see* **Note 34**) all the way *almost* up to the arteriotomy incision (you will be able to see the inflated balloon inside the carotid artery as it approaches the arteriotomy site). *Be very careful here* and do not withdraw the catheter too close to the arteriotomy hole. If the inflated balloon gets too close to this hole, it will slip out and an open arterial bleed will

ensue. This, of course, will cause significant blood loss if the common carotid is not immediately clamped or retracted (*see* **Note 31**). If this does happen, stop blood flow immediately with a clamp or suture retraction on the common carotid artery and re-group.

17. Deflate the balloon and re-insert the catheter all the way to the aortic arch again. Inflate the balloon at desired volume or pressure and withdraw the catheter with rotation. Perform this procedure a total of three times with rotation to ensure complete and reproducible removal of the endothelial lining and distension of the vessel wall (*see* **Note 35**).

18. Now for the removal of the catheter and closure of the arteriotomy site. Hold a 7S forcep in each hand and grab the ends of the loosely tied proximal suture on the external carotid artery [*see* **Figs 3** (denoted as 2) and **4**]. These will be pulled tight to tie the suture and to close off the arteriotomy following removal of the catheter. With the inflated balloon near the arteriotomy hole on the external carotid and the ends of the suture held in the forcep tips, quickly deflate the balloon, remove the catheter from the vessel, and tie the suture to close the arteriotomy hole (*see* **Note 36**). Place the catheter aside and tie this suture again. Quickly clean up pooled blood and inspect the surgical site for arterial bleeding (*see* **Note 37**). If the internal carotid artery is clamped or retracted, release it to restore blood flow. Again, check for any leaks. Release the retracted distal suture on the external carotid artery by cutting the suture (keep the knot tied) and remove the suture still looped on the common carotid artery. Make sure the common carotid artery is patent and pulsatile (tent it up) with luminal blood flow. Check for any other bleeding or unusual happenings (lack of pulsatility, no blood flow, apparent thrombosis, etc.). Add topical lidocaine and check for any leaks.

19. Swab away the lidocaine and any pooled blood and remove dried blood from the muscles, skin, and all body areas. Make sure all tissues are kept moist.

20. Remove all clamps and excess suture if still present. Place overlying tissues on top of the carotid vasculature in layers.

21. Close glandular tissue using 6.0 blue monofilament sutures and a running (continuous) suture (a.k.a., "mattress stitch"). Be sure to double-knot and tuck both ends.

22. Tent up the skin for the folds to meet and close the skin using either skin sutures or standard rodent wound staples and skin stapler. If using skin staples keep them closely approximated from end to end (*see* **Note 14**). Following skin closure, inspect the wound and check for any openings and correct if needed.

23. Swab an anti-septic/bactericide/virucide agent on all sides of wound (*see* **Note 38**) to reduce likelihood of infection.

24. Inject 3–5 ml supplemental fluids to amend for fluid loss, swab ophthalmic ointment over eyes, moisten mouth and tongue (*see* **Note 2**), and provide an appropriate analgesic for the animal according to institutional guidelines (*see* **Note 15**). Continue with post-operative animal care described below.

3.4. Post-Operative Procedures

1. Once adequate steps are taken to ensure animal comfort (*see* **Notes 2** and **4**), return the animal to his cage and place the animal prone on suitable bedding. Avoid using corncob bedding as this can get lodged in the wound area and be a source of discomfort or infection. It is recommended to use sterile gauze pads directly below the wound to maintain a degree of cleanliness during recovery. Tuck the hind legs under the body for support, prop up the upper thorax of the animal using rolled-up surgical blankets or towels and tuck the front legs under the neck/head (*see* **Note 16**). This aids breathing during recovery. Remember to keep the animal prone throughout recovery until ambulatory. It has been recommended to the author that placement of a heating blanket (set at a low-to-medium temperature) or a microwaveable heating pad inside the animal cage during recovery aids the animal during recovery (*see* **Note 11**). Also, during this time, it is important to monitor breathing rate and rhythm. If these become labored or slow, supplemental air or oxygen via a nasal cannula should be used.

2. Monitor the animal routinely during recovery until sternally recumbent and ambulatory. Make sure to keep the eyes, mouth, and tongue moist. Inspect the wound to make sure there is not bleeding or that it does not become dehiscent and open for infection.

3. Once animal is ambulatory, provide sufficient water to drink and food and return to animal care facility. A symptom commonly demonstrated by balloon-injured rats is partial ptosis of the left eye due to nerve manipulation associated with the surgery and ensuing malfunction of the eyelid elevator muscles.

4. Carefully remove the trocar from the balloon catheter sheath and thoroughly clean. This can be sterilized as well. Heparin is suggested to be used on the trocar to prevent formation of blood clots within the trocar lumen.

5. Thoroughly but carefully clean the balloon catheter, especially making sure to remove all clotted blood that might be around the balloon–catheter junction.

6. It is very important to check the degree of balloon inflation after every surgery to make sure that the balloon inflated the appropriate extent during the operation. With any given ballooning procedure, the balloon can become distended or otherwise lop-sided or even perforated, so it is imperative to make sure the balloon is still able to inflate to the expected volume or pressure. Return the balloon tip to an aliquot of sterilized water for storage (*see* **Note 39**).

7. Collect all instruments in a stainless steel surgical pan, add surgical instrument cleaner or detergent, thoroughly clean each instrument (*see* **Note 40**), and dry.

8. Clean and sterilize the entire surgery area, and swab the area with 70% alcohol.

9. Specific to the research design, plan next series of experiments accordingly to obtain relevant vascular tissues for use in histology, expression analyses, or other endpoints as described in Chapter 1. Several examples of photomicrographs from rat uninjured or balloon-injured carotid arteries are presented in **Figs 6** and **7**.

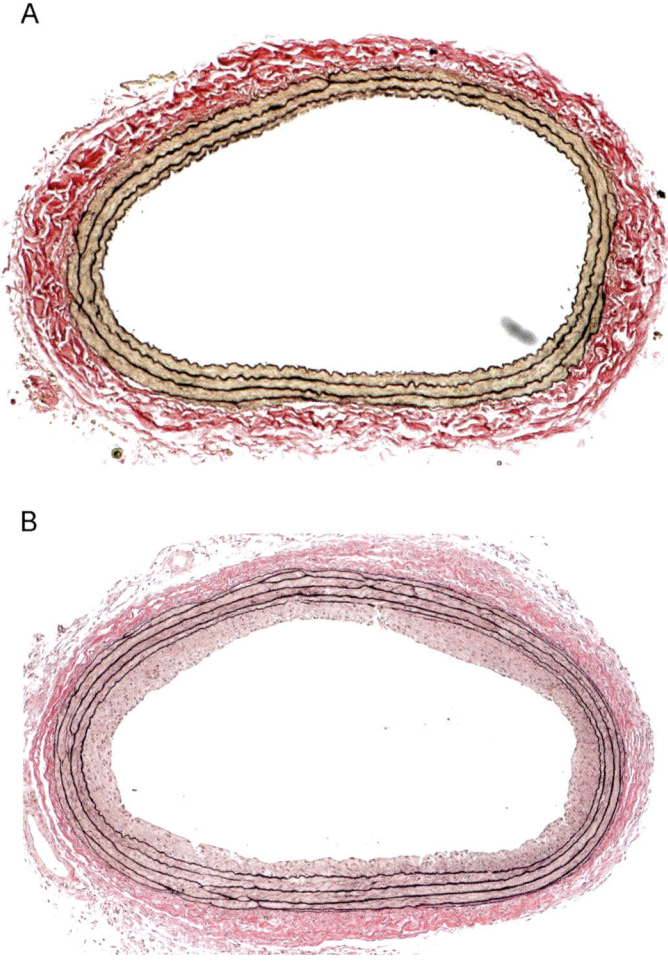

Fig. 6. Cross-sectional photomicrographs of rat uninjured (**A**) or balloon-injured (**B**) carotid arteries perfusion-fixed and treated with Verhoeff's elastin stain and Van Gieson counterstain. Injured arteries demonstrate robust and concentric neointima with luminal stenosis 2 weeks post-injury. Magnification for photos is ×100.

3.5. Animal Recovery

The investigator should be aware of the common signs of animal morbidity during the recovery period that can serve as indications for early euthanasia. General guidelines include immobility, huddled posture, inability to eat and/or

A

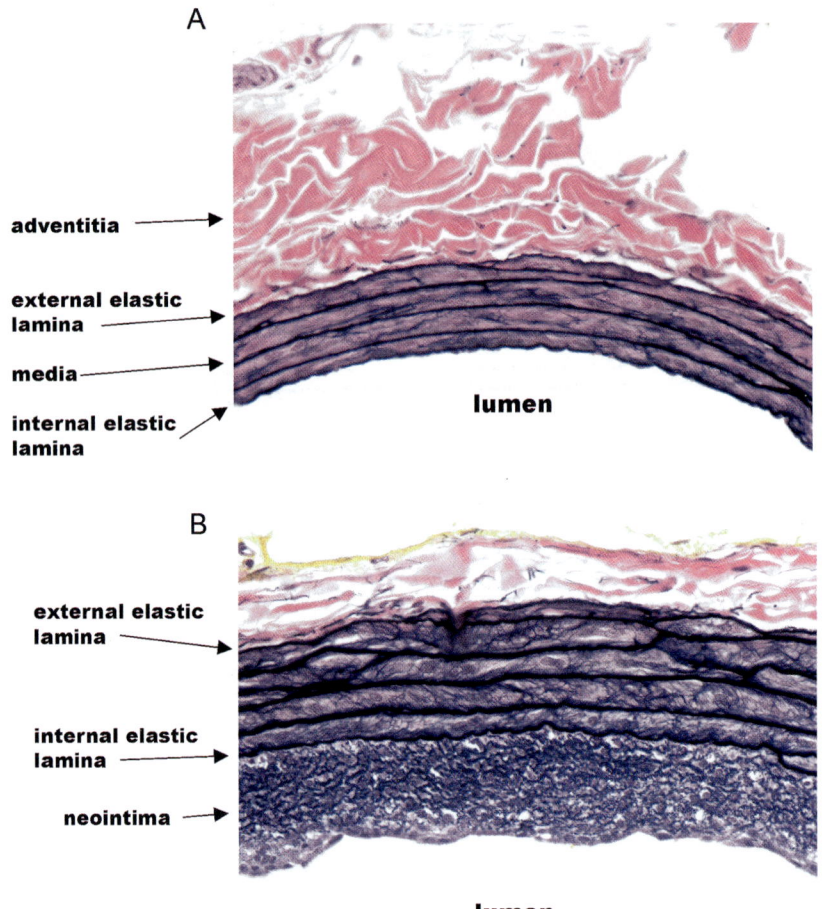

adventitia

external elastic
lamina

media

internal elastic
lamina

lumen

B

external elastic
lamina

internal elastic
lamina

neointima

lumen

Fig. 7. Detailed photomicrographs of rat uninjured (**A**) or balloon-injured (**B**) carotid artery sections demonstrating a matrix-rich neointima. Specifics of histological protocols and morphometric analyses germane to the rat carotid artery balloon injury model are discussed in Chapter 1. Magnification for photos is ×400.

drink, ruffled fur, self-mutilation, vocalization, wound dehiscence, hypothermia, and/or > 20% weight loss.

4. Conclusions

The rat carotid artery balloon injury model described in these methods presents a practical and highly useful in vivo animal model that mimics many biophysical, cellular, chemical, and molecular mechanisms found in humans with injured and/or diseased vasculature; however, knowledge of the caveats and limitations of this model is imperative for successful completion of such studies, scientifically valid interpretation of results, and legitimate comparisons to the human condition. Described herein are peri-operative protocols for the rat balloon injury model that have been used over many years with great success; however, modification of these methods may be required for individual needs and/or based on individual experiences or preferences. Histological and morphometric methods pertinent to the rat carotid artery balloon injury model and other animal vascular injury models are described in Chapter 1 and should be consulted before designing such research studies.

5. Notes

1. A primary caveat of the rat carotid artery balloon injury model is that this intervention uses a normal eutrophic blood vessel lacking pre-existing atherogenic or vasoproliferative pathology. This is in sharp contrast to clinical balloon angioplasty procedures performed on diseased vasculature in humans. Although the response of healthy vessels to balloon intervention involves many of the same cellular and molecular signals that are involved during the response in diseased vessels, the investigator must be aware that these are independent processes and should not be confused. Vascular SMCs are primarily responsible for the adaptive response to injury in the rat, whereas in diseased human vasculature a mixed population of vascular SMCs and endothelial cells, macrophages, and T cells interact in response to the inimical stimulus. Anatomical constraints of this model include lower percentage of medial wall elastin, a condensed subintimal layer, and lack of an existing vasa vasorum *(15)*. Studies using this model often report success in minimizing the extent of arterial remodeling using varied treatments that are not replicated in human studies, thus suggesting that this model may be a "ubiquitous responder" to wide-ranging experimental regimens. However, despite these inherent limitations this experimental approach remains a valuable tool with which to study many of the diverse mechanisms involved in the injury response.

2. In this method, several steps are taken by the investigator to ensure that appropriate animal care is provided and that the animal's welfare is considered seriously. In this regard, provision of ophthalmic solution for the eyes and water for the mouth and tongue is performed before surgery as well as during and following surgery to prevent drying of these tissues and discomfort to the animal.

3. An appropriate anesthetic should be chosen according to the specifics of the research project and in accordance with institutional guidelines for the care and use of animals. The anesthetic of choice should be placed adjacent to the surgical area along with a syringe and needle with an appropriate supplemental dose (10% original dose) already withdrawn in the syringe and ready to inject if needed.

4. Provision of supplemental fluids is essential during the surgery as the animals become dehydrated under anesthesia. Generally and unless significant blood loss occurs, 8–10 ml fluid per animal is provided via subcutaneous injection (using an 18–20 gauge needle and a 3–5 ml syringe). If substantial blood loss occurs during surgery, then additional fluids should be provided as well as supplemental nutrients (LR solution is suggested). Supplemental fluids can be given immediately before performing the surgery as well as through the duration of the surgery and during the recovery period.

5. One can also simply use long pieces of adhesive tape to hold down the rat appendages. It is suggested for the surgeon to loosely tie-down front and hind legs as well as to prepare a loop of suture (size 0, chromic gut) attached to a piece of tape and to use this to gently retract and hold down the head of the animal by looping around upper teeth. It is not recommended to restrain the tail of the animal as the rat uses the tail to regulate body temperature as well as the fact that the tail often flinches if the level of sedation becomes too light. The surgeon should regularly monitor tail movement as an indication that supplemental anesthesia might be needed. Keeping the tail unrestrained also allows easy access to the tail vasculature if the surgeon desires intravenous or intra-arterial intervention.

6. The author uses topical warm lidocaine hydrochloride (0.8 grams in 40 ml PBS) to provide mild anesthesia to the exposed tissues, to decrease incidence of muscular spasm of the carotid vasculature, and to provide a moderate degree of vasodilation to simplify insertion of the catheter through the arteriotomy incision.

7. Insertion of the balloon catheter into the arteriotomy incision on the external carotid artery is simplified if the surgeon uses an introducing guiding needle or catheter sheath as a trocar. Before the surgery, the balloon tip must be advanced through the trocar and the catheter with trocar should be readily available during surgery. The trocar allows precise insertion of the balloon tip into the arteriotomy hole without interference caused by the bulbous head of the balloon. The author recommends using a trocar suitable for 2 French catheters, namely 18 gauge thin-walled or ultrathin-walled needles. Caution must be practiced, however, when using a trocar such as a thin-walled needle. Immediately upon insertion of the needle bevel into the arteriotomy, carefully lower the luer-lock end of the needle in order to slightly raise the bevel tip. Advance the trocar bevel slightly more into the lumen and proceed with insertion of the balloon catheter through the trocar. Once the balloon catheter is inserted and advanced to the clamp on the common carotid artery, gently slide the trocar down the catheter sheath toward the syringe and out of the way. If during balloon advancement, the balloon tip does not enter the artery lumen but instead moves alongside the vessel (and therefore out of the arteriotomy), withdraw it slightly and gently advance and rotate the trocar bevel in the direction that the balloon tip is moving. This will help guide the balloon into the artery lumen and keep it from moving into the arteriotomy at an angle. One other note regarding use of a trocar necessitates mention here. If using a needle

as a trocar, the tip is extremely sharp and will easily penetrate the backside of the vessel ("backwall") if the trocar is inserted at too sharp of an angle. If this happens, the catheter will still be able to be advanced through the trocar; however, the balloon will not be inserted into the artery lumen and will usually be located underneath or adjacent to the common carotid artery. Remove the balloon catheter and the trocar and re-attempt insertion of the trocar into the arteriotomy hole and into the lumen. Once a "backwall" has happened, effort must be made to keep the catheter from going in that same direction upon repeated attempts.

8. A regulated manual (or automatic) inflation pump can be used to achieve consistent and reproducible inflation of the balloon at a given pressure (atmosphere). This is a good measure to use for calibration of the balloon and for verification of the extent of inflation. The surgeon must be aware, however, of the limitations and practical difficulties of using an inflation device during surgery. It is recommended that if an inflation device is to be used during surgery, a separate operator performs the inflation of the balloon while the surgeon performs manipulation of the catheter inside the vessel along with withdrawal of the inflated balloon to induce injury.

9. Before placing suture sections around various sites on the carotid artery vasculature, it expedites the process for the surgeon to have cut several lengths of suture (4-0 black braided silk and 6-0 Prolene blue monofilament) ahead of time (during the setup) and to have these readily available. Approximately 3 in. sections of both silk and blue monofilament sutures as well as 5–6 in. lengths of silk suture will be needed.

10. It is highly recommended that the surgeon use several independent sources of light or multi-directional fiber-optic lighting during the surgery in order to reduce incidence of shadows and to enhance visualization of the carotid vasculature.

11. A temperature-regulated (via anal probe) heating pad is the best choice to use during the surgery to avoid hyperthermia for the animal. If heating lamps are used, make sure to precisely control the temperature of the animal to avoid hyperthermia. It can be difficult to adequately control the temperature of the animal when using heat lamps.

12. A broad range of surgical tools can be used with complete confidence for a successful surgery; therefore, specific item numbers or manufacturing information is included for a majority of these items; yet, these are simply recommendations made by the author. The most important point concerning the surgical instruments is that the surgeon is comfortable and at ease with their use.

13. Pertinent to all retraction procedures performed in this method, the surgeon can choose between using a surgical retractor (commercially available) and using sutures to keep skin and/or tissues out of the way. The use of surgical retractors is straightforward. If the surgeon wishes to use suture, it is recommended to use 6-0 suture with suture needle attached. Once the needle has been passed through the tissue to be retracted, gently tie the suture thread, retract the suture away from the animal, and clamp the suture thread taut with a hemostat or tape it down to the operating table out of the way of the surgeon.

14. A skin staple remover is highly recommended to be available to the surgeon during post-operative closing procedures if the surgeon chooses to use skin staples to close the skin (instead of skin suture). A common occurrence during this procedure is the placement of unapposed skin staples that creates gaps between the skin folds and openings of the underlying tissues. This can lead to wound dehiscence and infection in the animal. Use of a manual skin staple remover makes removal of incorrectly placed staples simple and straightforward.

15. The author recommends use of an appropriate analgesic post-operatively if this does not interfere with the investigators research study and if it abides by the local institutional animal care and use guidelines. Buprenorphine is the analgesic of choice of the author for use in rats (0.1–0.5 mg/kg subcutaneous injection immediately following surgery and then every 12 h as needed, or according to institutional policy).

16. The author uses rolled-up surgical drapes or towels propped under the thorax of the animal during recovery to keep animal upright ($\sim 30°$ angle) in an effort to ease breathing until consciousness is attained.

17. The primary example of an emergency that can occur in this protocol is unexpected rupture of the carotid artery or accidental removal (slippage) of the balloon catheter from the artery. Items to keep readily available for such an occurrence include arterial clips or clamps of various sizes, large and small retractors, extra sections of tape and suture, and additional gauze and plenty of cotton-tipped applicators. Also, extra surgical tools (in case one drops and breaks) and extra balloon catheters along with trocar(s) should be available. The more one is prepared, the better chance of dealing with an unexpected situation successfully and the better prognosis for the animal in case of emergency.

18. When removing air bubbles from the stopcock and catheter end, the author recommends using a broken wooden stick from the cotton swabs to "sweep out" any lodged air bubbles. Otherwise, shaking these items will inadvertently remove the water and the surgeon will have to start over again. It is essential to create a closed system here with water filling the syringe and stopcock. The closed system will also include the lumen of the balloon catheter which will similarly be filled with water.

19. For sterilizing surgical tools, the author recommends using a glass bead sterilizer. This apparatus should be situated nearby and needs to be pre-heated for 30 min. The tips or other operating parts of the surgical tool needs to be immersed in the glass beads for 10–15 s (longer times can eventually damage the instrument), after which the tool should be placed on a sterile surgical blanket at an appropriate and convenient location in the surgical area. Keep in mind not to touch anything with the sterilized portion of each surgical tool after sterilization. In the case of surgical forceps (especially Pattern 7S), never lay them down on their tips! Always lay forceps with tips up to maintain sterility and to keep from damaging the tips.

20. When using the clippers to remove hair, be careful as the neck area is very pliable and will depress with minimal pressure thus causing difficulty in breathing for the

animal. At the sternum, it is easy for the clippers to nick the skin and to cause minor cuts that could bleed and cause discomfort to the animal. It might be better to use scissors for hair removal near the chest area.

21. For scissors to be used to cut the skin, the author recommends using large straight, sharp/blunt scissors with one serrated edge on the sharp side. When cutting through the skin, use the sharp serrated side of the scissors to penetrate the skin, and remember to keep the tip up during the skin incision to avoid penetrating underlying tissue.

22. Do not be tempted to use a scalpel to perform the neck incision. Even with the use of brand new and sharp scalpels one must exert sufficient pressure on the blade in order to make the cut (this tissue is very pliable). This will depress the trachea and impede the ability of the animal to breathe and should not be attempted.

23. These methods are for left carotid artery injury; however, injury can be performed on the right carotid artery just as well. The left carotid artery presents a longer section for intervention than does the right carotid and is the choice for this author and many other investigators. If one chooses the right carotid artery for injury, remember that the right carotid artery branches off the innominate (brachio-cephalic) artery after bifurcation from the aortic root, hence shortening the available section for injury. The innominate artery is not suitable for inclusion in the injury protocol due to larger caliber compared to the right common carotid.

24. If the surgeon is using instruments in his/her right hand and performing this surgery on the left carotid artery, then caution must be practiced to ensure that the surgical instruments in use are not impinging upon the trachea and causing difficulty in breathing for the animal. This inadvertent act is easy to do when focusing on blunt dissection and the carotid vasculature. Be aware of the breathing rate and rhythm of the animal and the location of the surgical instrument(s) during this and all steps of the surgery.

25. As one moves proximally along the common carotid artery toward the sternum, the vagus nerve travels alongside the artery in parallel fashion. However, in many animals just before the carotid artery moves under the rib cage, the vagus nerve moves across the artery in a medial direction where it continues to travel parallel to the artery. This creates inconvenience for the surgeon who must avoid contact with and manipulation of the vagus nerve. Additionally, oftentimes, a nerve plexus exists overlying the carotid vasculature that should not be manipulated or cut. This is especially imperative at the carotid bifurcation, a site for blood pressure control. If necessary, these vital nerve components should be carefully moved aside and out of the way of the carotid artery. Do not attempt to cut these essential nerves.

26. Considerable variation can exist in the exact anatomical location of the common carotid artery bifurcation. In most animals, the branch point for the internal and external carotid arteries occurs on the distal common carotid at a site that provides easy access to the external branch for surgical intervention. However, in some animals, the bifurcation occurs more distally toward the head, thus making a

shorter segment on the external branch for vascular access. Make a note for each animal as to the exact location on the common carotid artery the bifurcation is located (i.e., 5 mm from hyoid cartilage, 15 mm from sternum, etc.) using a ruler or micro-caliper.

27. Hemostasis in the main carotid artery should be induced and all retrograde blood flow controlled before insertion of the catheter into the lumen of the carotid artery. If not, the surgery can still proceed but there will be bleeding once the arteriotomy is made. This can lead to dramatic blood loss and difficulty in visualization of the surgical area. It is highly recommended by the author that temporary cessation of blood flow (via retraction and/or clamping) takes place as an essential component of this method. If blood flow is halted for a few moments before making the arteriotomy incision and entry of the catheter, then once the arteriotomy hole is made it will be clearly seen and insertion of the catheter will be greatly simplified. Once the catheter is inserted into the arteriotomy and advanced a bit, then the retractors and/or clamps can be released. This will restore blood flow only partially though because the catheter will now be inside the vessel lumen which will halt most blood flow, even in an uninflated state (some leaking is still expected with an uninflated balloon).

28. In this step, all adjacent tissues must have been separated from the external carotid artery as distally up the vasculature as possible toward the head. This will create a workable length of the external carotid branch thus providing a longer section with which to make the arteriotomy hole and to insert the balloon catheter. If a problem exists with the first arteriotomy incision (which is made as distally as possible), then that site will be closed and the surgeon will move proximally if there exists a sufficient length of external carotid artery branch remaining with which to work. However, caution must be practiced if the arteriotomy hole is made *too close* to the most distal suture knot [*see* **Figs 3** (denoted as 3) and **4**]. In that case, the suture knot itself could impede insertion of the balloon tip, and the surgeon may have to make a second arteriotomy at a more proximal site.

29. Up to this point, no vascular intervention or blood flow alteration has occurred; yet once these small vessels are ligated, then the surgeon must work quickly but in as scientifically and medically sound fashion as possible. This is a critical time during the surgery, and prolonged time under blood flow cessation may cause undesired side effects. From the time when the first vessel is ligated through when the catheter is removed and the arteriotomy incision is closed is the most critical and significant portion of the entire method.

30. As mentioned before, retraction and stoppage of blood flow can take place either through the use of an arterial clamp or with a suture looped around the vessel and retracted. For the common carotid artery, the author recommends using an arterial clamp to ensure that all blood flow is stopped through this vessel. If suture is used here, it must be significantly retracted to a large degree in order to stop blood flow, and the normal geometry of the carotid artery is dramatically perturbed; therefore,

the author recommends against using suture to halt common carotid artery blood flow in this step.

31. It is recommended to keep the suture in place here even though an arterial clamp has been placed on the common carotid artery to stop blood flow. The reason being is that this clamp will be removed during the balloon injury, and if a major arterial leak occurs from any number of reasons (accidental removal of the balloon tip during the catheter withdrawal procedure being a primary example), then this suture is already looped around the common carotid artery and can easily and quickly be retracted to stop common carotid artery blood flow. However, if during advancement of the balloon catheter the balloon tip gets "caught" or "hung-up" on the suture that is looped around the common carotid artery and therefore cannot be advanced, then either gently loosen that suture or remove it all together. If removed, be aware that if a carotid artery leak should occur then hemostasis will be more difficult without that suture already in place.

32. If the arteriotomy incision is made too deep or if the entire vessel is cut through, do not panic! Because this initial arteriotomy site was far up the external carotid artery on the most distal portion of that vessel (hopefully), ample length of external carotid artery should still exist to try this again. But first, if the vessel is cut deeply but is still intact, try to perform insertion of the balloon catheter and the balloon injury as normal but be especially prudent of the amount of pressure applied to the vessel by the catheter. If too much pressure is applied, then this will surely cause the arteriotomy incision to expand and the vessel to break. If the vessel breaks (either from the initial arteriotomy attempt or from too much pressure from the catheter), the surgeon has several options. If the catheter was already inserted into the artery, then the balloon injury itself could still take place if the external artery tissue is held fast by forceps or hemostats. This is often too difficult to perform and can lead to tears and more breaks in the artery. Alternatively and the preferred choice of the author, the surgeon can attach the broken proximal portion of the external artery to the broken distal portion via sutures and pull these two sections together. Joining the two broken portions of the external artery will then provide additional vessel length with which to attempt another arteriotomy incision, this time proximal to the initial cut and to the re-attached vessel, and will also importantly restore normal vascular geometry. Remember, the entire external carotid artery will be ligated after injury, so this does not cause further complication to the animal or to the scientific integrity of the experiment.

33. Another alternative for the surgeon if the vessel breaks is to gently grab the lip of the proximal section with fine forceps (7S) and then to try to insert the catheter into the vessel by holding or tenting it up. This is similar to putting on a sock and using only two fingers on one hand! Generally, once the catheter is inserted into the common carotid, it will advance without further complication.

34. The author recommends manual withdrawal of the balloon catheter with rotation as this is simpler to perform than when using forceps. Rotation of the balloon catheter

is essential in order to ensure concentric injury and to maintain consistency in the injury along a cross-section of the vessel. This is especially important if the balloon does not inflate perfectly round but instead becomes lopsided inside the blood vessel. An alternative to full rotation of the balloon catheter is to partially rotate the catheter using a side-to-side motion.

35. The degree of balloon inflation will directly determine the extent of vascular injury and ensuing cellular and molecular mechanisms responsible for neointima development and medial wall remodeling *(16)*; therefore, maintaining consistency and reproducibility of balloon inflation within and between animals is essential.

36. For removal of the deflated balloon catheter and closure of the arteriotomy site, it is recommended that the surgeon have an assistant. If the surgeon performs this technique alone, then successful removal of the catheter can be achieved without significant blood loss by holding the catheter (near the syringe) in his/her mouth and gently moving their head back to remove the catheter from the vessel. The surgeon should try to maintain his/her head at the same level as that of the animal. Once the catheter is removed, quickly pull back on the forceps in each hand to tie the suture and close the arteriotomy hole. This approach was devised by the author and is used routinely in his laboratory with much success.

37. Visualization of an arterial bleed is made easier by adding lidocaine to the vasculature. This will serve to dilate the artery and will enhance blood flow thus augmenting existing leaks in the vessels. Generally speaking, if leaks are present they are usually located around the arteriotomy incision and are due to inadequate closure of that suture. If this is the case, tie another suture around the arteriotomy at that site [*see* **Figs 3** (denoted as 2) and **4**] to stop the bleeding.

38. Do not to apply the topical anti-septic/bactericide/virucide agent directly on top of the wound incision but instead spread it all around the periphery of the wound using cotton-tipped applicators.

39. Following surgery and cleaning of the balloon catheter, always store the balloon tip in sterilized water or other suitable liquid medium. If not, the balloon will quickly desiccate and will not be able to be used again. With proper care, a single balloon can be used repeatedly for these animal surgeries as long as the balloon remains intact and not perforated, inflates the appropriate degree, and is completely circular or concentric during inflation. An estimate by the author for use of a single balloon catheter is approximately 15 surgeries without complications.

40. When cleaning the surgical tools, remember to open all scissors and hemostats and to clean both the inside and outside edges of the blades. Open arterial clamps and clean their insides as well. Be very careful with the instruments during cleaning, as this is a time when most accidents occur that can ruin expensive tools!

Acknowledgments

Work in preparation of this chapter was supported by NHLBI grants HL-59868 and HL-081720 and by a grant from the American Heart Association.

The author would like to apologize to investigators whose works were not cited in this methodological report due to space limitations and the personal perspective with which this chapter was prepared.

References

1. Clowes, A.W., Reidy, M.A., and Clowes, M.M. (1983) Mechanisms of stenosis after arterial injury. *Lab. Invest.* **49**, 208–215.
2. Clowes, A.W., Reidy, M.A., and Clowes, M.M. (1983) Kinetics of cellular proliferation after arterial injury. I. Smooth muscle growth in the absence of endothelium. *Lab. Invest.* **49**, 327–333.
3. Clowes, A.W. and Clowes, M.M. (1985) Kinetics of cellular proliferation after arterial injury. II. Inhibition of smooth muscle growth by heparin. *Lab. Invest.* **52**, 611–616.
4. Clowes, A.W., Clowes, M.M., and Reidy, M.A. (1986) Kinetics of cellular proliferation after arterial injury. III. Endothelial and smooth muscle growth in chronically denuded vessels. *Lab. Invest.* **54**, 295–303.
5. Clowes, A.W., and Reidy, M.A. (1991) Prevention of stenosis after vascular reconstruction: pharmacologic control of intimal hyperplasia – A review. *J. Vasc. Surg.* **13**, 885–891.
6. Majesky, M.W. (1994) Neointima formation after acute vascular injury. Role of counteradhevise extracellular matrix proteins. *Tex. Heart Inst. J.* **21**, 78–85.
7. Schwartz, S.M., deBlois, D., and O'Brien, E.R.M. (1995) The intima: soil for atherosclerosis and restenosis. *Circ. Res.* **77**, 445–465.
8. Zubilewicz, T., Wronski, J., Bourriez, A., Terlecki, P., Guinault, A.-M., Muscatelli-Groux, B., Michalak, J., Melliere, D., Becquemin, J.P., and Allaire, E (2001) Injury in vascular surgery – the intimal hyperplastic response. *Med. Sci. Monit.* **7**, 316–324.
9. Sapru, H.N. and Krieger, A.J. (1977) Carotid and aortic chemoreceptor function in the rat. *J. Appl. Physiol.* **42**, 344–348.
10. Gabeler, E.E.E., van Hillegersberg, R., Statius van Eps, R.G., Sluiter, W., Gussenhoven, E.J., Mulder, P., and van Urk, H. (2002) A comparison of balloon injury models of endovascular lesions in rat arteries. *BMC Cardiovasc. Disord.* **2**, 16–28.
11. Majesky, M.W., Schwartz, S.M., Clowes, M.M., and Clowes, A.W. (1987) Heparin regulates smooth muscle S phase entry in the injured rat carotid artery. *Circ. Res.* **61**, 296–300.
12. West, J.L., and Hubbell, J.A. (1996) Separation of the arterial wall from blood contact using hydrogel barriers reduces intimal thickening after balloon injury in the rat: the roles of medial and luminal factors in arterial healing. *Proc. Natl. Acad. Sci. U.S.A.* **93**, 13188–13193.
13. Wallner, K., Sharifi, B.G., Shah, P.K., Noguchi, S., DeLeon, H., and Wilcox, J.N. (2001) Adventitial remodeling after angioplasty is associated with expression of tenascin mRNA by adventitial myofibroblasts. *J. Am. Coll. Cardiol.* **37**, 655–661.

14. Li, G., Chen, S-J., Oparil, S., Chen, Y-F., and Thompson, J.A. (2000) Direct in vivo evidence demonstrating neointimal migration of adventitial fibroblasts after balloon injury of rat carotid arteries. *Circulation* **101**, 1362–1365.
15. Sims, F.H. (1989) A comparison of structural features of the walls of coronary arteries from 10 different species. *Pathology* **21**, 115–124.
16. Indolfi, C., Esposito, G., Di Lorenzo, E., Rapacciuolo, A., Feliciello, A., Porcellini, A., Avvedimento, V.E., Condorelli, M., and Chiariello, M. (1995) Smooth muscle cell proliferation is proportional to the degree of balloon injury in a rat model of angioplasty. *Circulation* **92**, 1230–1235.

2

Histological and Morphometric Analyses for Rat Carotid Balloon Injury Model

David A. Tulis

Summary

Experiments aimed at analyzing the response of blood vessels to mechanical injury and ensuing remodeling responses often employ the highly characterized carotid artery balloon injury model in laboratory rats. This approach utilizes luminal insertion of a balloon embolectomy catheter into the common carotid artery with inflation and withdrawal resulting in an injury characterized by vascular endothelial cell (EC) denudation and medial wall distension. The adaptive response to this injury is typified by robust vascular smooth muscle cell (SMC) replication and migration, SMC apoptosis and necrosis, enhanced synthesis and deposition of extracellular matrix (ECM) components, partial vascular EC regeneration from the border zones, luminal narrowing, and establishment of a neointima in time-dependent fashion. Evaluation of these adaptive responses to blood vessel injury can include acute and longer term qualitative and quantitative measures including expression analyses, activity assays, immunostaining for a plethora of factors and signals, and morphometry of neointima formation and gross mural remodeling. This chapter presents a logical continuation of **Chapter 1** that offers details for performing the rat carotid artery balloon injury model in a standard laboratory setting by providing commonly used protocols for performing histological and morphometric analyses in such studies. Moreover, procedures, caveats, and considerations included in this chapter are highly relevant for alternative animal vascular physiology/pathophysiology studies and in particular those related to mechanisms of vascular injury and repair. Included in this chapter are specifics for in situ perfusion-fixation, tissue harvesting and processing for both snap-frozen and paraffin-embedded protocols, specimen embedding and sectioning, slide preparation, several standard histological staining steps, and routine morphological assessment.

From: *Methods in Molecular Medicine, Vol. 139: Vascular Biology Protocols*
Edited by: N. Sreejayan and J. Ren © Humana Press Inc., Totowa, NJ

Key Words: Balloon injury; Embedding; Harvesting; Histology; Medial wall; Microscopy; Morphometry; Neointima; Perfusion-fixation; Processing; Rat carotid artery; Remodeling; Sectioning; Staining.

1. Introduction

Use of histological and morphometric techniques is essential for the examination of disease processes as well as for the investigation of many of the mechanisms that contribute to a wide variety of tissue pathologies. As a logical continuation of varied animal survival surgeries, these techniques involve tissue fixation and harvesting from the animal post-mortem, tissue processing, embedding, and sectioning, microscope slide preparation, tissue staining for a multitude of cellular and/or molecular components in the specimen, and analysis through microscopy by the investigator(s). Germane to animal vascular studies and specifically the rat carotid artery balloon injury model (*see* **Chapter 1**), this chapter details commonly used "histo-techniques" for routine performance of several of these key steps. Particular emphases are placed on protocols for performing in situ perfusion-fixation, tissue harvesting for both fixed and fresh samples, several commonly employed staining procedures, and standardized morphometric analyses. Based on the breadth of histological and morphometric protocols currently available, many of which are suitable for vascular tissues, this chapter is not intended to serve as a comprehensive summation of all techniques and strategies but rather as a general guide for continuance of studies employing the rat carotid artery balloon injury model. Thus, only brief outlines of protocols are provided for several of these steps. For interested readers, excellent comprehensive resources are available for consultation on a wide variety of histological and morphometric techniques *(1,2)*, including a summary of procedures specific to animal vascular studies *(3)*. In conclusion of this chapter, **Subheading 4** provides important considerations for practical utility of these protocols.

2. Materials

Materials needed to perform these procedures are listed below in subheadings for the various steps. Where appropriate, brief mention of their use is included along with item numbers, brand names, and/or manufacturer preferred by the author.

2.1. In Situ Perfusion-Fixation

2.1.1. Solutions

1. Anesthetic of choice (*see* **Note 1**).
2. Fixative of choice (*see* **Notes 2** and **3**).

3. Pre-warned phosphate-buffered saline (PBS) or vasodilator solution (*see* **Note 4**).
4. Absolute alcohol (for cleaning surgical area).
5. Water (for flushing system).
6. Disinfectant (for flushing system).
7. Heparin (for treating over-the-needle catheter).

2.1.2. Supplies

1. Ample lighting.
2. Rack and clamps (for perfusion-fixation apparatus).
3. Three-way stopcocks with luer-lock end (used for attachment of tubing sections).
4. Two-way stopcocks with luer-lock end (used for attachment of short section of tubing to the over-the-needle catheter).
5. Tubing (3/16 in. inside diameter, 5/16 in. outside diameter vinyl).
6. Needles (26 gauge for anesthetic).
7. Syringes (1 ml for anesthetic and for washing lumen ex vivo; 50–60 ml capacity for perfusion-fixation apparatus).
8. Intravascular over-the-needle catheter (18–20 gauge, 1.5–2 in.; *see* **Note 5**) or similar guiding needle.
9. Collecting pan (for perfusion-fixation and animal waste).
10. Gauze.
11. Cotton-tipped swabs (have plenty of these available).
12. Face mask or shield, protective goggles, and/or gloves (per institutional guidelines).
13. Tape (cut into sections).

2.1.3. Surgical Instruments

1. Scissors: large, medium, and fine (Roboz Surgical Instrument Company, Inc., Gaithersburg, MD, USA).
2. Forceps: large (heart-holding or other large serrated-edge and medium), medium, and small curved.
3. Microcaliper (Roboz) or small metric ruler (to measure vessel lengths).

2.2. Tissue Harvesting

2.2.1. For Perfusion-Fixed Tissues

1. Fixative of choice (*see* **Notes 2** and **3**).
2. Alcohol (70%).
3. Petri dish (for washing lumen ex vivo).
4. Gauze.
5. Cotton-tipped swabs.
6. Suture (size 4–0 black braided silk; for tying on arteries to maintain orientation).
7. Labeled tubes.

2.2.2. For Fresh Tissues

1. Ice and ice cooler (for cleaning vessel ex vivo).
2. Liquid nitrogen (for snap-freezing).
3. Dry ice and methanol (for snap-freezing).
4. Freezing medium (Fisherbrand Super Friendly Freeze' It (Fisher Health Care, Curtin Matheson Scientific, Houston, TX, USA), or other suitable cytological fixative).
5. Petri dish (for cleaning vessel ex vivo).
6. Gauze and cotton-tipped swabs.

2.3. Tissue Processing

2.3.1. Solutions

1. Ethanol of varying percentages (70–95%, absolute).
2. Xylene (or other clearing agent, *see* **Note 6**).
3. Paraffin (or other embedding medium, *see* **Subheading 2.4.** and **Note 7**).

2.3.2. Supplies

1. Processing cassettes with lids (Unisette, Omnisette, or Histosette II, Fisherbrand).
2. Kimwipes or other fine-grade porous tissues (mainly used for small and/or highly valuable tissues).
3. Automated tissue processor.
4. Coplin jars or glass-staining dishes with covers (if manual processing is performed).
5. Pencil (for labeling embedding blocks/cassettes).

2.4. Embedding

2.4.1. For Paraffin-Embedded Tissues

1. Paraffin (TissuePrep or Tissue Path Paraplast, Fisherbrand, or Paramat).
2. Paraffin repellent.
3. Embedding blocks or rings (Tissue Path, Fisherbrand).
4. Stainless-steel base molds (HistoPrep, Fisherbrand).
5. Paraffin embedder with cold plate (an automated machine is preferred).
6. Forceps heater (highly recommended).

2.4.2. For Frozen Tissues

1. Tissue-freezing medium (TFM or O.C.T., TBS).
2. Embedding blocks or rings (Tissue Path, Fisherbrand).
3. Stainless-steel base molds (HistoPrep, Fisherbrand).

2.5. Sectioning

2.5.1. For Paraffin-Embedded Tissues

1. Rotary microtome (or other microsectioning tool).
2. Microtome stainless-steel blades (disposable).
3. Flotation bath (TissuePrep circular flotation bath, Fisherbrand).
4. Granular gelation for flotation bath (to increase surface tension of water but is not essential).
5. Fine forceps.
6. Small brush.
7. Water [diethyl pyrocarbonate (DEPC)-treated; *see* **Subheading 3.5.** and **Note 8**].

2.5.2. For Frozen Tissues

1. Cryostat (or other microsectioning tool used for frozen specimens).
2. Cryostat blades.

2.6. Slide Preparation

1. Incubator or slide warmer.
2. Microscope slides (Superfrost, Fisherbrand).

2.7. Routine Staining Procedures for Histomorphometry

1. Cover-slips/cover glasses.
2. Mounting medium (Permount, Fisherbrand).
3. Light microscope (for visual inspection of tissue samples and staining efficacy).
4. Glass jars with lids (to hold staining solutions).
5. Staining solutions (individual recipes; *see* **Subheading 3.7.**).

2.8. Microscopic and Morphological Evaluation

1. Ultra-fine point permanent marker.
2. Light microscope with $\times 10$ and $\times 40$ objectives (and other objectives based on personal preference).
3. Complete image analysis system with camera, interface, software of choice, and computer.
4. Graphical and statistical software of choice.

2.9. Endothelial Cell Regeneration

1. Evans blue dye (*see* **Note 9**).
2. Large beaker.
3. Hot water.
4. Syringe (1 ml) with needle (26–28 g).
5. Small microscissors (McPherson-Vannas, Roboz).

6. Small curved forceps (Microdissecting tweezers, Roboz).
7. Dissecting dish with silicone pad and pins (Electron Microscopy Services, Hatfield, PA, USA).
8. Small ruler or other measuring device (caliper).

2.10. Hematoxylin and Eosin Staining Method

1. Acid alcohol, 1% (1% HCl in 70% ethanol).
2. Ammonia water, 0.05%.
3. Eosin, 1%.

2.11. Verhoeff's Elastic Tissue Stain with Van Gieson Counterstain

1. Verhoeff's solution: 4% alcoholic hematoxylin (20 ml), 10% aqueous ferric chloride (8 ml), 2 g Lugol's iodine, 4 g potassium iodine, and 100 ml distilled water.
2. Van Gieson solution: 1% aqueous solution of acid fuchsin (10 ml) and saturated aqueous solution of picric acid (200 ml).

2.12. Masson's Trichrome

1. Solution A: acid fuchsin (0.5 g), glacial acetic acid (0.5 ml), and distilled water (100 ml).
2. Solution B: phosphomolybdic acid (1 g) and distilled water (100 ml).
3. Solution C: methyl blue (2 g), glacial acetic acid (2.5 ml), and distilled water (100 ml).

3. Methods

Histological preparation of animal tissues following an experimental procedure or intervention generally encompasses the following steps performed in succession: tissue fixation, processing, embedding, sectioning and slide preparation, staining, and microscopic analysis.

3.1. In Situ Perfusion-Fixation

The primary purpose for perfusing the animal carcass immediately following euthanasia is to rid the tissues of interest of resident blood components that can cause interference in ensuing microscopic examination. Once tissues are removed from a body however, they can rapidly undergo autolysis, putrefaction, and environmental degradation. Fixation is a complex series of chemical events that "fixes" or retains tissues as close to their living state as possible and preserves tissue integrity, from correct anatomical orientation to ultrastructure, without rearrangement or loss of vital cellular and/or molecular components. These processes are performed in situ at mean arterial pressure (MAP) in an effort to maintain the vasculature at normal distending pressures.

3.1.1. Setup

Preparation of all reagents and solutions, catheters, supplies, and so on should be accomplished ahead of the time when performing the perfusion-fixation protocol. This is especially important if acute studies are being conducted when timing is critical. As is the case when performing animal survival surgery, the initial steps of this protocol involve manipulations on a living animal, and therefore, preparedness and expediency are essential. Prepare PBS or saline (or vasodilator) solutions ahead of time and set in a 37 °C incubator until temperature is reached. Prepare fixative at desired concentration or percentage. It is useful to have individual aliquots (60–100 ml, depending on protocol) of each solution readily available to simplify pouring the solutions into the syringes of the perfusion-fixation apparatus.

The perfusion-fixation apparatus (*see* **Figs 1** and **2**) must be constructed ahead of time if it is to be used (instead of a perfusion pump, which will not be discussed) for perfusion-fixation. By using a tall rack stand situated on a laboratory bench, attach two clamps to the stand at a level that results in a column of fluid approximately equal to 100 mmHg (*see* **Note 10**). This is calculated using the fact that 1.36 cm of water (or fluid of similar viscosity) in a column equals 133 Pa ($1333 \, dyn/cm^2$), which also equals 1 mmHg. A mean pressure of 100 mmHg therefore equates to a column of fluid 136 cm, or approximately 53.5 inches, high (similarly, a mean pressure of 120 mmHg equates to a column of fluid with height 64.25 inches). This is the height for the column of PBS (or vasodilator solution) and for fixative, respectively, each in individual syringes with individual tubing (up to a point). Remove the plungers and protective caps from two 50-ml to 60-ml syringes and attach the ends of the tubing cut to appropriate lengths (to make a column 53.5 inches high for 100 mmHg pressure) to the "needle ends" of the syringes (held upside down). Next, attach the other two ends of the tubing each to one of the available ends of the three-way stopcock (*not* the luer-lock end) (*see* **Figs 1** and **2**). Now, by using an additional short piece of tubing (8–12 inches), attach it to the luer-lock end of the three-way stopcock and then attach a two-way stopcock to the other end of this short piece of tubing. To this two-way stopcock, the luer-lock hub of an over-the-needle catheter (used for cardiac puncture and perfusion) will be fastened. Attach the two syringes in an inverted fashion on the clamps already fixed to the rack and let the tubing drape to the floor (*see* **Figs 1** and **2** and **Note 11**). Clearly label one syringe "PBS" or "vasodilator" and the other one "fixative."

When choosing tubing, pick a size that can fit relatively easily but snugly onto the ends of the syringes, the stopcocks, and the luer-lock hub of the

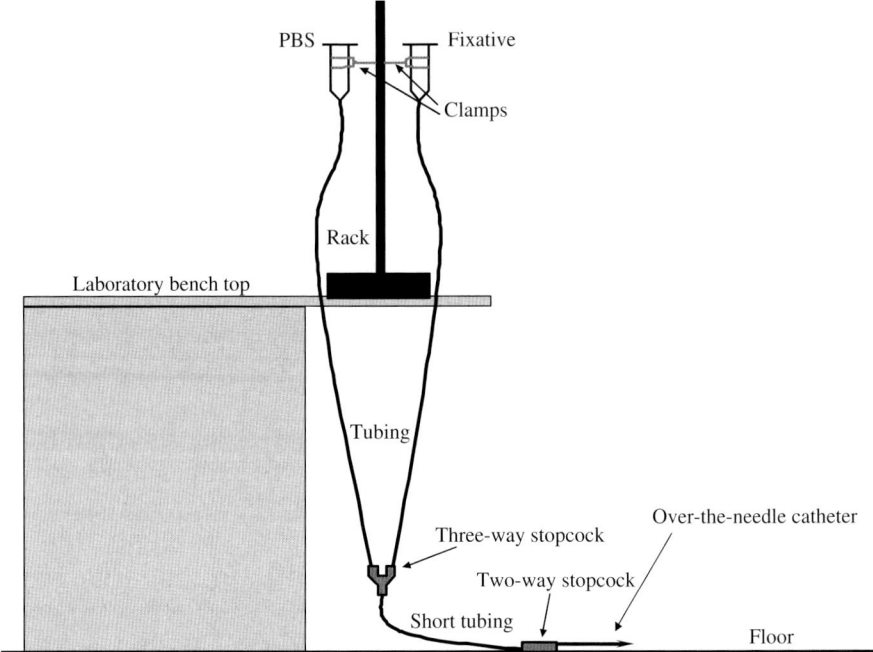

Fig. 1. A diagram of the apparatus used for performing in situ perfusion-fixation on rat carotid arteries. With the use of a laboratory bench rack, two clamps and syringes, tubing, stopcocks, and an over-the-needle catheter, this apparatus can be easily constructed for assistance in performing histomorphometric studies in a variety of animal tissues. This apparatus utilizes two independent columns of fluid that approximate pressures of 100 mmHg (see text for detailed description) and is used for transcardial perfusion and fixation of rat carotid arteries with phosphate-buffered saline (PBS) (or vasodilator solution). Tissues are then harvested for processing, embedding, sectioning, and staining with subsequent analytical evaluation.

over-the-needle catheter. As pressure equals force (acceleration of gravity) per unit area, then the cross-sectional area of the column (or tubing) will not affect the pressure calculation (*see* **Note 10**). The author uses 3/16 in. (inside diameter) tubing in his laboratory setup with much success; however, slightly smaller or larger tubing can also be used as long as it fits tightly onto both the stopcocks and the over-the-needle catheter. Become familiar with opening and closing the stopcocks so that this maneuver can be performed with one hand. Practice using water in both syringes and opening and closing the two in-series stopcocks to see the magnitude and speed of flow from the two-way

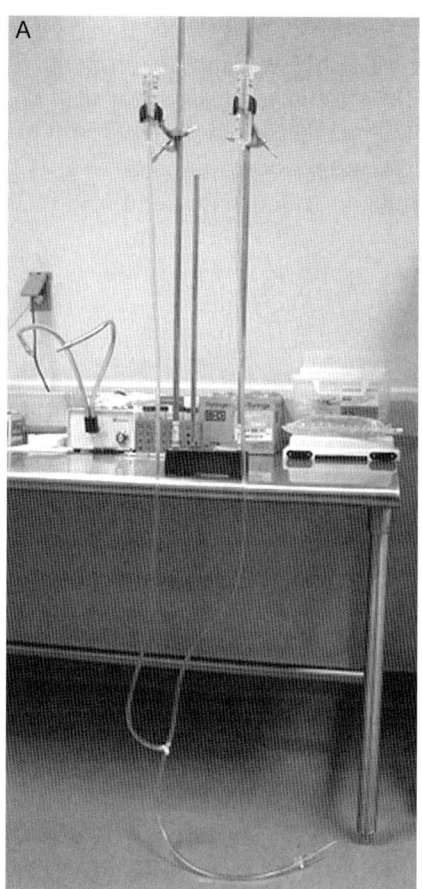

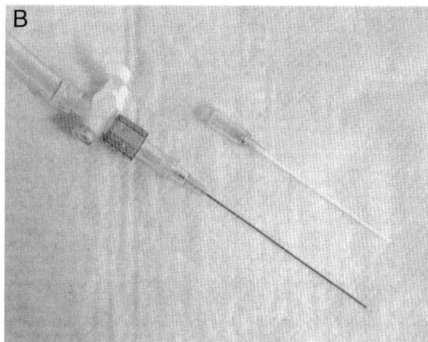

Fig. 2. Photographs of an in situ perfusion-fixation apparatus. (**A**) Complete apparatus; (**B**) close-up photograph of the over-the-needle catheter (with catheter removed) attached to stopcock and tubing.

stopcock or catheter end (*see* **Note 12**). Make sure that the apparatus is in proper working condition and that the stopcocks are firmly attached to the tubing and no leaks are apparent prior to performing the protocol. Ahead of the time (~15–20 min) for performing the perfusion-fixation and with both stopcocks in closed position, fill the appropriate (labeled) syringe with approximately 60 ml warmed PBS (or vasodilator solution) and the other one with approximately 60 ml fixative (*see* **Note 13**). Now, *importantly* open the stopcocks so that *only* fixative will flow through the tubing, through the first stopcock and short segment of "common" tubing, and through the distal stopcock and out. Close both stopcocks and remove any trapped air bubbles in the stopcocks and tubing. Next, open the stopcocks to allow *only* PBS (or dilator solution) to run through the tubing and stopcocks and out. Close the stopcocks to stop flow and purge all air bubbles. The order of performing this preparatory step is very important and critical for adequate perfusion and fixation of tissues (*see* **Note 14**). Refill the syringes with PBS and fixative solutions to account for fluid loss in this preparatory step.

Several other steps need to be completed prior to the perfusion protocol. Prepare all surgical instruments with thorough cleaning and proper convenient placement in the surgical area. Make sure adequate numbers of cotton-tipped swabs and pieces of gauze are available and located nearby. Cut sections of black 4–0 suture and have these nearby as well. Prepare labeled tubes (using alcohol-proof markers) filled with either fixative (for post-fixation of ~4 h) or 70% ethanol (to hold tissues following post-fixation and until ready for processing). For each tissue, two tubes will be made available, one with fixative and the other with 70% ethanol.

3.1.2. Protocol

Once again, *before* anesthetizing the animal, make sure that everything is ready for the protocol so that delays do not occur after the animal has been sedated. If recommended by the institutional animal care and use committee and based on specifics of the research study, the surgeon should choose to wear a face mask or shield, protective goggles, and/or gloves as appropriate. Also, it is recommended to perform this protocol with the perfusion-fixation apparatus situated in a fume hood so that fumes from the fixative do not affect the surgeon or other personnel in the immediate area. Weigh the rat and provide anesthetic (as routine) to achieve a surgical plane of anesthesia. Wait until the animal is fully sedated and then follow with an overdose of anesthetic (recommended 2.5× original dose or according to institutional guidelines). Monitor breathing and heart rate patterns. The animal should be extremely sedated so that overdose is imminent. Needless to say, make sure that the anesthetic does not cause unwanted bias or interfere with the scientific aims of the study.

Place the animal on its back on a surgical tray with the head toward you and tape down the front and hind legs. The surgeon may also want to loosely wrap a suture around the upper teeth of the rat and tape down the ends (to hold the head still). Working quickly, by using large serrated-edge scissors, make a midline incision starting at the sub-sternal notch through the sternum (and ribcage), moving in a caudal direction. Keep the tips of the scissors up to avoid puncture of heart or underlying tissues. Keep cutting through the sternum and ribcage all the way down to the diaphragm. It is important to work rapidly here once the initial cut is made, considering that euthanasia is being induced through pneumothorax (followed by overdose) and the fact that expediency will usually provide more accurate data related to the living condition. Once the incision has been made through the entire ribcage, by using large forceps or hemostats, gently retract the ribs apart from each other to expose underlying tissues. Do not pull too tightly at this stage, as that could crush the tissues of interest (which lie directly below this opening).

Now, the thoracic aorta will be clamped to perfuse only the upper thorax, neck, and head of the animal (*see* **Note 13**); however, if for scientific or other reasons the investigator chooses to perfuse the entire animal, then this next step can be avoided. Rotate the surgical tray 90° to the right so that now the animal is lying horizontally on the surgical table with head facing the right. Still by using the large scissors, at the bottom of the ribcage where the initial cut ended, make a longitudinal cut along the diaphragm toward the back on the left side of the animal. The left side of the animal is used in this step because access to the thoracic aorta is straightforward. Separate the ribs a bit to expose the lungs and gently move those aside to expose the underlying thoracic aorta. This appears as a whitish band running caudally on top of the vertebral column. Clamp the thoracic aorta with large hemostat and lay the hemostat to the side of the animal carcass and out of the way.

Next, rotate the surgical tray another 90° to the right (now the head of the animal is directed away from the surgeon), and by using large forceps, carefully retract the ribs to further expose the underlying carotid vasculature and the heart and aorta. At this point, the surgeon may need to make an additional cut through the ribs to gain visual access to the heart and aorta. If needed to adequately visualize the aortic root, carefully remove any fatty tissues lying on top of the heart. By using heart-holding or blunt forceps, gently grasp the heart and insert the over-the-needle catheter tip into the apex and advance it fully into the left ventricle (*see* **Note 5**). Retract *only* the needle leaving the catheter sheath indwelling and advance it straight-inward until it appears in the aortic root immediately as it leaves the heart. If resistance is met, pull back on the catheter and attempt advancing it to the aorta again. Once in place, gently attach the end of the catheter onto the luer-lock end of the two-way stopcock (at end of tubing) and carefully tape down the catheter end and/or tubing to keep it secure.

At this point, the surgeon needs to make sure that the animal carcass (along with the catheter in place and tubing attached) is at a point where perfusion will occur at MAP (*see* **Figs 1** and **2**). If following the steps used by the author, this will involve carefully moving the surgical tray containing the animal to the floor just below the perfusion-fixation apparatus. When the animal carcass is in place, by using small scissors, carefully cut the right atrium to make an outflow and carefully open the stopcocks to allow flow-through of *only* the warm PBS (or vasodilator solution) through the system. If adequate perfusion takes place, the surgeon will be able to immediately see blood leaving the incised right atrium and exiting the body (*see* **Note 15**). The color of the outflow will gradually become clearer as blood is removed from the upper thorax, neck, and head regions of the animal. Monitor this outflow to ensure adequate perfusion. Also, *importantly* watch the level of PBS (or dilator) in both the syringe and the tubing to make sure that air does not enter the animal (*see* **Note 16**).

When the fluid leaving the right atrium is clear and after an adequate volume of PBS (or dilator solution) has perfused the animal, carefully switch the three-way stopcock to now allow fixative to flow through the system and to enter the animal. Again, monitor the level of the fixative and make sure that air does not enter the animal. As the fixative flows through the body, sometimes (but not always) the muscles will twitch in tetany as indication that adequate tissue fixation is occurring. Once the tissues are completely fixed, "rigor" will set in the upper body and tissues will be hard. At this point, remove the catheter from the heart and place the animal carcass on an absorbant pad on the operating table for tissue harvesting (*see* **Note 17**). Quickly run liberal volumes of clean water along with disinfectant through both syringes and the entire perfusion-fixation apparatus including the catheter and needle and collect in a pan. The surgeon may want to use heparin on both the needle and the over-the-needle catheter to dissolve any existing blood clots that have occurred. The surgeon will now want to proceed with tissue harvesting, but at a convenient later time, clean all instruments and clean and sterilize the surgical area with alcohol.

3.2. Tissue Harvesting

3.2.1. For Perfusion-Fixed Tissues

The animal carcass should now be on the surgical table with the upper thorax, neck, and head completely fixed and with the carcass situated so that the head of the animal is directed toward the surgeon. The author suggests wearing a face mask at this point. By using forceps and small scissors, remove all tissue and muscular fascia surrounding the left carotid artery (LCA) and expose the entire left carotid vasculature from the bifurcation to the aorta. Carefully free the LCA from all adjacent tissues and especially the underlying connective fascia. By using fine scissors, cut the distal end of the carotid artery at the bifurcation, gently lift it up, and separate the entire length of the artery from extraneous tissues in a caudal direction to the aorta. Cut the carotid artery from the aorta and place it in a Petri dish filled with fixative, making sure to *maintain proper orientation* of the vessel as it was inside the animal. It helps to have a small piece of gauze soaked in fixative inside the Petri dish. Gently clean extraneous tissue still attached to the vessel and make sure that the lumen is free of blood clots. If blood components are still resident inside the lumen, *gently* flush the lumen with fixative (through syringe and needle) to remove these. To ensure proper vessel orientation throughout the processing steps, at this point attach a small loop of suture around the most distal end of the artery and make a note of this. Place the LCA in an appropriately labeled vial containing the same fixative, incubate (post-fix) for an additional 4 h (*see* **Note 18**), and then transfer the tissues to vials containing 70% alcohol until ready for processing. Next, perform the exact same protocol for the right carotid artery (RCA). Once vessels are in fixative, make all relevant notes about the time of incubation of the samples in fixative, the efficiency/efficacy of the perfusion-fixation protocol, or any other unique observations that might impact the results and/or data interpretation. Place the animal carcass in a necropsy bag, close securely, and place in an appropriate freezer or other container according to institutional guidelines.

The author will now discuss a method that can be used for tissue harvesting if the investigator has incorporated a protocol whereby a certain treatment has been applied to the exposed distal portion of the LCA vasculature immediately following balloon injury (of course, this is based on specifics of the research design but is a commonly used method). In this case, the entire length of the LCA will have been injured, yet only the distal portion (approximately half) will have been treated with a drug, blocking antibody, or other agent (*see* **Fig. 3**) *(4)*. During harvesting, proceed as usual until the entire length of the common carotid is exposed and freed from underlying and adjacent tissues. Remove the entire length of the LCA and place in a Petri dish filled with fixative making sure to maintain proper orientation of the vessel. Once the vessel is cleaned, cut the vessel approximately in half and tie a suture around the distal end of each vessel section. Place each vessel section into individually labeled tubes (labeled "distal LCA" or "proximal LCA" or similar) containing fixative, post-fix for 4 h (*see* **Note 18**), and then place tissues in 70% alcohol until ready for processing. Harvest the RCA as routine.

3.2.2. For Fresh Tissues

When using specific antibodies for immunostaining that work optimally on fresh tissues or when performing certain protocols such as Western blotting,

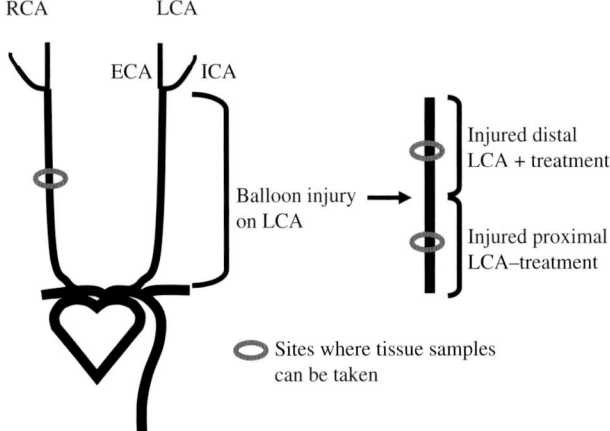

Fig. 3. A scheme for treating the balloon-injured distal section of a carotid artery for comparison with the injured untreated proximal section from the same animal (redrawn from **ref.** *4*). The entire length of the left carotid artery (LCA) is balloon injured, yet only the distal portion is subjected to a treatment of choice (as determined by the investigator). The right carotid artery (RCA) serves as an uninjured intra-animal control. This scheme can be used for designing appropriate research experiments as well as for histological procedures utilizing inter-animal comparisons. ECA, external carotid artery; ICA, internal carotid artery.

fresh unfixed tissues must be removed from the animal expediently and snap-frozen. Two protocols are pertinent here, one using fresh tissues for cryostat sectioning and for making slides and one utilizing fresh tissues for Western blotting or similar experiments that incorporate whole unsectioned tissues.

Prior to performing this method, collect ice in a small cooler and pack it to the top. Lay a Petri dish on the ice, place a small piece of gauze inside the dish, and fill with PBS. Fill a container with liquid nitrogen and cover. If liquid nitrogen is not available, then partially fill a beaker with methanol, carefully add chunks of dry ice, and allow time for the liquid to cool. Label vials accordingly using an alcohol-proof marker (or label the inside of the vial top and/or mark accordingly with tape). Once the animal is completely sedated and placed supine on a surgical tray with legs retracted and with the head toward the surgeon, working quickly cut from just above the sternum through the ribs to the diaphragm (with scissor tips up) and expose the carotid and aortic vasculature and the heart (the same as performed for perfusion-fixation). Carefully separate the ribs to expose the underlying tissues. Control bleeding with ample use of gauze and/or cotton swabs. By using blunt dissection, carefully remove the overlying muscles and all adjacent tissues to expose and isolate the LCA. Cut the most distal end of the carotid artery at the bifurcation and hold the cut end gently with fine forceps. By delicately holding the end of the vessel with the forceps, cut the remainder of the carotid artery free from associated tissues caudally all the way to the aortic arch and remove the vessel from the animal. Immediately place the artery in ice-cold PBS (in the Petri dish on ice) and promptly clean the artery of blood and extraneous tissue. If tissue is to be used for cryostat sectioning and to make slides, then tie a suture around the distal end of the artery, hold the artery upright inside an embedding block (or ring) situated in a stainless-steel base mold on sitting on dry ice in a cooler, and fill the embedding block with tissue-freezing medium until the vessel is covered. Keep the block on dry ice until completely frozen and then transport to a deep freezer ($-80\,^{\circ}$C). If the research plan calls for separate analyses of the distal and proximal sections of the carotid artery (if the distal end has been treated), then similar steps must be taken to appropriately define the individual sections and to maintain their proper orientation. If the tissue is to be used for Western blotting or another protocol that requires fresh whole tissue, once the vessel is clean simply place it in a pre-labeled empty vial and drop it into liquid nitrogen (or ethanol solution with dry ice) to ensure instant freezing. These procedures do not require placing suture around the vessel as the entire vessel section will be used in the ensuing method. Transfer the vials containing vessels to deep freeze until ready for use in a specific protocol. These steps are repeated for the RCA. Follow-up procedures include collecting and thoroughly cleaning all instruments, cleaning and sterilizing the surgery area with 70% alcohol, making relevant notes on the technique and any other important observations, and proper disposal of the animal carcass according to institutional guidelines.

3.3. Tissue Processing

For cryostat sectioning as the frozen tissues will go directly from harvesting into embedding in an appropriate freezing medium, no steps are included here specific for processing; therefore, the following descriptions are relevant to perfusion-fixed samples only. After the tissues have been fixed, they must be dehydrated and cleared in order for adequate paraffin infiltration, embedding, and sectioning. This step is termed "tissue processing" and involves tissues incubated in a series of graded alcohols, usually 70 through 100%, to remove water. A first step in 30% alcohol can be used for delicate, highly valuable, or susceptible animal tissues. The next step is clearing which removes traces of dehydrant alcohol and uses an agent miscible with the embedding medium.

Xylene is the most common clearing agent used for tissue processing for these tissues (*see* **Note 6**). The final step of processing is infiltration of the tissues in liquid paraffin or other embedding medium (*see* **Note 7**). For protocols involving rat vascular tissues, it is recommended that automated (mechanical) processing be used to ensure reproducibility and consistency from sample to sample. Automated processing involves movement of tissues (in tissue cassettes) through graded alcohols, xylenes, and liquid paraffin on a pre-set time scale. Newer processors have the ability to treat the samples with vacuum and/or heat during processing. These machines also allow individual protocols to be entered based on the desired incubation time in each of the solutions.

Processing of rat carotid arterial tissues starts following in situ fixation, post-fixation (4 h), and storage of the samples in 70% alcohol at room temperature. At this point, the orientation or the vessel will have been ensured with suture placement around the distal end. By using fine forceps, carefully remove the vessel from 70% alcohol and place it directly in a pre-labeled (with pencil) embedding cassette and snap the lid closed (*see* **Note 19**). A 12-step processing procedure is employed for these tissues suitable for either manual or automated processing (*see* **Table 1**). If manual processing is used, then the investigator must first prepare all solutions in glass-staining dishes (with lids). Volatile solutions should be placed in an appropriate fume hood. For automated processing, the timing clock on the apparatus must first be set according to the time schedule for incubations. The following incubation schedule has been used successfully by the author for a variety of animal vascular tissues including rat carotid arteries; however, individual variations (within reason) can be used with anticipated success. Of note, *all* incubation steps are performed under vacuum and pressure and at room temperature *except* the final two steps (liquid paraffin) that are performed under elevated temperature. The author routinely performs tissue processing during the overnight hours with the sample incubating in liquid paraffin (step 12) and ready for embedding the next morning.

3.4. Embedding

Frozen tissues will have been harvested fresh from the animal, cleaned while immersed in ice-cold PBS, suture tied around the distal end to ensure orientation, and quickly embedded in an appropriate medium, frozen, and ready for cryostat sectioning. Generally, frozen tissues are embedded in an unfixed state; however, certain procedures require use of a post-fixation step (generally

Table 1
A recommended 12-step tissue processing protocol for use in animal vascular injury studies. All incubation steps occur under vacuum and pressure and, unless otherwise noted, are performed at room temperature. The total time for processing according to this protocol is >13 hours which makes it ideal for overnight incubation.

Processing step	Bath	Time (min)
1	50% ethanol	120–180
2	80% ethanol	25
3	80% ethanol	25
4	95% ethanol	40
5	95% ethanol	40
6	100% ethanol	40
7	100% ethanol	40
8	100% ethanol	55
9	Xylene	60
10	Xylene	90
11	Paraffin, liquid (at 58–60 °C)	90
12	Paraffin, liquid (at 58–60 °C)	> 90

formol-calcium) to reduce chance of diffusion of labile substances (*1*). Paraffin-embedded tissues in holding cassettes will be immersed in liquid paraffin in the final stage of processing. The use of a paraffin embedder with a cold plate is highly suggested for this step, although manual embedding can be performed using melted paraffin in a suitable container over a hot plate and an ice bucket or other cold container for wax hardening.

When using a paraffin embedder, make sure that the embedder is pre-heated (at least 3° above the melting point of the wax), that all the paraffin pellets are in liquified form, and that the cold plate is pre-chilled. Run a test-embedding block to make sure that the machine is fully operational and that the paraffin hardens after a short time on the cold plate. Use of a forceps heater during the embedding procedure is highly recommended and will ease manipulation of the tissue sections. Remove the cassette from the processor (in liquid paraffin) and place it on the heating block of the embedder and gently open. Carefully remove the vessel by using fine forceps and place the vessel alone on the heating block (to keep the wax melted). Place a pre-labeled embedding block or ring in a stainless-steel base mold and also place this on the heating block. Now, with one hand hold the embedding block and base mold under the paraffin dispenser and with forceps in the other hand carefully pick up the vessel so that it can be embedded in proper orientation (according to the research design) inside the embedding block. The author normally embeds vascular tissues with the suture-end down first, thereby during sectioning the samples are taken first from the most distal portion of the vessel continuing in a proximal direction. Holding the vessel absolutely still in correct orientation and perfectly vertical (only if vessel cross-sections are desired; *see* **Note 20**) inside the embedding block, fill the block with liquid paraffin around the vessel, release the vessel from the forceps, and carefully move the filled block to the cold plate. Be careful when releasing the vessel from the forceps as the wax can cause them to stick together and move the vessel out of its vertical position. In short time, the block will start to harden. Once the paraffin is completely solidified, move the block from the cold plate, release the stainless-steel base mold, and store the block at room temperature until ready for sectioning.

3.5. Sectioning

3.5.1. For Paraffin-Embedded Tissues

After tissues have been embedded in paraffin and hardened, sections must be made and placed on a specimen slide for microscopic inspection and analysis. Prior to this step, make sure to pre-heat ($\sim$10 °C below the melting point of the wax) a flotation bath containing DEPC-treated water or other suitable agent (*see* **Note 8**). DEPC derivatizes histidine residues and is therefore an effective method to inactivate nucleases including RNAse. Have the flotation bath located adjacent to the microtome and next to a supply of microscope slides. Sectioning of paraffin-embedded tissues is performed with a microtome that allows precisely thin and standardized tissue cross-sections to be cut in repetitive manner. Sections in paraffin can be cut between 3 and 10 µm in thickness but usually range between 5 and 8 µm. The author uses 5-µm sections for rat carotid arteries to be used in standard histology procedures. Keeping a very sharp microtome knife blade is essential for proper cutting (*see* **Note 21**). First, place the embedding block on the specimen holder of the microtome with the vessel to be cut directed toward the investigator. Set the desired thickness of the section to be cut on the microtome control. Slowly rotate the microtome handle, and the tissue block will advance the desired distance (5 µm) with each rotation. Prior to the first cut, *make sure* that the arm holding the embedding block has not advanced a distance that will cut the entire embedding block in half. To be sure to avoid this, make sure that the arm is retracted a sufficient distance *prior* to placing the embedding block in it and *prior* to rotating the microtome handle. With each microtome cut, the block will move down and over the immovable microtome blade, and a thin section of wax (containing the tissue) will be placed on the blade itself. With adequate training and experience, "ribbons" or "sheets" of adjacent serial paraffin sections can be cut from a single block. Next, by using forceps or a small brush, carefully move the cut paraffin sections to the heated flotation bath nearby that will help to expand the wax and remove wrinkles or folds from the tissues. During this step, it is recommended that the trailing end of the wax ribbon or sheet make contact with the water first, thus producing a slight drag that will help remove these wrinkles and folds. Keep the wax sections floating on the water only for a short time (<30 s) before picking them up on the slides. Longer times spent floating on the heated water could drastically expand the wax and distort the tissue it contains. Several common problems associated with microtomy and tissue sectioning are described in **Note 22** as well as in available resources *(1,2)*.

3.5.2. Frozen Tissues

Frozen tissues are cut in a similar fashion and employ use of a cryostat, basically a microtome inside a refrigerated container. Temperatures within the cryostat normally range between −15 and −30 °C, thus keeping the sections frozen throughout the cutting procedure. In particular, cryostat sectioning of non-fatty unfixed tissues including rat carotid arteries works optimally at around −25 °C. Vascular tissues in the freezing embedding medium in the embedding blocks (prepared immediately following harvesting from the animal *post-mortem* with subsequent cleaning) are placed on the specimen holder of the cryostat, and the proper thickness is set on the control. Like microtomy, rotation of the cryostat handle will cut tissue sections at desired thickness. Tissues are *not* placed in a flotation bath while using the cryostat, but instead the cut tissue sections can be directly picked up on microscope slides, dried, and made ready for additional histological approaches.

3.6. Slide Preparation

At this step, paraffin-embedded sections are floating on DEPC-treated water in the flotation bath and are ready to be picked up on microscope slides. This procedure is straightforward, yet the investigator should attempt to capture the wax sections completely flat on the slide with the wax spread out evenly in all directions. Once the wax sections containing tissue have been picked up on the pre-labeled slides (*see* **Note 23**), blot excess water and rest the slide on an angle on absorbant towels until dry. Place the slides flat in a pre-warmed incubator or drying oven (with temperature set at melting point of the wax) or slide warmer for at least 30 min to help desiccate the section and to enhance adherence of the section to the slide. If heat might harm certain antigens essential for immunostaining, then avoid this step or simply lower the temperature of the incubator or slide warmer. For delicate or highly valuable tissues, it is recommended to incubate the slides at 37 °C overnight. Once dried, store the slides in an appropriate dry location until ready for further histological assays. Tissue sections cut on the cryostat are ready to be picked up by a microscope slide following sectioning. These should be rested on an angle on absorbant towels and, when completely dried, stored for further analyses.

3.7. Routine Staining Procedures for Histomorphometry

The slides are now ready to undergo one of a variety of histological staining procedures according to the research design and experimental endpoints. A listing of various histological stains useful for animal vascular studies is available *(3)*. In general, the initial steps of all staining procedures involve a reversal of the procedures taken through tissue processing and embedding. In other words, the paraffin wax that impregnated the tissue must now be extracted, and the tissue must be rehydrated to allow water-soluble dyes to penetrate the sections. So, before staining can be performed, the slides are de-paraffinized in xylenes and rehydrated through serial dilutions of ethanols (100, 95, 80, and 50%) to water. Once the wax is removed and the tissues hydrated, then protocols are followed pertinent to the stain desired. Once adequate staining is achieved, then the stained section on the microscope slide is cover-slipped with a thin glass or plastic sheet and with the use of an appropriate adhesive mounting medium. This protects and preserves the tissues and also provides enhanced optical resolution for microscopic viewing. Once the mounting media has dried, slides can be stored at room temperature for an extended period of time (*see* **Note 24**). Several commonly used staining procedures for rodent vascular tissues are briefly described here. These are *succinct and generalized* guidelines for the reader to follow, and the author recommends consulting more thorough references for details of these protocols before use (*see* **ref.** *1,2*).

Of all the dyes used for histomorphometry, a combination of hematoxylin and eosin (H&E) is the most common. Hematoxylin behaves like a basic dye and stains nucleic acids of the cell nucleus blue/black. However, hematoxylin itself is not a stain, and the major oxidization product is hematein, an anionic natural dye with color properties. Hematein, although, needs an enhancing agent that strengthens the tissue-dye bond in order for it to stain the tissue. This agent, termed a mordant, is a metal cation such as aluminum, iron, or tungsten. The type of mordant used strongly dictates the types of tissues stained and the final colored product. Eosin, an acidic xanthene dye, stains collagen pink and cellular cytoplasm red. A brief method for H&E staining of paraffin-embedded tissue sections is listed here based on a standardized protocol *(2)*.

3.7.1. Hematoxylin and Eosin Staining Method

1. De-paraffinize and rehydrate tissues to distilled water.
2. Remove fixative pigments (*see* **Note 25**).
3. Stain with a hematoxylin of choice for an appropriate length of time (*see* **Note 26**).
4. Wash in slow running tap water until sections become blue (<5 min) (*see* **Note 27**).
5. Differentiate in 1% acid alcohol for 5–10 s.
6. Wash in slow running tap water until sections again become blue (10–15 min) or dip slides into ammonia water and then wash in running tap water for 5 min (*see* **Note 27**).
7. Stain in 1% eosin for 10 min.
8. Wash in running tap water for 1–5 min (*see* **Note 27**).
9. Dehydrate through serial solutions of graded alcohols, clear with xylene, dry, and cover-slip.

Results from H&E staining are blue/black nuclei, red/pink cytoplasm, deep pink muscle fibers and fibrin, and red/orange red blood cells. An example of an H&E-stained rat balloon-injured carotid artery cross-section along with a close-up view in the inset is shown in **Fig. 4**.

The most commonly used and instructive staining procedure employed by the author for rat carotid artery balloon injury studies is a combination of a hematoxylin stain using an iron mordant along with an acid fuchsin/picric acid solution. This staining procedure is termed Verhoeff's elastic tissue stain with Van Gieson (VVG) counterstain and is used for routine histomorphometry and image analysis vital to animal vascular injury studies. Ferric chloride (10%) is included in the hematoxylin solution along with Lugol's iodine and a 2% ferric chloride differentiation step. Elastin fibers stain intensely black, and other cellular components are easily detectable, thus allowing for precise quantitation of routine morphometric parameters as discussed in **Subheading 3.8.**

3.7.2. Verhoeff's Elastic Tissue Stain with Van Gieson Counterstain (see **Note 28**)

1. Deparaffinize and rehydrate tissues to distilled water.
2. Stain for 20 min or until fibers are blue/black in Verhoeff's solution (*see* **Note 26**).
3. Wash in slow running warm tap water (*see* **Note 27**).
4. Differentiate in 2% ferric chloride until background is clear.
5. Wash in slow running warm tap water (*see* **Note 27**).
6. Place in 95% alcohol to remove iodine stain.
7. Wash in slow running warm tap water <5 min (*see* **Note 27**).
8. Counterstain in Van Gieson solution for 3–5 min.
9. Dehydrate through serial solutions of graded alcohols, clear with xylene, dry, and cover-slip.

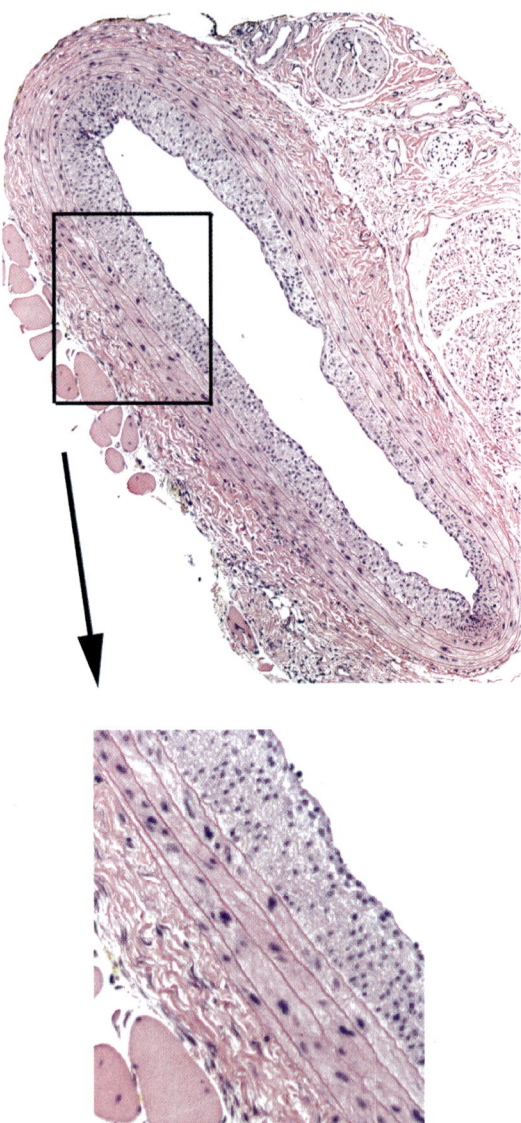

Fig. 4. A cross-section of a rat balloon-injured carotid artery 2 weeks post-injury stained with hematoxylin and eosin clearly depicts cellular-rich neointima development. Magnification is ×100 for the large photomicrograph.

Results from VVG staining are intensely blue/black elastic fibers, blue-to-black nuclei (based on variables for hematoxylin staining), and red collagen with associated tissues (fat and nerves) staining yellow. Several photomicrographs of VVG-stained cross-sections of a rat balloon-injured carotid arteries are shown in **Fig. 5** (*see* **Figs 6** and **7** in **Chapter 1** for additional VVG-stained photomicrographs of rat balloon-injured carotid arteries). **Figure 5** also illustrates several common problems associated with the rat carotid artery balloon injury model and associated histomorphometry.

Several highly used staining procedures preferentially color connective tissue in the vasculature, namely collagen, fibrin, and elastin. Connective tissue forms a scaffolding architecture around cellular components in the vessel wall and provides support to these tissues. Connective tissue can also house various cellular components including fibroblasts, mast cells, adipose cells, histiocytes, reticular cells, and bone cells *(1)*. A common technique used in analyzing collagen and extracellular matrix (ECM) components in rat carotid arteries is *Masson's trichrome stain*. This procedure colors nuclei blue/black, collagen and other sulfated muco-substances blue to blue/green, and cytoplasm, muscle, and various ECM components including fibrin red. An illustration of a Masson's trichrome-stained rat balloon-injured carotid artery is shown in **Fig. 6**, and a brief outline of this procedure is listed below *(1)*.

1. Deparaffinize and rehydrate tissues to distilled water.
2. Remove fixative pigments (*see* **Notes 25** and **29**).
3. Wash in slow running tap water.
4. Stain nuclei with celestine blue hematoxylin (*see* **Note 30**).
5. Differentiate with 1% acid alcohol.
6. Wash well in slow running tap water.
7. Stain with solution A for 5 min.
8. Rinse in distilled water.
9. Treat with solution B for 5 min.
10. Drain but do not rinse.
11. Stain with solution C for 3–5 min.
12. Rinse in distilled water.
13. Treat with 1% acetic acid for 2 min.
14. Dehydrate through serial solutions of graded alcohols, clear with xylene, dry, and cover-slip.

3.8. Microscopic and Morphological Evaluation

Following detailed and well-differentiated staining of vascular tissues, they are now ready to be carefully examined, qualified, and quantified under microscopy. Generally speaking, most of the histomorphometry protocols used

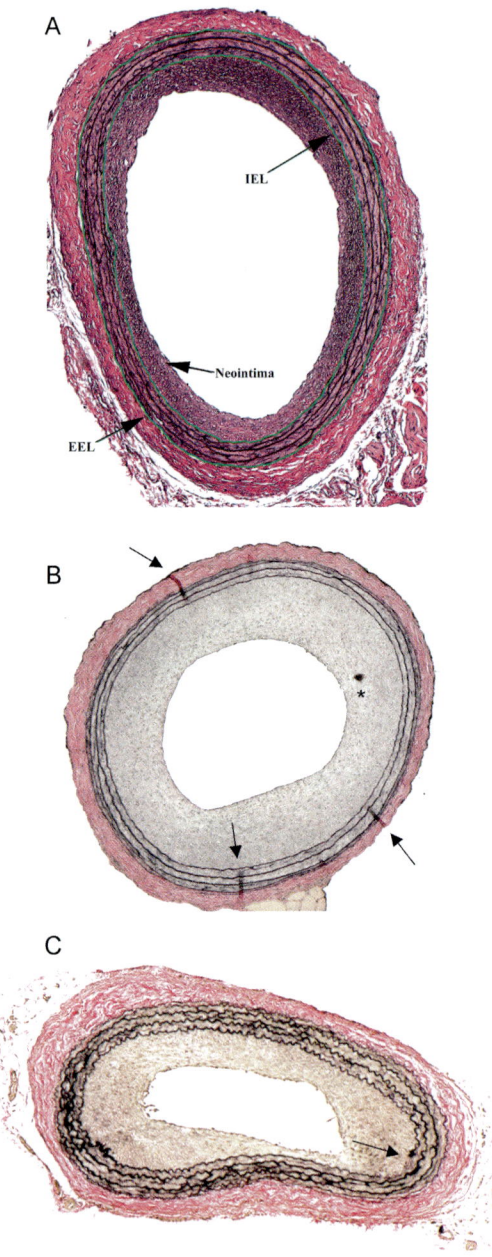

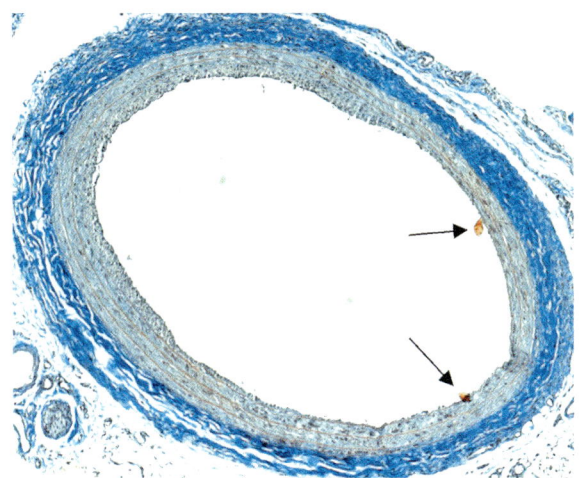

Fig. 6. A cross-section of a rat balloon-injured carotid artery 2 weeks post-injury stained with Masson's trichrome technique. Robust adventitial collagen is stained deep blue. Several histological artifacts (possibly floating tissue debris) are shown (arrows). Note the minimal neointima development (compared with **Figs 4** and **5**), implying an impaired injury response in this animal. Magnification is ×100.

for routine examination of rat arterial tissues use transillumination under light microscopy. Phase contrast, interference, polarized light, fluorescence, electron, and confocal microscopy are alternate more advanced means by which the evaluation of prepared tissue samples can take place. These procedures are highly dependent upon an operating image analysis system, complete with a

◀──

Fig. 5. Variations in Verhoeff–Van Gieson-stained cross-sections of rat balloon-injured carotid arteries 2 weeks post-injury. Panel **A** depicts a concentric, elastin-rich neointima with the internal elastic lamina (IEL) and external elastic lamina (EEL) digitally traced and clearly defined. A hearty adventitial layer is clearly present. Panel **B** shows a robust and concentric neointima containing artifacts (tissue folds and tissue debris) from improper histology (most likely problems associated with sectioning and/or slide preparation). Arrows indicate tissue folds; asterisk indicates tissue debris deposited on the slide. Panel **C** illustrates a non-concentric neointima (diminished in lower right corner) as a result of improper ballooning during surgery. Highly corrugated IEL and EEL suggest that perfuse-fixation was not performed at an adequate distending pressure of the animal. Also evident is a break in the elastin staining of the IEL (indicated by arrow). Magnification is ×100 for all photomicrographs.

A   B   C   D

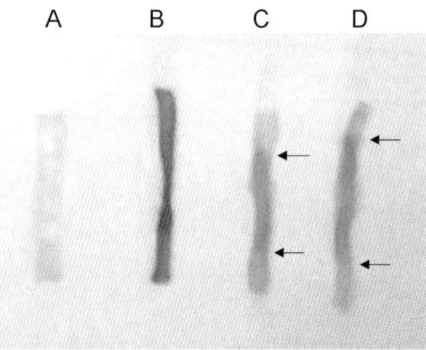

Fig. 7. Photographs of rat balloon-injured carotid arteries at various times post-injury treated with Evans blue dye (redrawn from **ref**. **7**). Tissues were treated with Evans blue (0.5 ml of a 5% solution) in situ 10 min prior to sacrifice, perfusion-fixed, harvested intact, split longitudinally, and pinned out on a silicon-padded dish for analysis. Panel **A** shows a contralateral uninjured right carotid artery 2 weeks following injury on the left carotid artery (LCA). Absence of Evans blue staining indicates an intact endothelial layer. Panel **B** shows an injured LCA 30 min post-injury, and complete loss of the endothelial lining is indicated by complete Evans blue staining of the sub-endothelial matrix along the entire length of the vessel. Panels **C** and **D** illustrate two carotid arteries with partial endothelial regrowth from the border zones 2 weeks post-injury. White (unstained) regions at the proximal and distal ends of these vessels are indicated by arrows and suggest that endothelial cells have partially regenerated into the central injured section at this time point.

light microscope, attached camera with an interface to a computer, and an image analysis software program. The basic goal of image analysis is to generate data, which describes aspects of the tissue specimen on the microscope slide. The objective here is to describe, in terms of both qualitative characteristics and precise quantitative measurements, anatomical parameters of each vessel cross-section. Specific parameters suitable for analysis for rat carotid artery injury studies include the perimeters (lengths) of the lumen, the internal elastic lamina (IEL), and the external elastic lamina (EEL); areas *inside* the lumen, IEL, and EEL (these are all measured directly); and areas *of* the medial wall and neointima (these are calculated). Additionally, thickness of the neointima and/or medial wall can be measured directly if so desired as an additional means to estimate arterial remodeling. These parameters can also be quantitated for the adventitial layer if desired.

For standard morphologic assessment of stained slides under microscopy, the author primarily employs VVG-stained cover-slipped sections. The microscope to be used should be prepared and "calibrated" ahead of time to center the light path and to achieve the best transillumination of the specimen with optimal resolution. The eyepieces should be adjusted to compensate for variation between individual users as well. Calibration of the image analysis program must also be completed ahead of time based on the instructions for each particular software program. This usually involves use of a stage micrometer to calibrate distances and lengths and is dependent upon the magnification of each separate objective used. By using an ultra-fine point marker, make a mark around each vessel directly on the cover-slip of each microscope slide to simplify location when the slide is on the microscope stage and under an objective. Initially, by using a low-powered objective, locate the vessel cross-section and then move to progressively higher powered objectives until suitable magnification is achieved. The author performs most of his analyses using a ×10 objective (total magnification ×100) for gross vessel inspection and a ×40 objective (total magnification ×400) for more detailed examination. The author also utilizes an oil immersion ×100 objective (total magnification ×1000) for meticulous scrutiny of stained specimens. First, it is recommended to scan all vessel cross-sections in a given batch to rule out specimens that cannot be used for a variety of reasons (broken or ruffled tissues, inadequately stained sections, artifacts that could directly hinder inter-pretation, etc.; *see* **Notes 20, 22, 25**, and **29**). Make detailed notes regarding all specimens that are removed from experimental analysis and the reason(s) why they will not be included. Digitally capture and save all remaining images ("image acquisition") that will be used for qualification/quantitation and locate these in appropriate file(s) on the computer. Once these images have been saved, then proceed with specimen qualification and/or data quantitation of the morpholo-gical variables according to the instructions pertinent to the software of choice.

For tissue specimen analyses, both qualitative subjective interpretation and quantitative objective measurements should be performed. Notes should be made for characteristics of each vessel cross-section including the nature of the neointima (concentricity, degree, consistency, elastic versus cellular compo-sition, staining intensity of a particular dye, integrity of luminal lining, etc.), the medial wall (disrupted laminal layers, broken or fragmented tissue, and staining), and the adventitia as well as the presence of any artifacts that might be present (*see* **Notes 20, 22, 25**, and **29**). As mentioned, specific quantifiable parameters suitable for analysis of rat arterial tissues include perimeters of and areas inside the lumen, IEL, and EEL. Conversely, if only perimeter lengths are measured, then areas can be calculated: area $= \pi[(\text{perimeter}/2\pi)^2]$ (**5**). Based on these data, parameters for the medial wall and neointima can then be directly determined (i.e., area inside the

EEL – area inside the IEL = medial wall area; area inside the IEL – lumen area = neointimal area). Measurements of neointimal, medial, and/or adventitial thickness can also be performed. As noted, the exact subjective and/or objective measurements to be performed for each group of tissues are entirely dependent upon the goals of the research study and are at the discretion of the investigator(s).

3.9. Endothelial Cell Regeneration

This section stands alone as a separate protocol that can be used to measure the extent of endothelial cell (EC) regrowth from the border zones of the injured area (to estimate endothelial regeneration) and can also serve as an indication of the success of the injury in the rat carotid artery if performed shortly after injury *(5–7)*. This method utilizes pre-treatment of the animal with Evans blue dye, a sub-endothelial matrix-specific stain and basement membrane-specific stain, a short time prior to sacrifice (10–60 min) followed by perfusion-fixation of the animal as described.

Prepare the Evans blue dye ahead of time according to concentration desired (*see* **Note 9**). Intravenous injections in rat tail vessels (or other suitable site for vascular access if so desired) are required for this protocol, so assistance from an individual experienced in performing such techniques is recommended. At the desired time for the examination of EC function (as determined by the investigator), anesthetize the animal as routine. Once a surgical plane of anesthesia is achieved, immerse the entire tail of the rat (all the way up to the testis) in very hot water for several minutes. This serves to vasodilate the tail vein making intravenous access easier. Quickly dry the tail and lay the animal supine on the operating table. The area for injection will appear darkened underneath the skin as the vein has become dilated. With a slight angle to a needle (attached to a syringe containing Evans blue dye), make an injection through the skin into the tail vein and gently withdraw the plunger to make sure venous access is achieved (blood will draw into the syringe). *Slowly* inject the desired quantity of Evans blue dye (*see* **Note 9**). The entire body of the animal will gradually turn a dark blue, thus ensuring success in the administration. After the desired time (10–60 min following injection), perfuse-fix the animal as described in Section 3.1. Immediately following harvesting of the *entire lengths* of both the LCA (injured) and the RCA all the way from the bifurcation to the aortic arch (for the left carotid) and from the bifurcation to the innominate artery (for the right carotid), maintaining proper orientation lay them on a dissecting dish (coated with silicone and containing a small volume of fixative) and cut each in half *longitudinally* by using fine forceps and microscissors. Pin out the entire lengths of the cut arteries on the silicone pad and perform quantitation of the following areas: unstained proximal and distal portions (indicating an intact endothelial layer), the stained central portion of the vessel (the denuded section), and the entire vessel. **Figure 7** illustrates results from an Evans blue EC regrowth assay on rat carotid arteries at both acute and 2-week time points (redrawn from ref. 7).

4. Conclusions

For animal vascular injury studies that employ histologic and morphometric analyses, investigators must first prepare their work plan according to their desired research goals keeping in mind specifics and individual requirements for tissue fixation, processing, embedding, sectioning, staining, and microscopy. Each of these inherent steps has many variables that can directly impact the success of otherwise routine procedures. It is suggested that the investigators utilize control samples of tissue to optimize these protocols for any given project before using valuable experimental specimens. The rat carotid artery balloon injury model is a useful procedure that allows investigation into many cellular, molecular, physical, and chemical mechanisms involved in the vascular injury response. Success of this model, however, is largely dependent upon consistent, reproducible, and scientifically accurate histomorphometry, with the ultimate goal of discerning valuable scientific data from control and/or treated animal tissues. The background, protocols, and special considerations included in this chapter aim to provide a solid basis for successful and routine performance of many of these procedures; nevertheless, individual preferences, practical experiences, and research goals may dictate alternate courses of action.

5. Notes

1. An anesthetic to sedate the animal prior to sacrifice should be chosen consistent with the scientific goals of the experimental plan and in accordance with the guidelines of the institutional animal care and use committee. The choice of anesthetic should not cause physiological interference with specific endpoints in the study as determined by the investigator(s).
2. A variety of fixatives can be used as determined by the investigator(s) that account for differences in tissue architecture and composition. As in **Note 1**, these should not interfere with specific endpoints that will be analyzed in the study. Fixatives can generally be classified as *coagulative* or *cross-linking*. Coagulative fixatives work by establishing a penetrable network in tissues thus allowing for the penetration of processing solutions and embedding media. These include mercuric chloride and picric acid (a common reagent in many stains and dyes). Cross-linking fixatives

operate by linking and stabilizing proteins in the tissue, thus creating a "gel" that limits denaturation of the tissue and enhances solution penetration. These include the commonly used fixatives formaldehyde, glutaraldehyde, and potassium dichromate. A comprehensive list of suitable fixatives for use in various animal vascular studies is available *(3)*. Good choices for general fixatives for use in histomorphometry in rat arterial injury studies are formaldehyde or 10% neutral-buffered formalin (pH 6.8). These agents fix by cross-linking proteins, particularly lysine residues, but do so in a manner that maintains antigenicity and therefore are consistent with immunostaining techniques. These agents work slowly but penetrate tissues well and are recommended for use by the author. Zinc formalin (formaldehyde) maintains good tissue morphometry and preserves epitopes for prolonged periods of time, thus allowing immunostaining protocols to be performed at a later date. Zinc formalin also eliminates the need for laborious and risky antigen recovery protocols. A 2% solution of buffered glutaraldehyde penetrates tissues poorly but fixes them quickly, gives good nuclear and cytoplasmic differentiation, and can be used for electron microscopy analysis; however, glutaraldehyde causes deformation of α-helix protein structures and is not recommended for immunostaining protocols. Alternate choices for fixatives include mercurials such as B-5 or Zenker's solution, alcohols, picrates including Bouin's solution, or oxidizing agents.

3. A number of factors will affect the fixation process for vascular tissues. Fixation is optimized at *neutral pH* between 6 and 8. Prolonged duration of surgical techniques or other interventions that cause tissue hypoxia prior to fixation will lower the pH of those tissues, so adequate buffering capacity of the fixative will prevent excessive acidity. *Penetration* of tissues depends on the diffusibility for each individual fixative and on the thickness of the tissue to be fixed. Formalin and alcohols penetrate the best while glutaraldehyde penetrates poorly. The *ratio of volume of fixative to tissue weight* is important for optimum fixation and is recommended at 10:1 [although 20:1 has also been suggested *(3)*]. If large amounts of tissues are being fixed, then replacing used or "de-fixed" fixative with new solution will minimize the volume needed. Mild agitation of the tissue in fixative will also enhance fixation. *Temperature* of the preparation is directly proportional to the degree of fixation as long as the increasing temperatures do not interfere or disrupt the tissue specimens. Heated formalin is often the initial step in automated tissue-processing procedures. The *concentration* of the fixative should be adjusted to the lowest level achievable where fixation is still possible, as higher concentrations can adversely affect tissues. Examples include formalin at 10% and glutaraldehyde between 0.25 and 4%. Finally, the *duration of time* between obtaining the tissue and fixation is very important. Fresh tissues will fix better than ones that have dried too much following removal from the carcass. With increasing time away from the carcass, artifacts can be introduced, cellular organelles and molecular/cellular signals can be lost, epitopes can be masked, and

nuclear shrinkage and condensation can occur. These potential biases are controlled for when using in situ perfusion-fixation procedures included here as long as the time from sacrifice to perfusing the tissues with fixative is kept brief.

4. To remove blood and other circulating factors from the tissues of interest, use of a warmed saline (PBS) solution perfused at MAP is recommended. Alternately, if the investigator chooses to dilate the vasculature, a suitable vasodilator solution (that does not interfere with the scientific questions being asked) can be used for perfusion. In the laboratory of the author, warmed PBS solution perfused at MAP completely removes blood from the vessel lumen and completely dilates the carotid arteries (noticed as a "non-corrugated" appearance of the elastic laminae).

5. The author recommends using an intravascular over-the-needle Teflon catheter (20 gauge × 2 in. Terumo Surflo I.V. catheter, National Health Resources, LLC) for trans-cardial perfusion-fixation instead of a similar-sized needle for a very important reason. By using only a needle, the surgeon takes a chance that once cardiac puncture is achieved, the sharp bevel of the needle will advance without resistance through any tissues it encounters. Based on the angle and direction of the needle being inserted, this can often miss the aortic valve and aorta and puncture through adjacent areas of the base of the heart, likely the left atrium. When this happens, the needle must be withdrawn and additional attempts made to advance the needle directly into the aorta. However, this will leave puncture holes in the heart through which the perfusion fluids will travel, thus decreasing the pressure and the amount of fluid perfusing the carotid vasculature. By using an over-the-needle catheter, once cardiac puncture is achieved, the surgeon carefully withdraws the needle leaving the flexible catheter in place inside the heart. This then can be advanced easily through the aortic valve and into the aorta. If resistance is met with the catheter tip, it will not penetrate that tissue and can be withdrawn slightly and insertion into the aorta can be attempted again. Once the catheter is inserted into the aorta (evidenced visually through the aortic wall), it can be attached to the luer-lock end of the two-way stopcock and perfusion can continue.

6. Clearing of tissues is vital for adequate removal of remnant dehydrant (alcohol, acetone, etc.) and for proper infiltration of the embedding medium into the tissue. The most common clearing agent for vascular tissues is xylene, but toluene also works well but may be cost-limited. Other clearing agents that could be used for these tissues include chloroform (slow action and health hazard), methyl salicylate (oil of wintergreen and expensive), limolene derivatives, or complexes of long-chain aliphatic hydrocarbons (less forgiving with poorly fixed, severely dehydrated, or sectioned tissues).

7. The most common embedding agent used for histology is paraffin, a waxy crystalline substance composed of a complex mixture of hydrocarbons with a density similar to most normal tissue. Paraffins can differ in their melting points (40–70 °C), a consideration for hardness in relation to tissue type (a high melting point makes a harder paraffin block than one with a low melting point). In general,

paraffin is inexpensive, easy to handle, and straightforward in terms of embedding and sectioning. During processing, vacuum and pressure are applied to the tissues to enhance penetration of the paraffin-embedding medium. Several commonly used paraffins include TissuePrep, Paramat, or Paraplast (contains plasticizers for ease in sectioning). Other embedding agents that could be used include various plastic resins (methyl or glycol methacrylate, araldite, and epon). These agents solidify particularly hard and are especially useful for cutting very thin sections; however, they require specialized (i.e., expensive) reagents for dehydration and clearing and specialized microtomy procedures. For practical purposes, use of plastic resins for embedding may be "overkill" for routine rodent vascular studies.

8. Alcohol, small quantities of detergent, or other suitable agents can be added to the flotation bath to reduce surface tension and to allow the cut wax sections to flatten out with ease.

9. Various concentrations and/or doses of Evans blue dye are reported in the literature for performing EC regeneration assays in the rat balloon-injured carotid artery. Original articles reported 50–60 mg/kg dose of Evans blue dye given intravenously 30–60 min prior to perfusion-fixation *(5,6)*, yet the author has successfully used 0.5 ml of a 5% Evans blue solution in saline 10 min prior to sacrifice (*see* **Fig. 7**) *(7)*.

10. For purposes of the perfusion-fixation protocol, in standard normotensive laboratory rats MAP is estimated to be approximately 100 mmHg. A measured value for rat diastolic blood pressure (DBP) is 90 mmHg, whereas a range for rat systolic blood pressure (SBP) is 116–180 mmHg *(8)*. In this protocol, perfusion is performed at approximate MAP to replicate distending pressure under normal conditions and to fix the tissues at this normal pressure. If hypertensive animals are to be used and if the investigator(s) wishes to perfuse at their respective elevated pressures, or if for some reason the investigator chooses to perfuse at a pressure different than the MAP, then the calculation of MAP on an animal-to-animal basis should be performed $\{MAP = [(2 \times DBP) + SBP]/3\}$.

11. The author chooses to set up this apparatus on a laboratory bench top with the tubing draped to the floor (for a total height of ~53.5 inches; *see* **Figs 1** and **2**). In this manner, when the perfusion step is conducted, the investigator places the animal carcass on the floor (in a surgical pan) to maintain columns of fluid equivalent to an MAP of 100 mmHg. However, if the investigator wishes this apparatus can be set up on a top shelf in a laboratory (with the animal carcass on a lower bench top), whereby columns of fluid at the appropriate heights can still be maintained for perfusion at MAP.

12. Be cautious with the three-way stopcock during opening and closing. When the switch in a three-way stopcock is turned in a specific direction, both columns of fluid will become open to each other and the levels of fluid in each section of tubing will equilibrate. This will result in mixing of the fixative and PBS (or dilator solution) and will render those solutions useless. Practice with opening and

closing the three-way stopcock to ascertain proper switching to open *either* the PBS *or* the fixative (but *not* both at the same time).

13. These volumes of fluid are suggested when performing a perfusion-fixation protocol for *only* the carotid vasculature as outlined in this chapter. This method involves clamping the thoracic aorta to prevent perfusion of the lower body, and thus only the upper thorax, neck, and head regions are perfused in this protocol. If an investigator chooses to perfuse the entire body, then larger volumes of fluid will be needed (estimated ~120–150 ml) to achieve adequate perfusion and fixation of all tissues.

14. In this step, it is imperative that PBS (or vasodilator solution) flow through the tissues first, followed by fixative. This will result in adequate removal of blood from the vasculature and will keep the vessels expanded at approximate MAP. Followed immediately by fixative, the vessels will be fixed at an expanded open caliber at MAP without interference of blood components in the lumen. To achieve this, make sure that the fluid in the short section of "common" tubing is filled with PBS (or vasodilator solution) and that the first stopcock that will be opened for perfusion allows *only* PBS to flow through the system, followed then by fixative.

15. If adequate perfusion does not take place (estimated by lack of flow through the system and/or improper outflow), the surgeon needs to re-evaluate the adequacy of the catheter insertion to make sure that it is inside the aortic lumen and also to make sure that the tip of the catheter is not pushed up against the inside wall of the aorta, thus preventing fluid movement through it. The surgeon should also check the outflow incision in the right atrium, and if this is not adequate, make additional cuts in the tissue to increase the area for outflow. One other factor that could exist to reduce or prevent outflow is the formation of a large thrombus at the incision site on the right atrium. Gently swab the area around the right atrium to remove any clots possibly located there.

16. If air bubbles enter the vasculature of the animal during perfusion, they can get lodged at certain sites that could cause problems during the sectioning procedures. More often than not, air bubbles trapped in the carotid arteries will be removed during processing; however, if air should remain trapped inside the lumen of a carotid artery, then during sectioning this could cause crushing or disintegration of the tissue by the microtome blade.

17. Immediately following perfusion-fixation, gently remove the carcass from the surgical tray and place nearby on an absorbent pad or towels. Carefully empty contents of the surgical tray (mostly PBS or vasodilator solution, fixative, and blood) into an appropriately labeled waste container and cap the container. Wash the tray to remove residual fixative.

18. It has been suggested by others that to ensure complete fixation for animal vascular tissues, excised specimens should be post-fixed for at least 24 h prior to transferring the samples to 70% alcohol *(3)*. This author routinely post-fixes rat carotid artery samples for 4 h with success; however, longer post-fixation times can be used at

the discretion of the investigator(s). The investigator(s) although must be aware of potential problems associated with prolonged fixation times *(1)*.

19. For a routine rat carotid artery sample, it can be placed directly in a pre-labeled embedding cassette for processing. However, if the investigator wishes (especially important for small samples like mouse vessels), the sample can be first placed inside a fine porous paper (Kimwipe) that is folded several times and then placed inside the cassette.

20. If vessel cross-sections are desired to be analyzed under microscopy, then it is imperative that the vessel be in perfect vertical orientation during the embedding procedure. If the vessel is embedded slanted or otherwise not vertical, then during microtome cutting the vessel cross-sections will appear skewed and asymmetrical. These *cannot* be correctly analyzed for morphometry because of inherent biases in the tissue. If this is noted for a particular embedded vessel, then the investigator must re-embed the tissue and prepare a new block. To do this, first lay the embedding block (containing the vessel) on the heating block of the paraffin embedder to melt away most of the wax. When the wax has sufficiently melted away from the tissue, either carefully re-orient the tissue so it is vertical or completely remove the tissue from the block and prepare a new block. Either way, once the vessel is properly aligned inside the embedding block, refill the block with liquid paraffin and let harden as routine.

21. A dull blade during microtome sectioning can destroy the tissue and render it useless even if it is re-embedded in a new block. In fact, tissues can even be pulled out of a paraffin block if the blade is too dull. Replacing dull or heavily used blades is essential for proper sectioning and preservation of important experimental tissues. Knives can be either of a standard thick metal variety that needs custom sharpening (and are expensive too) or of the disposable kind (inexpensive and suitable for most histology specimens).

22. It is important to have properly fixed, cleared, and embedded tissues or artifacts that can be introduced during the sectioning step. Several common artifacts that become apparent during sectioning include tearing or ripping of the wax section after cutting, the appearance of holes in the sections, and folding or wrinkling of the tissues and/or wax section. Insufficient dehydration prior to clearing (during processing), inadequate infiltration with paraffin (during processing and embedding), and/or the presence of air bubbles in the tissues can all lead to problems evidenced during sectioning. Tissue-processing cycles should allow sufficient time for tissue dehydration, with at least one step involving incubation in 100% ethanol. Covering or sealing all solutions from the ambient air also helps to avoid tissue rehydration, especially in humid environments. During in situ perfusion-fixation and tissue harvesting, make all necessary steps to avoid the introduction of air into the vasculature as the presence of air inside the vessel lumen will surely cause destruction of the tissue during sectioning.

23. For the studies described in this chapter, universal 76×25 mm slides with thickness 1–1.2 mm are recommended. These are cost-effective and do not break easily. More expensive coated or frosted slides or slides with polished edges can be used if desired by the investigator.

24. Following staining and cover-slipping, slides can be stored for long periods of time unless a decaying-type of stain has been used. Most light stains will slowly deteriorate over time, thus decreasing clarity of the stain and ease of visualization. Fluorescent stains or radioactive treatments have inherent limits to the length of time that they can be analyzed, as their treatments involve a natural decay that will lessen the stain with time.

25. Commonly used fixative pigments including those associated with formalin, if not adequately removed from the tissues during the processing and/or embedding steps, can render artifacts in the samples under acidic conditions that can be observed during microscopic evaluation of the stained tissues. Formalin residues commonly appear as brown or brownish-black anisotropic (birefringent) deposits in tissues that have been fixed in acidic hematoxylin. To remove these pigment, residues use of a saturated alcoholic picric acid solution on unstained tissues is recommended. This is included in the protocol described for Verhoeff's elastic tissue stain with Van Gieson counterstain in this chapter.

26. Determining optimal staining times for hematoxylin in routine procedures is complex and dependent upon several variables. These include the type of hematoxylin used (Mayer's 10–20 min; Ehrlich's 20–45 min), the age of the stain (intensity decreases with age), the intensity of use of the stain, progressive versus regressive procedures (based on staining depth), pre-treatment and post-treatment of tissues including factors for processing, clearing, embedding, and staining, and personal preference of the investigator(s). It is recommended by the author that control samples of tissues be used with various incubation times until an optimal staining time is achieved.

27. A weak alkali solution such as standard tap water or 0.05% ammonia water is used to differentiate these stains in a process termed "bluing." The use of warm water is suggested to further enhance the intensity of elastin fiber staining. This is an essential step in the differentiation of the dye and in conversion of the initial stain to its final colored product.

28. Pre-treatment of tissues with 1% potassium permanganate for 5 min followed by oxalic acid improves the sharpness and intensity of elastin fiber staining in this protocol.

29. Use of a Lugol's iodine/sodium thiosulfate sequence has been suggested to remove mercury pigment artifacts in tissues fixed with a mercury-based fixative (*1*). Mercury residue is usually evidenced as a brown/black extracellular crystal. Iodine is the classic method for removing this pigment, followed by bleaching in a weak sodium thiosulfate (hypo) solution.

30. The celestine blue aluminum hematoxylin nuclear stain is recommended for use in conjunction with Masson's trichrome technique. This method uses deparaffinized and rehydrated tissues that are stained with celestine blue solution [celestine blue B (2.5 g), ferric ammonium sulfate (25 g), glycerin (70 ml), and distilled water (500 ml)] for 5 min followed by rinsing in tap water. Tissues are then stained in an alum hematoxylin of choice for 5 min and washed in tap water until blue color sets. The remaining staining procedure is then followed.

Acknowledgments

Work in preparation of this chapter was supported by NHLBI grants HL-59868 and HL-081720 and by a grant from the American Heart Association. The author would like to apologize to investigators whose works were not cited in this report due to space limitations and the personal context in which this chapter was prepared.

References

1. Bancroft, J.D., and Gamble, M., Eds. *Theory and Practice of Histological Techniques*, 5th edition. Churchill Livingstone, Harcourt Publishers Limited, 2002.
2. Junqueira, L.C., Carneiro, J., and Kelley, R.O. *Basic Histology*, 8th edition. Appleton & Lange, A Simon & Schuster Company, 1995.
3. Seifert, P., Rogers, C., and Edelman, E.R. Histological and immunohistological methods for vascular animal model studies. In: *Contemporary Cardiology: Vascular Disease and Injury: Preclinical Research*, Simon, D., and Rogers, C., Eds. Humana Press, 2001.
4. Tulis, D.A., Durante, W., Peyton, K.J., Chapman, G.B., Evans, A.J., and Schafer, A.I. (2000) YC-1, a benzyl indazole derivative, stimulates vascular cGMP and inhibits neointima formation. *Biochem. Biophys. Res. Commun.* **279**, 646–652.
5. Clowes, A.W., Reidy, M.A., and Clowes, M.M. (1983) Mechanisms of stenosis after arterial injury. *Lab. Invest.* **49**, 208–215.
6. Clowes, A.W., Clowes, M.M., and Reidy, M.A. (1986) Kinetics of cellular proliferation after arterial injury. III. Endothelial and smooth muscle growth in chronically denuded vessels. *Lab. Invest.* **54**, 295–303.
7. Tulis, D.A., Durante, W., Peyton, K.J., Evans, A.J., and Schafer, A.I. (2001) Heme oxygenase-1 attenuates vascular remodeling following balloon injury in rat carotid arteries. *Atherosclerosis* **155**, 113–122.
8. Waynforth, H.B., and Flecknell, P.A., Eds. *Experimental and Surgical Technique in the Rat*, 2nd edition. Academic Press, A Harcourt Science and Technology Company, 2001.

3

Plaque Rupture Model in Mice

Takeshi Sasaki, Kae Nakamura, and Masafumi Kuzuya

Summary

It is widely believed that rupture of a vulnerable atherosclerotic plaque leads to acute coronary events and stroke. However, the exact mechanisms involved in the plaque rupture remain unknown. Pathological animal models are valuable in the research on human disease mechanism, their therapy, and drug development. Recently, we have proposed a novel murine model of atherosclerotic plaque rupture associated with not only intraplaque hemorrhage but also luminal thrombus using combination treatments of ligation and perivascular cuff placement on the carotid artery.

Key Words: Animal model; Plaque rupture; Thrombus; Ligation; Cuff placement.

1. Introduction

It is widely believed that rupture of a vulnerable atherosclerotic plaque leads to acute coronary events and stroke. The vulnerable plaque is generally composed of an atrophic fibrous cap, a lipid-rich necrotic core, the accumulation of inflammatory cells *(1,2)*, and imbalance between extracellular matrix synthesis and degradation resulting in decreased extracellular matrix protein content *(3–5)*. However, the exact mechanisms involved in the plaque rupture remain unknown. Pathological animal models are valuable in the research on human disease mechanism, their therapy, and drug development. Although several plaque rupture models using the apolipoprotein E (ApoE)-deficient mouse have been proposed, in most of the models plaque rupture has been seen less frequently even in old mice after prolonged feeding with very high cholesterol diets *(6,7)*. Furthermore, there is no convincing evidence of the

From: *Methods in Molecular Medicine, Vol. 139: Vascular Biology Protocols*
Edited by: N. Sreejayan and J. Ren © Humana Press Inc., Totowa, NJ

formation of occlusive thrombus at the site of presumed rupture *(6–8)*, which is characteristic of the human plaque rupture leading to coronary heart disease and stroke.

Recently, we have developed a novel murine model of atherosclerotic plaque rupture *(9)* by modification of occlusive thrombosis model in the wild-type mice *(10)*. The left common carotid arteries of male ApoE-deficient mice (9 weeks old) were ligated just proximal to their bifurcations. Following 4 weeks on a standard diet, the mice received polyethylene cuff placement just proximal to the ligated site. The cuff placement evoked decrease in collagen content, intraplaque hemorrhage, and plaque rupture with luminal thrombus in this region within a few days after cuff placement. This model would help us not only to understand the mechanism of human plaque rupture but also to assess various agents in the future.

2. Materials

2.1. Animal

1. The male ApoE-deficient mice (9 weeks old) from Jackson Laboratory (Bar Harbor, ME, USA).
2. Mice are housed under a 12L/12D cycle (light on at 08:00 h and light off at 20:00 h) and provided with a standard diet (Oriental Yeast, Osaka, Japan) and tap water ad libitum throughout the experimental period.

2.2. Ligation and Cuff Placement

1. Silk surgical suture from Kono (5-0, Chiba, Japan) for ligation.
2. Polyethylene cuff: Length 2 mm and inside and outside diameter 0.580 mm and 0.965 mm, respectively (Becton Dickinson, Franklin Lakes, NJ, USA), incised unilaterally (*see* **Fig. 1A**) (*see* **Note 1**).
3. Stereoscopic microscope from Olympus (Tokyo, Japan) for animal operation.
4. Pentobarbital sodium from Dainippon Pharmaceutical (Osaka, Japan) for animal anesthesia.

2.3. Tissue Collection and Processing

1. Phosphate-buffered saline (PBS): Prepare $10\times$ stock solution with 1.37 M NaCl, 80 mM Na_2HPO_4, and 147 mM KH_2PO_2 (adjust to pH 7.4). Working solution by dilution of one part with nine parts of water (*see* **Note 2**).
2. Fixative solution: Paraformaldehyde (Merck, Darmstadt, Germany). Prepare 4% (w/v) solution in PBS fresh for each experiment (*see* **Note 3**).
3. Optimal cutting temperature (OCT) compound (Sakura Finetechnical, Tokyo, Japan).

A

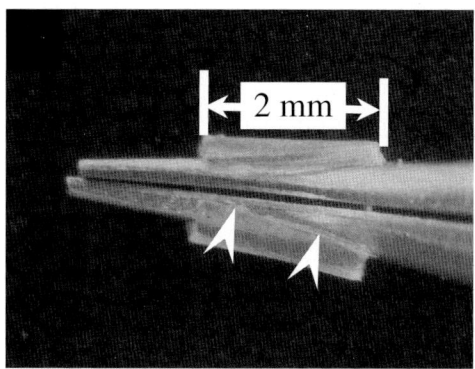

B

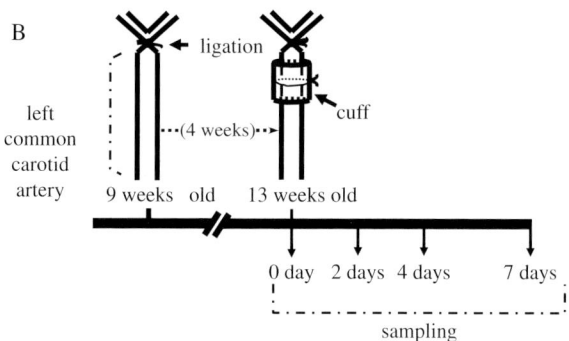

Fig. 1. (**A**) Photograph of the polyethylene cuff. White arrow heads indicate the unilateral incision line in the cuff. (**B**) Schematic diagram of ligation and cuff placement in the mouse left carotid artery. Briefly, the left common carotid arteries of 9-week-old male apolipoprotein E-deficient mice are ligated just proximal to their bifurcations. Following 4 weeks (13 weeks old), the mice receive polyethylene cuff placement just proximal to the ligated site. The carotid arteries are collected just before 0, 2, 4, or 7 days after the cuff placement and studied histologically.

2.4. Histological and Immunohistochemical Staining

1. Hematoxylin: Mayer's Hematoxylin solution from Muto Pure Chemicals (Tokyo, Japan).
2. Eosin: Prepare eosin solution with 0.5 g Eosin Y (Merck) and 20 μL acetic acid in 100 mL distilled water.
3. Picro-sirius red: Prepare picro-sirius red solution with 0.5 g Sirius red F3B (Polysciences, Warrington, PA) in 500 mL saturated aqueous picric acid.
4. Acetic acid water: Prepare 50 mL acetic acid and glacial with 110 mL distilled water.

5. Analyzer and polarizer for polarized microscopic observation of picro-sirius red staining from Olympus.
6. Normal serum buffer: Dilute 15 μL normal goat serum (Vector Laboratories, Burlingame, CA, USA) with 1 mL PBS.
7. Primary antibodies: alkaline phosphatase-conjugated monoclonal antibody against alpha smooth muscle cell (α-SMC) actin (Sigma-Aldrich, St Louis, MO, USA) and rabbit polyclonal antibody against fibrin(ogen) (DAKO, Carpinteria, CA, USA). Dilute antibody against α-SMC actin (1:50) and fibrin(ogen) (1:200) with normal serum buffer just before the use.
8. Secondary antibodies: Biotinylated anti-rabbit IgG from Vector Laboratories. Dilute the biotinylated anti-rabbit IgG with distilled water (1.5 mg/mL). Just before the use, secondary antibody is diluted (1:200) with normal serum buffer.
9. Vectastain ABC-AP solution: Vectastain ABC-AP kit (AK-5000) from Vector Laboratories. Vectastain ABC-AP kit contains reagents A and B. Vectastain ABC-AP solution is made from the mixture of 10 μL reagent A and 10 μL reagent B in 1 mL PBS (*see* **Note 4**).
10. Tris–HCl buffer: 100 mM Tris–HCl, pH 8.2–8.5.
11. Alkaline phosphatase substrate solution: Vector Red Substrate kit (SK-5100) and Levamisole solution from Vector Laboratories. Vector Red Substrate kit contains reagents 1, 2, and 3. Alkaline phosphatase substrate solution is made from the mixture of 20 μL reagent 1, 20 μL reagent 2, 20 μL reagent 3, and 10 μL levamisole solution in 1 mL Tris–HCl buffer (*see* **Note 5**).
12. Entellan New from Merck.

3. Methods

3.1. Animal Surgery

Experimental design is presented in schematic diagram (*see* **Fig. 1B**).

1. Perform the ligation and cuff placement using a stereoscopic microscope.
2. Anesthetize the animal (9-week-old mouse) with an intraperitoneal (i.p.) injection of pentobarbital sodium (50 mg/kg).
3. Ligation: Perform a neck incision (*see* **Fig. 2A**) and carefully expose the left common carotid artery and ligated proximal to their bifurcations (*see* **Fig. 2B and C**) using 5-0 silk surgical suture.
4. Suture the surgery wound (*see* **Fig. 2D**) and allow the mouse to recover.
5. Four weeks after ligation, reoperate to put a polyethylene cuff around the ligated carotid artery.
6. Cuff placement: Anesthetize the mouse by i.p. pentobarbital injection. After reopening neck in mice, remove the connective tissue and carefully expose the ligated common carotid artery (*see* **Fig. 2E**). Apply a polyethylene cuff just proximal to the ligated site (*see* **Note 6**).
7. After cuff placement around carotid artery, constrain the cuff by 5-0 silk suture (*see* **Fig. 2F**).

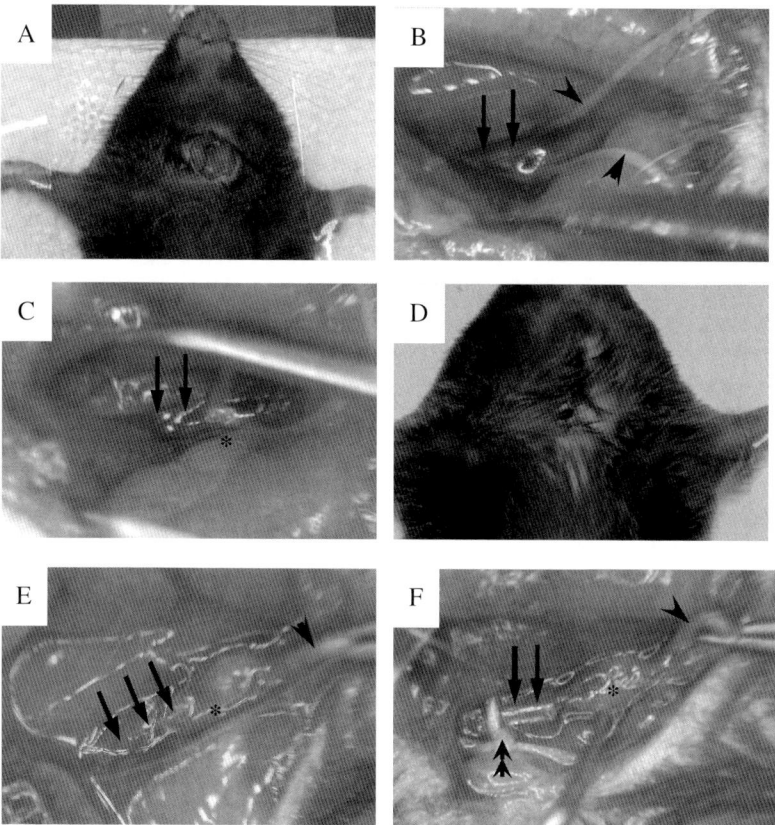

Fig. 2. Actual photographs of ligation and cuff placement in the mouse left carotid artery. (**A**) Incision of neck in the mouse; (**B**) passing the 5-0 silk surgical suture under left common carotid artery; (**C**) ligation of the common carotid artery proximal to bifurcations of external and internal carotid artery; (**D**) suturing the incision of neck; (**E**) left common carotid artery 4 weeks after ligation; and (**F**) cuff placement just proximal to the ligated site. Arrows indicate left common carotid artery, arrow heads indicate the silk surgical suture for ligation, asterisks indicate knot of ligation, and double arrow head indicates constraint by 5-0 silk suture.

3.2. Tissue Collection and Processing

1. Just before 0, 2, 4, or 7 days after the cuff placement, anesthetize the mice by i.p. pentobarbital injection.
2. Perfuse the mice through left cardiac ventricle with isotonic saline (15 mL) and fixative solution under physiological pressure.

3. Remove the left carotid arteries and immerse in fixative solution for 16 h (4 °C).
4. Separate the left carotid arteries into three parts: the distal, internal, and proximal regions of the cuff. After the removal of the cuff, embed each vessel in OCT compound, snap-freeze in liquid nitrogen, and store at −20 °C until use (*see* **Note 7**).

3.3. Histological and Immunohistochemical Staining

1. Prepare 6 μm cross cryosections from the carotid artery using the cryostat (Leica, Wetzlar, Germany) for every 60-μm intervals.
2. Dry each cryosection under a stream of cool air for 10 min and store at −20 °C until use.

3.3.1. Hematoxylin and Eosin (H&E) Staining

1. Place the sections in distilled water for 10 min and stain with hematoxylin (5 min) and eosin (5 min).
2. Dehydrate the stained sections with a graded series of ethanols, clear with xylene, and mount with Entellan New (*see* **Fig. 3A–D**).

3.3.2. Picro-sirius Red Staining

1. Place the sections in distilled water for 10 min and incubate it with picro-sirius red solution for 1 h.
2. Wash with two changes of acetic acid water.
3. Dehydrate routinely with a graded series of ethanol, clear with xylene, and mount with Entellan New.
4. Observe the sections using a polarized microscope (*see* **Fig. 4A–C**) (*see* **Note 8**).

3.3.3. Immunohistochemical Staining

The procedure using unconjugated primary antibody (antifibrinogen) is as follows (*see* **steps 1–11**) and using biotinylated primary antibody (anti-α-SMC actin) is the abbreviated protocol (*see* **steps 1–3** and **8–11**).

1. Place the sections in distilled water for 10 min and wash twice in PBS for 5 min.
2. Incubate sections for 20 min with normal goat serum for blocking.
3. Discard the normal goat serum on sections and incubate sections for overnight (18 h) with diluted primary antiserum (4 °C).
4. Wash sections three times with PBS for 5 min.
5. Incubate sections for 1 h with diluted biotinylated secondary antibody solution (4 °C).
6. Wash sections three times in PBS for 5 min.
7. Incubate sections for 1 h with Vectastain ABC-AP solution (4 °C).
8. Wash sections twice in PBS and distilled water for 5 min.

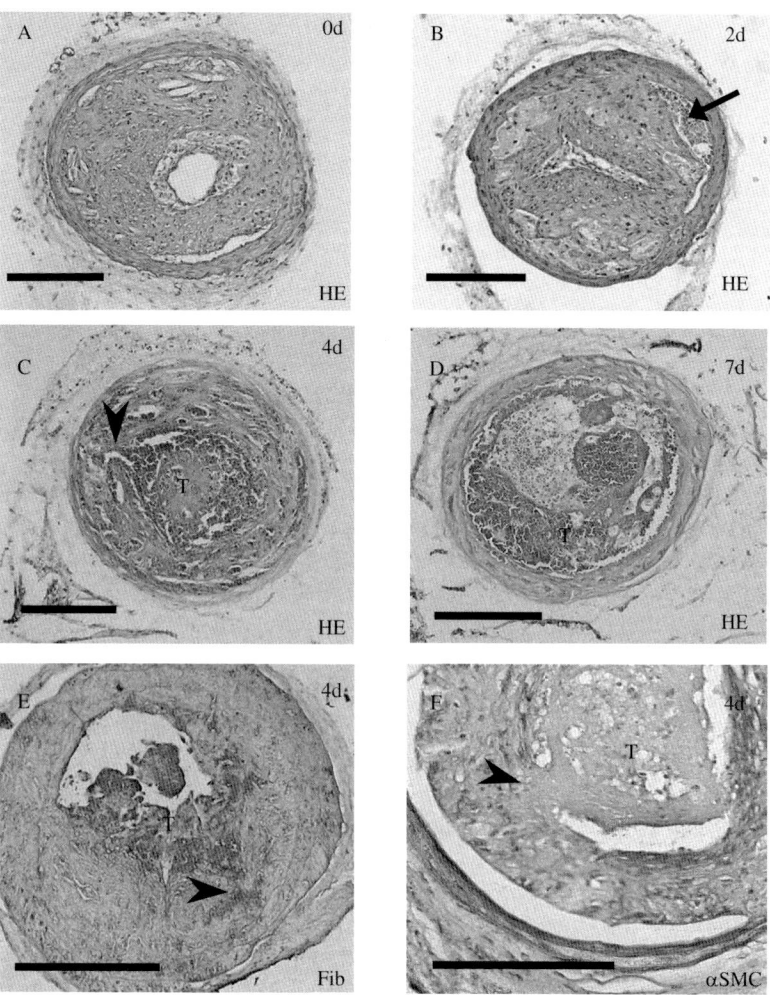

Fig. 3. Plaque rupture at the intracuff region in the left carotid arteries of apolipoprotein E-knockout mice. (**A–D**) H&E staining of cross-sections of the left carotid artery before ligation (**A**), at 2 days (**B**), at 4 days (**C**), and at 7 days (**D**) after cuff placement. (**E and F**) immunostaining for (**E**) fibrin(ogen) (Fib) and (**F**) alpha smooth muscle cell (α-SMC) actin of the carotid artery at 4 days after cuff placement. Bars = $100\,\mu$m and T, thrombus. Arrows indicate intraplaque hemorrhage, and arrow heads indicate the cracks of fibrous caps.

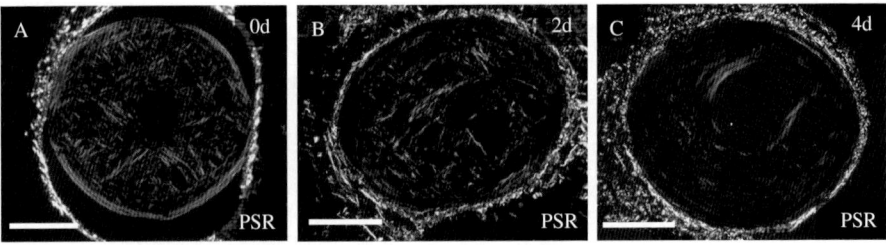

Fig. 4. Collagen content in neointima. (**A–C**) Picro-sirius red (PSR) staining viewed under polarized light before cuff placement (**A**), at 2 days (**B**), and at 4 days (**C**) after cuff placement. Bars = 100 μm.

9. Incubate sections with alkaline phosphatase substrate solution (Vector Red Substrate kit) until desired stain intensity develops.
10. Wash sections in distilled water for 5 min and counterstain with hematoxylin (30 s–1 min).
11. Dehydrate routinely with a graded series of ethanol, clear with xylene, and mount with Entellan New (*see* **Fig. 3E and F**).

4. Notes

1. The polyethylene cuff is nicked using a scalpel and acuminate forceps under a stereoscopic microscope. The nick should be made linear along a major axis to make it easy to place the cuff around carotid artery.
2. This PBS is able to be used in immunohistochemical staining procedure.
3. Paraformaldehyde solution is prepared just before use.
4. This mixture solution has to be prepared 30 min before use.
5. This mixture solution is prepared just before application. The incubation time in this solution is beyond 45 min, which will not increase sensitivity unless freshly made solution is reapplied to sections.
6. At the placement of the cuff around a carotid artery, the nick of the cuff is expanded from the inner side using acuminate forceps (*see* **Fig. 1A**).
7. Plaque rupture is mainly observed at the internal region of the cuff. Representative results (*see* **Figs 3** and **4**) are sections at intracuff region of carotid arteries.
8. Polarized microscopic system is combination of polarized analyzer and polarizer with a standard light microscopy.

Acknowledgments

This work was supported by a research grant from the Scientific Research Fund of the Ministry of Education, Science, and Cultures, Japan (no. 13671182 and 17700377).

References

1. Zaman A. G., Helft G., Worthley S. G., and Badimon J. J. (2000) The role of plaque rupture and thrombosis in coronary artery disease. *Atherosclerosis* **149**, 251–266.
2. Rekhter M. D. (2002) How to evaluate plaque vulnerability in animal models of atherosclerosis? *Cardiovasc. Res.* **54**, 36–41.
3. Sukhova G. K., Schonbeck U., Rabkin E., Schoen F. J., Poole A. R., Billinghurst R. 'C., et al. (1999) Evidence for increased collagenolysis by interstitial collagenases-1 and -3 in vulnerable human atheromatous plaques. *Circulation* **99**, 2503–2509.
4. Loftus I. M., Naylor A. R., Goodall S., Crowther M., Jones L., Bell P. R., et al. (2000) Increased matrix metalloproteinase-9 activity in unstable carotid plaques. A potential role in acute plaque disruption. *Stroke* **31**, 40–47.
5. Galis Z. S., Sukhova G. K., Lark M. W., and Libby P. (1994) Increased expression of matrix metalloproteinases and matrix degrading activity in vulnerable regions of human atherosclerotic plaques. *J. Clin. Invest.* **94**, 2493–2503.
6. Rosenfeld M. E., Polinsky P., Virmani R., Kauser K., Rubanyi G., and Schwartz S. M. (2000) Advanced atherosclerotic lesions in the innominate artery of the ApoE knockout mouse. *Arterioscler. Thromb. Vasc. Biol.* **20**, 2587–2592.
7. Calara F., Silvestre M., Casanada F., Yuan N., Napoli C., and Palinski W. (2001) Spontaneous plaque rupture and secondary thrombosis in apolipoprotein E-deficient and LDL receptor-deficient mice. *J. Pathol.* **195**, 257–263.
8. Johnson J., Carson K., Williams H., Karanam S., Newby A., Angelini G., et al. (2005) Plaque rupture after short periods of fat feeding in the apolipoprotein E-knockout mouse: model characterization and effects of pravastatin treatment. *Circulation* **111**, 1422–1430.
9. Sasaki T., Kuzuya M., Nakamura K., Cheng X. W., Shibata T., Sato K., et al. (2006) A simple method of plaque rupture induction in apolipoprotein E-deficient mice. *Arterioscler. Thromb. Vasc. Biol.* **26**, 1304–1309.
10. Sasaki T., Kuzuya M., Cheng X. W., Nakamura K., Tamaya-Mori N., Maeda K., et al. (2004) A novel model of occlusive thrombus formation in mice. *Lab. Invest.* **84**, 1526–1532.

4

Immunostaining of Mouse Atherosclerotic Lesions

Hong Lu, Debra L. Rateri, and Alan Daugherty

Summary

Atherosclerotic lesions develop through interactions with diverse cell types whose functions are determined by a complex array of regulators. Immunostaining is now a commonly applied technique to identify these numerous cell types and regulators in lesions. The principle of the technique is that an antibody is incubated with the tissue under conditions that favor a specific interaction with its antigen. This is subsequently visualized most commonly by a chromogenic substrate that produces a colored precipitate at the location of the antigen–antibody interaction. When appropriately applied, it is a powerful technique to provide mechanistic insight into the atherogenic process. However, the complexity of atherosclerotic tissue can provide challenges to ensuring that development of the chromogen is due to specific antigen–antibody interaction. Thus, the determination of specific interactions frequently requires the judicious use of appropriate control experiments.

Key Words: Atherosclerosis; Immunostaining; Antigen–Antibody Interactions; Chromogens; Vascular Cells.

1. Introduction

Atherosclerosis is a chronic disease that is characterized by complex cellular changes that vary at progressive stages of lesion development. Early stages are characterized by dysfunctional endothelium that leads to the attraction and adherence of monocytes. These cells undergo diapedesis through this intact monolayer to become residence macrophages in the intima. These macrophages progressively acquire lipids and become greatly hypertrophied. With continued progression, smooth muscle cells become considerably more prevalent. At the terminal stages of the disease, thrombotic events associated with lesions can

From: *Methods in Molecular Medicine, Vol. 139: Vascular Biology Protocols*
Edited by: N. Sreejayan and J. Ren © Humana Press Inc., Totowa, NJ

occlude arteries and lead to acute cardiovascular syndromes. In addition to the cell types that are most abundant, there are many other cell types that may be less prevalent, but still exert a role in lesion development. These include dendritic cells, mast cells, natural killer cells, B lymphocytes, and many subtypes of T lymphocytes *(1–3)*.

The complexity of the cellular changes that characterize development of atherosclerotic lesions resulted in the implication of many regulators in recruitment and change in function of these different cell types. Many of these regulators are proteins, but fats and carbohydrates may also have atherogenic roles. Adducts of these are also hypothesized to be involved in the atherogenic process. The detection of these biological entities by immunostaining of atherosclerotic lesions assists in validating their role in the disease. The determination of the spatial distribution of a regulator within a lesion can provide biological insight. For example, a regulator that is involved in plaque rupture would be expected to be most abundant at the lesion margins where this event occurs.

Immunostaining is a widely applied technique for the detection of lesional cell types and regulators of the atherogenic process. Its application to detection and spatial distribution of antigens can provide valuable associative evidence of a role in atherogenesis. The recognition that lesion size is not an important predicator of acute cardiovascular events led to the increasing use of cellular immunostaining to determine the propensity that a lesion may evolve into a thrombogenic nidus.

There has been a great emphasis on animal models of atherosclerosis to glean insight into mechanisms of lesion formation. Most of the contemporary studies use mouse models of the disease. The most commonly used are apolipoprotein E (apoE$^{-/-}$) and low-density lipoprotein (LDL) receptor$^{-/-}$ mice, although there are many other genetically manipulated mice that either develop atherosclerotic lesions spontaneously or are susceptible to diets that induce lesions *(4)*. The technical details of determining the size of atherosclerotic lesions in mice has been published previously *(5)*. There are many benefits of using mice in atherosclerosis studies including the extensive control of strain background, the relative ease of genetic manipulation, and economical benefits relative to large animals. For a perspective of immunostaining studies on atherosclerotic lesions, this species also benefits from the large number of commercially available antibodies that react against cells and proteins of mice.

1.1. Use of Immunostaining in Atherosclerosis

Immunostaining is a potentially powerful technique for the detection of the antigens in tissue. It is most commonly applied to detection of proteins, but can also be applied to other antigenic determinants. However, the technique is also fraught with issues of specificity that need to be considered in the interpretation of staining. The potential problems of lack of specificity need to be addressed with particular care in atherosclerotic tissues in which the complex nature of the tissue enhances the propensity of chromogen color development that is unrelated to the antigen–antibody interaction. Therefore, before giving an example of a specific protocol, we will discuss the many variables that need to be taken into account in experimental design to establish that chromogen development is reflective of specific antigen–antibody interactions.

1.2. Staining Systems

A common system for immunostaining involves the application of reagents in a "puddle" over the tissue sample and overnight incubations with primary and secondary antibodies at 4 °C. The major shortcoming to this approach is the length of time needed for completion of the protocol which could stretch to several days.

An alternative to this is the use of the MicroProbe system (FisherBiotech). The basic principle of this system is that antibodies will come to a binding equilibrium rapidly at physiological temperatures. The system uses slides that are placed face-to-face to form a small gap that permits reagents to be introduced by capillary flow to bathe tissue sections with solutions. The small volume of fluids used to cover the tissue permits rapid heating in the ovens of this system. Consequently, incubation times with the primary and secondary are usually shortened to approximately 15 min. Even the longest protocol using paraffin-embedded sections can be completed within 2 h.

1.3. Tissue Collection and Processing

Atherosclerotic lesions in mice have mostly been quantified in the aortic sinus and on the intimal surface of the entire aortic tree (6). However, as measurement of lesion size in the aortic sinus requires the acquisition of tissue sections, this is the most common region used in immunostaining studies. More recently, lesions that develop in the innominate artery have been recognized to exhibit properties of more advanced atherosclerosis, compared to that which is present in the sinus (7). Thus, it is likely there will be an increasing number of studies that deal with immunostaining performed on this region.

The decision on tissue preparation is frequently a balance of the desire to obtain optimal tissue integrity, without modifying or sterically hindering the epitope(s) of the antigen that will be detected by immunostaining. The most native state of proteins is presumed to be in tissues that are embedded in a support media [Optimal cutting temperature (OCT) embedding media (Cat no. 4583, Sakura) that is placed with the tissue in a plastic mold], frozen, and cut in a cryostat. Sections can be stored frozen at $-20\,^\circ$C, preferable in a non–frost-free freezer so that tissues will not desiccate.

To prevent tissue putrefaction during and after the immunostaining process, sections must be incubated with some form of fixative. A common mode of fixation in frozen tissues is by incubation with chilled acetone or ethanol to cause protein precipitation with lipid removal. Tissue sections can also be fixed by exposure to a cross-linking agent, such as paraformaldehyde. These reagents function by cross-linking the primary amines in protein. The extent of the cross-linking will vary depending on the type and duration of fixation. While this process preserves tissue, it also has the potential to attenuate or ablate antigen–antibody interactions.

Although use of frozen tissues enhances the probability of maintaining antigen–antibody interactions, it does not optimally maintain tissue architecture. Also, the practical limit of tissue section thinness in fibrous tissues such as arteries is usually about 8 μm. This leads to lack of distinction of individual cells when the sections are viewed under a microscope. An alternative mode to use frozen tissues is to fix tissues with a cross-linking agent and embed them in paraffin. This form of tissue processing maintains tissue structure while the support enables tissue sections to be cut in the 2-μm range that facilitates viewing the cellular elements of lesions with a microscope. However, the fixation, hydration, and heating involved in this technique can adversely affect epitope recognition by an antibody.

1.4. Antibody Type

Most of the mammalian antibodies used in immunostaining are immunoglobulin G (IgGs), with a lesser number of IgMs. With the development of better secondary antibodies to chicken Igs, there is an increased use of IgY antibodies for immunostaining.

There are several solutions in which antibodies can be supplied for immunostaining. The least desirable solution of obtaining an antibody is as a serum from an immunized animal. Because the specific antibody will only be a small fraction of the entire solution, many other constituents of antisera may enhance the likelihood of non-specific interactions. However, antiserum has been used successfully especially with highly expressed antigens (8).

Antibodies are also frequently supplied as unfractionated IgGs in which the Ig fraction has been obtained either by salt precipitation or by affinity chromatography using a material that interacts with the crystallized fragment

(Fc) portion of the Ig. This purification step substantially reduces the propensity to develop non-specific interactions compared to antiserum. However, the specific antibody will probably still represent a relatively small fraction of the total Ig that will be incubated with the tissue.

The preferred use of antibodies is after affinity purification. This isolation is usually achieved by the conjugation of the immunogen to a Sephadex column and the use of high salt conditions to displace the bound antibody. For monoclonal antibodies, a similar level of purity is obtained without the need for affinity purification.

Another factor that needs to be considered is the species used to develop the antibody. For example, monoclonal antibodies are frequently developed in mice. The most common procedure for visualizing immunostaining is by the use of a secondary antibody that recognizes the Fc portion of the primary antibody. However, if a mouse antibody is used on mouse tissue, it is not possible for the secondary antibody to distinguish between the exogenously applied and the endogenous Ig. Mouse antibodies may be applied on mouse tissue if a conjugated material is used that negates the use of a secondary antibody. There are also kits available that will mask the epitopes of the endogenous Igs before incubation with the secondary antibody (Mouse on Mouse kits, Vector Laboratories).

Some examples of antisera and antibodies used to detect specific cell types in mouse atherosclerotic lesions are listed in **Table 1**.

1.5. Detection of Antigen–Antibody Interactions

There are many different techniques in which antigen–antibody interactions can be detected. One of the most rapid technique is to use antibodies that are directly linked to a fluorescent probe. These forms of conjugated antibody are widely used in cell biology, but have only been used in a relatively small number of atherosclerosis studies. The lesser use is partially related to the difficulties of providing a counterstain that permits the visualization of the positive immunoreactivity, while viewing its placement relative to the tissue structures. This may be partially overcome by the combination of nuclei staining and autofluorescence of elastin fibers when represented in merged photographs *(9)*. However, the autofluorescence of some arterial tissues is also a hindrance to this technique of visualization.

The most frequent form of visualizing antibody reactivity in atherosclerosis studies is using chromogens that are substrates for either peroxidase or alkaline phosphatase. The relative merits of different substrates will be discussed below (Table 2). Some primary antibodies are available with these enzymes directly

Table 1
Examples of Primary Antibodies used to Characterize Cellular Elements of Mouse Atherosclerotic Lesions by Immunostaining

Cell type	Examples of commonly used antibodies	Antigen	Suggested working dilutions	References
Endothelial cells	Biotinylated CD31 mAb (cat. no. 553371, BD Pharmingen)	CD31 (PECAM-1)	1 µg/ml	*(14,15)*
Leukocytes	Rat CD45 mAb (cat no. 550539, BD Pharmingen)	CD45	1 µg/ml	*(16)*
	Rat CD11b mAb (cat. no. ab6332, ABCAM)	Mac-1	1:25–1:200	*(14)*
Lymphocytes	Rat CD90.2 mAb—Thy-1.2 (cat. no. 553009, BD Pharmingen)	CD90.2	0.5 µg/ml	*(8,15,17)*
	Rat CD4—GK 1.5 (cat. no. CBL1305B, Chemicon)	CD4		*(18,19)*
	Rat CD8 mAb—YTS 105.18 (cat. no. MCA1108G, Serotec)	Alpha chain of CD8		*(19)*
	Rat CD19 mAb (cat. no. 553783, BD Pharmingen)	CD19	1:100	*(19,15)*

(Continued)

Table 1
(Continued)

Cell type	Examples of commonly used antibodies	Antigen	Suggested working dilutions	References
Macrophages	Rabbit macrophage antiserum (cat. no. AI-AD 31240, Accurate Chemicals)	Multiple	1:1000	(8,15,17)
	Rat CD68 mAb—FA-11 (cat. no. MCA1957, Serotec)	CD68	5 µg/ml	(20)
	Rat MOMA-2 mAb (cat. no. MCA 519G, Serotec)	Unknown		(21)
	Rat MAC-2 mAb (cat. no. ACL8942AP, Accurate Chemical)	Unknown		(22)
	Rat F4/80 mAb (cat. no. MCA497R, Serotec)	Unknown		(23)
Smooth muscle cells	Mouse anti-actin mAb alkaline phosphatase conjugated—1A4 (cat. no. A5691, Sigma)	Alpha actin		(7,21)
	Rabbit polyclonal (cat. no. ab5694-100, ABCAM)		2.5 µg/ml	

conjugated to the Fc portion of the antibody. These antibodies permit rapid development of immunostaining and require a limited number of controls. A common example is the alpha actin mouse monoclonal antibody, 1A4, conjugated to alkaline phosphatase that is used to detect smooth muscle cells. Direct enzyme-conjugated antibodies facilitate the rapid immunostaining of tissues when detecting a highly abundant antigen.

The use of an antibody conjugated to a single peroxidase or alkaline phosphatase molecule may not provide sufficient localization of enzymatic activity for detection of less abundant antigens. A common method of increasing the enzymatic activity in the vicinity of an antigen–antibody interaction is to use an amplification system such as the Vectastain Avidin–Biotin Complex (ABC) method (Vector Laboratories). This is a mixture of avidin and biotinylated enzyme that forms a complex. Once formed, this complex is incubated with the tissue in which the biotinylated antibody has reacted with the antigen. This results in the localization of many molecules of peroxidase or alkaline phosphatase at the site of antigen–antibody interactions. In addition to facilitating the detection of less abundant antigens, this also permits the use of more dilute concentrations of the primary antibody which will lessen the probability of non-specific interactions and have cost-saving benefits.

1.6. Visualization of Antigen–Antibody Interactions

There are several substrates that can be used as chromogens for peroxidase and alkaline phosphatase. They share the common property that interaction with the appropriate enzyme leads to a colored product that is precipitated on the tissue. The choice of the substrate is largely dependent on the esthetic decision of the operator's color preference. For example, while early studies commonly used the brown chromogen, 3, 3′ diaminobenzidine, more colorful alternatives are now more common, such as the red 3-amino-9-ethylcarbazole (AEC). The most commonly used substrates are listed in **Table 2**.

In addition to the visual appeal, the decision to use a specific chromogen has an impact on the subsequent processing of the tissue. For example, AEC is soluble in organic solvents. Therefore, the counterstain and mounting media that is used in conjunction with AEC have to be aqueous based. Also, if the tissue is to be immunostained for multiple antigens, the stability of the chromogen during these processing steps must be considered.

1.7. Antibody Concentrations

Determining the optimal dilutions of antisera or concentrations of antibodies are important parameters to define. Most commercially available antibodies are

Table 2
Common Peroxidase and Alkaline Phosphatase Substrates used as Chromogens in Immunostaining

Enzyme	Substrate	Color	Counterstain	Mounting medium
Peroxidase	DAB	Brown	Organic or aqueous	Permount
	DAB with nickel enhancement (DAB/Nickel)	Gray/black	Organic or aqueous	Permount
	3-amino 9-ethylcarbazole	Red	Aqueous only	Glycerol Gelatin
Alkaline phosphatase	Naphthol-AS-B1-phosphate/fast red TR	Red	Aqueous only	Glycerol Gelatin
	Naphthol-AS-MX-phosphate/fast red TR	Red	Aqueous only	Glycerol Gelatin
	Naphthol-AS-B1-phosphate/new fuschin	Red/violet	Aqueous only	Glycerol Gelatin
	Bromochloroindolylphosphate/Nitro Blue Tetrazolium	Purple	Organic or Aqueous	Permount

DAB, diaminobenzidine.

supplied with manufacturer's specification sheets containing suggested dilutions of concentrations. For newly developed antibodies or antigens that have not previously been detected in atherosclerotic tissues, the optimal dilutions and concentrations will have to be defined empirically. This is usually done by incubating tissue with a logarithmic progression of dilutions or concentrations (e.g., dilutions of 1:10, 1:30, 1:100, and 1:300 for antisera and concentrations of 10, 3, 1, and 0.3 μg/ml for antibodies). The aim is to define a dilution or concentration that provides the strongest immunostaining with the least non-specific background. Thus, these antisera and antibodies are usually compared to an appropriate non-immune control for the reasons described below (*see* **Note 1**).

1.8. Controls

1.8.1. Positive Controls

The need for a positive control tissue can vary according to the experimental aims. For example, if an operator fails to demonstrate immunostaining for a specific protein in atherosclerotic tissue, the simultaneous staining of a tissue that is known to express the protein would provide a positive control for determining the optimal conditions such as concentrations of primary and secondary antibodies. However, the ability to immunostain antigens is dependent on the localized antigen concentration exposed to the antibody. Thus, if the antigen is more diffusely distributed, or present at lower concentrations, it may not be detected under the same conditions used in the positive controls. Given the limitations of immunostaining, the inability to detect an antigen may be because it is present below the limits of detection, rather than being absent.

1.8.2. Negative Controls

As mentioned earlier, the complexity of atherosclerotic tissue enhances the potential for chromogen development due to non-specific interactions. Thus, the use of appropriate negative controls is a requirement for the meaningful interpretation of immunostaining. The underlying premise of negative controls is to determine whether chromogen development could be attributable to interactions other than the recognition of the antibody with its specific antigen.

The general premise of negative controls is to remove each component of the solutions used to perform immunostaining to determine whether the chromogen developed decreases accordingly (*See* **Fig. 1**). For example, in the full protocol that will be subsequently described in detail, any color development during the incubation of substrate only is indicative of insufficient

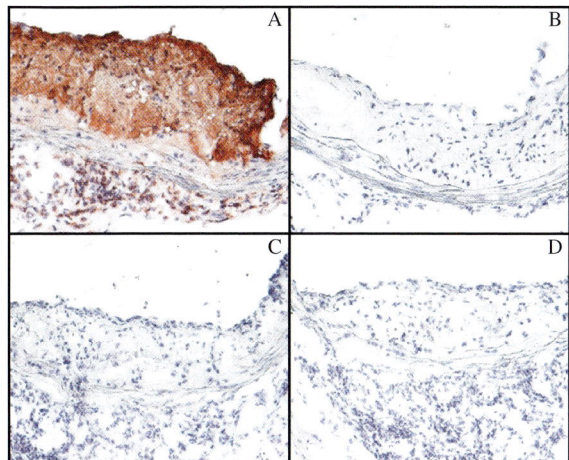

Fig. 1. Examples of immunostaining in serial sections of the aortic root for macrophage infiltration in atherosclerotic lesions. Examples of rat anti-mouse CD68 immunoglobulin G (IgG) is used as primary antibody. (**A**) Rat anti-mouse CD68 IgG, (**B**) non-immune rat IgG, (**C**) secondary antibody rabbit anti-rat IgG only, and (**D**) no primary and secondary antibodies immunostaining in the sections of aortic root from a fat-fed low-density lipoprotein receptor-deficient male mouse.

quenching of endogenous peroxidase or alkaline phosphatase. If this slide is negative, but color develops during incubation with just the ABC and substrate, this is indicative of direct interaction of the complex with the tissue. This occurs particularly in tissues with high endogenous concentrations of biotin. This may be a problem when considering potential positive control tissues such as renal and adrenal, but is not usually an issue in atherosclerotic tissue. If these two controls are negative, but color develops when just the primary antibody is excluded, this is because of direct interaction of the secondary antibody with the tissue. This may be overcome by reducing the concentration of the secondary antibody. Alternatively, another secondary antibody generated in a different species may be needed.

1.8.3. Non-Specific Interactions of the Primary Antibody

Defining possible non-specific interactions of the primary antibody can be more challenging. For antibodies used as antiserum, one control is the use of an equivalent dilution of sera from a non-immunized animal of the same species. A caveat to this approach is that the sera are usually derived from an animal that is different from the one in which the antisera were derived. Thus, it cannot be unequivocally defined that any differences seen between the antisera and non-immune sera is attributable solely to the absence of the specific antibody. The use of pre-immune sera from the same animal used to develop the antisera lessens these concerns. However, pre-immune serum from the same animal is seldom available in commercial antisera.

For antibodies supplied as unfractionated IgG, a minimal control is the use of an equivalent concentration of unfractionated IgG from a non-immunized animal of the same species. As with antisera, there is usually the concern that the IgG was isolated from a different animal used to generate the antibody.

With increased purification of the antibody, the concerns of non-specific interaction of the antibody are reduced. For example, the affinity-purified antibody can be compared to an equivalent concentration of the antibody that did not bind to the affinity column. In the case of monoclonal antibodies, the most common control is an isotype match that is developed against an antigen that would not be present in mammalian tissues.

A common control for the primary antibody is to perform a pre-incubation with the immunizing antigen. This can be an easily implemented approach in antibodies developed to peptide sequences. Ablation of immunostaining by pre-incubation with the immunogen is an excellent support for specificity. However, the immunogen could adhere directly to the atherosclerotic tissue resulting in an enhancement of chromogen development. We noted this previously when attempting this control of immunostaining of myeloperoxidase in which pre-incubation with the immunizing protein greatly enhanced chromogen development *(10)*. Thus, the inability of preabsorption to ablate immunostaining does not nullify the validity of the finding.

The use of tissues from a mouse in which genetic manipulation has deleted the protein, compared to tissues from mice that express the protein, may provide the ultimate test of specificity of the immunostaining procedure. However, there should be a note of caution if immunostaining is seen in genetically deficient mice. For example, all the known antibodies to the LDL receptor recognize the first cysteine repeat in the receptor-binding domain *(11)*. However, the LDL receptor-deficient mice were generated by truncation of the fourth exon *(12)*. Thus, while this mouse does not express functional LDL receptors, there is a potential for it to express immunostainable epitopes in a truncated form.

2. Materials

Subheading 1 provides an insight into some of the variables that need to be considered in the development of an immunostaining protocol. It will be apparent from this section that there are many potential variables, and thus it is not possible to provide a single protocol that would accommodate all immunostaining needs of atherosclerotic lesions. Instead, there is a need to understand the principle of the technique to develop a specific protocol to ensure authentic visualization of an antigen. The protocol described in **Subheading 2.1** is an example that may need to be tailored to specific needs.

The materials and reagents described in this chapter can be used provided that this technique will be performed using the MicroProbe system (*see* **Note 2**). Also, all reagents are listed to perform immunostaining on paraffin-embedded or frozen tissue sections.

2.1. Equipment and Supplies

1. MicroProbe staining system (Fisher Scientific).
2. MicroProbe slides (Fisher Scientific).
3. Coverslips: The thickness of the coverslips used depends on the properties of the microscope objectives.
4. Dewaxing agent—Limonene:xylene. Mix in a 3:1 ratio.
5. Hydrogen peroxide (30% vol/vol). Dilute in methanol (final concentration 1% vol/vol), which should be made fresh before each use. H_2O_2 in the original container should be stored in the refrigerator and protected from sunlight in order to slow its thermal decomposition.
6. Automation buffer stock solution ($10\times$, Biomeda Corp). Dilute to working solution by adding 100 ml stock solution to 900 ml distilled water and store at room temperature (RT).
7. Redusol (Biomeda Corp). This is a weak chromic acid solution that cleans the glass and facilitates capillary delivery of solutions.
8. Avidin–biotin blocking kit (Vector Laboratories). This kit is not commonly used in immunostaining of atherosclerotic lesions unless there is a lot of adventitia. However, some tissues (e.g., adrenal glands, liver, kidney, or brain tissues) contain endogenous biotin or biotin-binding proteins. These tissue sections are required to be pre-treated by avidin–biotin blocking before the addition of blocking agent and primary antibody.
9. Blocking serum diluted in phosphate buffered saline (PBS) (pH 7.4). Place normal serum (15 µl) in PBS (1 ml) before each use. Blocking serum must be from the same species used to generate the secondary antibody.
10. Primary antibody and control antibody (*see* **Table 1**) diluted in primary antibody diluting buffer (Biomeda Corp) before each use (*see* **Note 1**).
11. Secondary antibody diluted in PBS before each use. The secondary antibody must be ABC Kit (Vector Laboratories). The working solution should be incubated for 30 min at RT before use.

12. Peroxidase enhancer when using AEC (Biomeda Corp). This is a solution of AEC that increases the concentration of the substrate when the chromogen is subsequently incubated with the tissue in the presence of hydrogen peroxide.
13. Aqueous hematoxylin is used for counterstain of nuclei (*see* **Note 3**).
14. Mounting medium (*see* **Table 2**).

3. Methods

3.1. Immunostaining

As noted in **Subheading 2.1**, given the many variables, there is no protocol that will accommodate all immunostaining needs. The following protocol is based on the use of a primary and secondary antibody system using a peroxidase-based chromogen that is soluble in organic solvents. This may need to be adapted depending on the aim of the study and the availability of reagents.

3.1.1. For Paraffin-Embedded Sections

1. Heat for 5 min at 60 °C.
2. Incubate in dewaxing agent for 1 min at RT.
3. Blot the slides to remove dewaxing agent. We use reinforced wipers (Kimberly-Clark).
4. Incubate in dewaxing agent for 1 min at RT.
5. Blot the slides to remove dewaxing agent.
6. Heat for 2 min at 60 °C.
7. Incubate in dewaxing agent for 1 min at RT.
8. Blot the slides to remove dewaxing agent.
9. Incubate in absolute alcohol for 1 min at RT.
10. Blot the slides and wash in absolute alcohol twice at RT.

3.1.2. For Frozen Sections

1. Fix the slides using pre-chilled acetone or absolute alcohol for 5 min on ice.
2. Blot the slides to remove fixation solution at RT.

3.1.3. Subsequent Steps

1. Incubate in Redusol for 2 min at 40 °C.
2. Blot the slides to remove Redusol.
3. Incubate in automation buffer for 1 min at RT.
4. Rinse the slides four times in automation buffer at RT.
5. Incubate for 2 min at 40°C in 1% H_2O_2 in methanol to quench endogenous peroxidase activity.
6. Blot and rinse the slides in automation buffer four times at RT.
7. Incubate with blocking solution containing serum form the host of the secondary antibody for 10 min at 40 °C and rinse once.

8. Incubate with primary antibody for 15 min at 40 °C (*see* **Note 1**).
9. Blot and rinse five times in automation buffer at RT.
10. Incubate with secondary antibody for 15 min at 40 °C (*see* **Note 1**).
11. Blot and rinse five times in automation buffer at RT.
12. Incubate with ABC solution (detector complex) for 10 min at 40 °C.
13. Blot and rinse five times in automation buffer at RT.
14. Rinse with peroxidase enhancer solution once at RT and blot.
15. Incubate with chromogen reaction solution for 10 min at 40 °C.
16. Blot to remove chromogen reaction solution.
17. Incubate with chromogen reaction solution for 10 additional minutes at 40 °C.
18. Blot and rinse with distilled water twice at RT.
19. Incubate with aqueous hematoxylin for 10 s at RT. This step is used to assist in tissue visualization (*see* **Note 3**).
20. Rinse three times in automation buffer at RT.
21. Rinse with distilled water once at RT.
22. Remove the slides from the MicroProbe holder. Allow the slides to air dry.
23. Melt glycerol gelatin in hot water.
24. Drop glycerol gelatin onto slide and apply coverslip.
25. Allow the mounting media to completely set at RT. Clean with Windex and check the slides under microscope.

3.2. Assessment of Immunostaining

Subsection 1.8.2 has described some controls that can be performed to assess the potential of the authenticity of the chromogen development to specific antigen–antibody interactions. Once the immunostaining is complete, there can be additional indications gleaned from the viewing of the tissue sections under the microscope. For instance, it is usually encouraged to cut multiple atherosclerotic sections as described previously *(5,13)*. By viewing serially cut tissue sections, its may be assessed whether the chromogen development has been in a consistent region. This can be used to rule out artifacts that are attributable to factors such as incomplete washing of a specific section.

Immunostaining has been used classically as a method to qualitatively determine the presence and spatial distribution of antigens. However, more recently, it has been used as a quantitative technique in atherosclerosis research. The most common approach is to determine the percent of the area of an atherosclerotic tissue section that immunostains for a specific cellular antigen. For example, the percent of an area of lesions that immunostains for the macrophage antigen, CD68. This technique commonly uses a thresholding approach in which the image software will be set to measure the area above a certain hue intensity in a region of interest. Unfortunately, immunostaining tends to produce a continuum of hues and therefore the threshold will be user dependent. This potential shortcoming can be overcome by using multiple operators that are blinded to the experimental design. Another issue is that the antigen is unlikely to be uniformly present throughout the cell. For example, macrophages become so grossly hypertrophied in atherosclerotic lesions that an antigen such as CD68 will only be localized in a portion of each individual cell. Therefore, descriptions of the "percent area of macrophages in an atherosclerotic lesion" should be interpreted with caution.

4. Notes

1. There can be some variance in the affinity and specificity of antibodies among different lots. This is particularly the case for polyclonal antibodies that are used as either primary or secondary antibodies as they tend to be made in larger animals in which genetic homogeneity cannot be the same as in rodent colonies. Therefore, it is preferable to record the lot number of the reagents used in each experiment.
2. While the MicroProbe system permits rapid immunostaining, the bathing of the tissues by capillary action sometimes leads to air bubbles and subsequent incomplete coverage of the tissue during processing. Therefore, each slide should be carefully examined to ensure it is completely covered by fluid at each step of the procedure.
3. The duration of hematoxylin staining may alter between different types of atherosclerotic tissues. Incubation of hematoxylin with a highly cellular atherosclerotic lesion can lead to intense counterstaining that may obscure the immunostaining. The use of hematoxylin counterstaining may need some prior study to determine the optimal incubation time with tissue sections.

Acknowledgments

We thank members of the laboratory for providing feedback on this manuscript. The author's laboratory is supported by grants from the National Institutes of Health (HL80100 and HL70231). Hong Lu was supported from a postdoctoral fellowship from the Ohio Valley Affiliate of the American Heart Association.

References

1. Daugherty, A., and Rateri, D. L. (1993) Pathogenesis of atherosclerotic lesions. *Cardiol. Rev.* **1**, 157–166.
2. Ross R (1999) Atherosclerosis–an inflammatory disease. *N. Engl. J. Med.* **340**, 115–126.
3. Lusis, A. J. (2000) Atherosclerosis. *Nature* **407**, 233–241.
4. Daugherty, A. (2002) Mouse models of atherosclerosis. *Am. J. Med. Sci.* **323**, 3–10.
5. Daugherty, A., and Whitman, S. C. (2003) Quantification of atherosclerosis in mice. *Methods Mol. Biol.* **209**, 293–309.

6. Tangirala, R. K., Rubin, E. M., and Palinski, W. (1995) Quantitation of atherosclerosis in murine models: correlation between lesions in the aortic origin and in the entire aorta, and differences in the extent of lesions between sexes in LDL receptor-deficient and apolipoprotein E-deficient mice. *J. Lipid Res.* **36**, 2320–2328.

7. Rosenfeld, M. E., Polinsky, P., Virmani, R., Kauser, K., Rubanyi, G., and Schwartz, S. M. (2000) Advanced atherosclerotic lesions in the innominate artery of the ApoE knockout mouse. *Arterioscler. Thromb. Vasc. Biol.* **20**, 2587–2592.

8. Daugherty, A., Rateri, D. L., Lu, H., Inagami, T., and Cassis, L. A. (2004) Hypercholesterolemia stimulates angiotensin peptide synthesis and contributes to atherosclerosis through the AT1A receptor. *Circulation* **110**, 3849–3857.

9. Zhao, L., Moos, M. P., Grabner, R., Pedrono, F., Fan, J., Kaiser, B., John, N., Schmidt, S., Spanbroek, R., Lotzer, K., Huang, L., Cui, J., Rader, D. J., Evans, J. F., Habenicht, A. J., and Funk, C. D. (2004) The 5-lipoxygenase pathway promotes pathogenesis of hyperlipidemia-dependent aortic aneurysm. *Nat Med.* **10**, 966–973.

10. Daugherty, A., Dunn, J. L., Rateri, D. L., and Heinecke, J. W. (1994) Myeloperoxidase, a catalyst for lipoprotein oxidation, is expressed in human atherosclerotic lesions. *J. Clin. Invest.* **94**, 437–444.

11. van Driel, I. R., Goldstein, J. L., Sudhof, T. C., and Brown, M. S. (1987) First cysteine-rich repeat in ligand-binding domain of low density lipoprotein receptor binds Ca2+ and monoclonal antibodies, but not lipoproteins. *J. Biol. Chem.* **262**, 17443–17449.

12. Ishibashi, S., Brown, M. S., Goldstein, J. L., Gerard, R. D., Hammer, R. E., and Herz, J. (1993) Hypercholesterolemia in low density lipoprotein receptor knockout mice and its reversal by adenovirus-mediated gene delivery. *J. Clin. Invest.* **92**, 883–893.

13. Daugherty, A., and Rateri, D. L. (2005) Development of experimental designs for atherosclerosis studies in mice. *Methods* **36**, 129–138.

14. Hu, Y., Zhang, Z., Torsney, E., Afzal, A. R., Davison, F., Metzler, B., and Xu, Q. (2004) Abundant progenitor cells in the adventitia contribute to atherosclerosis of vein grafts in ApoE-deficient mice. *J. Clin. Invest.* **113**, 1258–1265.

15. Saraff, K., Babamusta, F., Cassis, L. A., and Daugherty, A. (2003) Aortic dissection precedes formation of aneurysms and atherosclerosis in angiotensin II-infused, apolipoprotein E-deficient mice. *Arterioscler. Thromb. Vasc. Biol.* **23**, 1621–1626.

16. Lutgens, E., Gijbels, M., Smook, M., Heeringa, P., Gotwals, P., Koteliansky, V. E., and Daemen, M. J. A. P. (2002) Transforming growth factor-beta mediates balance between inflammation and fibrosis during plaque progression. *Arterioscler. Thromb. Vasc. Biol.* **22**, 975–982.

17. Daugherty, A., Manning, M. W., and Cassis, L. A. (2000) Angiotensin II promotes atherosclerotic lesions and aneurysms in apolipoprotein E-deficient mice. *J. Clin. Invest.* **105**, 1605–1612.

18. Zhou, X. H., Stemme, S., and Hansson, G. K. (1996) Evidence for a local immune response in atherosclerosis: CD4(+) T cells infiltrate lesions of apolipoprotein-E-deficient mice. *Am. J. Pathol.* **149**, 359–366.

19. Roselaar, S. E., Kakkanathu, P. X., and Daugherty, A. (1996) Lymphocyte populations in atherosclerotic lesions of apoE − / − and LDL receptor−/− mice. Decreasing density with disease progression. *Arterioscler. Thromb. Vasc. Biol.* **16**, 1013–1018.

20. Trogan, E., Choudhury, R. P., Dansky, H. M., Rong, J. X., Breslow, J. L., and Fisher, E. A. (2002) Laser capture microdissection analysis of gene expression in macrophages from atherosclerotic lesions of apolipoprotein E-deficient mice. *Proc. Natl. Acad. Sci. U. S. A.* **99**, 2234–2239.

21. Boisvert, W. A., Santiago, R., Curtiss, L. K., and Terkeltaub, R. A. (1998) A leukocyte homologue of the IL-8 receptor CXCR-2 mediates the accumulation of macrophages in atherosclerotic lesions of LDL receptor-deficient mice. *J. Clin. Invest.* **101**, 353–363.

22. Rattazzi, M., Bennett, B. J., Bea, F., Kirk, E. A., Ricks, J. L., Speer, M., Schwartz, S. M., Giachelli, C. M., and Rosenfeld, M. E. (2005) Calcification of advanced atherosclerotic lesions in the innominate arteries of ApoE-deficient mice: potential role of chondrocyte-like cells. *Arterioscler. Thromb. Vasc. Biol.* **25**, 1420–1425.

23. Napoli, C., Ackah, E., deNigris, F., DelSoldato, P., DArmiento, F. P., Crimi, E., Condorelli, M., and Sessa, W. C. (2002) Chronic treatment with nitric oxide-releasing aspirin reduces plasma low-density lipoprotein oxidation and oxidative stress, arterial oxidation-specific epitopes, and atherogenesis in hypercholesterolemic mice. *Proc. Natl. Acad. Sci. U. S. A.* **99**, 12467–12470.

5

Surgical Animal Model of Ventricular Hypertrophy

Giuseppe Marano and Alberto U. Ferrari

Summary

In response to an increased afterload, the myocardium undergoes a complex adaptation by which wall stress is normalized and cardiac output is maintained. Although the consensus suggests that the increase of the myocardial mass is a necessary adaptive process to accommodate the increased workload, there is growing evidence that hypertrophy ultimately results in pathological remodeling and deterioration of cardiac function. Despite intense investigation, our understanding of the cellular mechanisms that are responsible for the initiation and the maintenance of this adaptation is largely incomplete and preventing or regressing left ventricular hypertrophy (LVH) is a major challenge. This chapter provides a detailed description of the procedures necessary to induce LVH by coarctation of the transverse aorta and to analyze the effects of the increased hemodynamic load on cardiac mass, cardiomyocyte size, and cardiac performance.

Key Words: Cardiac hypertrophy; Pressure overload; Cardiac surgery; Aorta banding.

1. Introduction

The availability of animal models mimicking human diseases is critical to understand the underlying pathogenetic mechanisms of many diseases, and cardiac hypertrophy/dysfunction/failure makes no exception. The aim of this chapter is to provide a detailed description of the techniques required to create a surgical model of pressure overload-induced cardiac hypertrophy through coarctation of the transverse portion of the aorta. Although initially implemented in the rat species, this has, in the beginning of 1990th years, shifted to the mouse species for a number of reasons, including small size, short gestation period, large litter size, and relatively low maintenance costs. Moreover, the mouse

From: *Methods in Molecular Medicine, Vol. 139: Vascular Biology Protocols*
Edited by: N. Sreejayan and J. Ren © Humana Press Inc., Totowa, NJ

genome has been largely characterized, and gene-targeted "knockout" (KO) and transgenic overexpressing mice have been generated. Thus, the techniques described herein formally relate to the mouse, but it should be appreciated that they are readily applicable to the rat as well.

With the advent of the DNA era, the mouse has become a very appealing animal species in biomedical research. In recent years, application of pressure overload in genetically engineered mice has allowed to elucidate the role of certain proteins and their related pathways in the development and maintenance of pressure overload-induced cardiac hypertrophy. In response to pressure overload, left ventricular hypertrophy (LVH) is thought to develop as an adaptive response by which cardiac output is maintained. Although this adaptation was thought to be beneficial at least at its initial stages, it is increasingly appreciated that it ultimately results in pathological remodeling and deterioration of cardiac function.

Recent attention has focused on the role of chronic activation of adrenergic receptor (AR)-mediated signaling. Nine different AR subtypes have been identified. In the heart, β_1-AR is the predominant receptor subtype approaching 75–80% of the total β-ARs. The specific function of individual AR subtypes has recently been addressed using genetically engineered mouse models. Overexpression of β_1-ARs and β_2-ARs (but not β_3-ARs), Gsα, and protein kinase A is known to cause myocardial hypertrophy and fibrosis (1–5), suggesting that chronic activation of the β-AR system in the heart is sufficient to induce a maladaptive ventricular remodeling. Mice overexpressing β_2-ARs, when challenged with transverse aortic constriction (TAC), also have a much higher incidence of progression to heart failure and death (6). Additionally, experiments in KO mice, in which epinephrine and norepinephrine are lacking (Dbh KO mice) (7), indicate that catecholamines are necessary for the increase in heart size after pressure overload. However, when hearts of β_1-AR and β_2-AR double KO mice, in which the predominant cardiac β-AR subtypes are lacking, are subjected to chronic pressure overload by transverse aortic banding, cardiac hypertrophy, cardiomyocyte enlargement, and increased gene expression of atrial natriuretic peptide (ANP) and β-isoform of myosin heavy chain (β-MHC) at 4 weeks after aortic banding occur (8). At this time, whether β-ARs are critical for the development of cardiac hypertrophy in response to pressure overload is still a matter of debate. Expression of a constitutively active mutant of α_{1B}-AR in the heart leads to ventricular hypertrophy with increased ANF expression (9), supporting the sufficiency of this receptor subtype for hypertrophy development. Aortic banding markedly reduces survival in mice deficient in α_{2A}-ARs or α_{2C}-ARs and induces a much higher incidence of heart failure along with enhanced LVH and fibrosis and elevated circulating catecholamines *(10)*. Cardiac pressure overload in mice with double KO of α_{1A}-ARs and α_{1B}-ARs causes a similar increase of heart weight and myocyte cross-sectional area in KO and wild-type mice 2 weeks after surgery. However, KO mice show a reduced survival and systolic function. In addition, KO hearts after TAC have increased interstitial fibrosis, increased apoptosis, and failed induction of the fetal hypertrophic genes *(11)*. These results indicate that α-AR signaling is required for cardiac adaptation in response to pressure overload. On the contrary, recent evidence in aortic banded rats indicates that α-AR signaling also contributes to the genesis of LVH and fibrosis as well as to disease progression and death *(12,13)*. Thus, much more physiological and molecular research is needed to obtain a clear picture about the role of ARs in the genesis and progression of hypertensive heart disease.

The aortic banding rodent model represents a convenient and relatively simple approach to reproducibly obtain animals with pressure overload-induced LVH; the model lends itself to countless applications for physiological, pharmacological, and molecular studies about a condition whose burden for human disease is extremely heavy.

2. Materials

2.1. Anesthesia and Tracheal Intubation

1. Plexiglas anesthesia box (13 cm height × 24 cm width × 13 cm diameter).
2. Anesthesia machine (Excel 210, Ohmeda, Madison, USA).
3. Anesthetic vaporizer for isoflurane.
4. Isoflurane (Forane, Abbott, Rome, Italy) and oxygen.
5. Heated rodent operating table (Harvard Apparatus, BS4 50–1247, Holliston, USA).
6. Arthroscope, straight bore, outer diameter (OD) 1.7 mm, 0° (straight), length 58 mm (Olympus, model A7002, Milan, Italy).
7. Light source (Olympus, model CLH 2).
8. Video camera (Olympus, model OTV SC).
9. Soft-tip, stiff guidewire (ARROW INT., Reading, PA, USA).
10. Lidocaine 1%.
11. Intubation tube: a 3-cm-long polyethylene tube (PE) catheter (OD = 0.7 mm).
12. PE cannula (OD = 1.1 mm, length = 50 mm).
13. PE catheter (OD = 0.7 mm), 3 cm long.
14. Y-shaped connector.

2.2. Pulmonary Ventilation and Aortic Banding, Evaluation of Hemodynamic Response, and Assessment of Cardiac Hypertrophy

1. Rodent volume controlled ventilator (Harvard Apparatus, 687 series).
2. Surgical microscope (Zeiss, OPMI-1FC, Arese, Italy).

3. Small animal clipper (Harvard Apparatus, size 50 clipper blade).
4. Curved forceps (Harvard Apparatus, BS4 52–34307).
5. Curved and straight scissors (Harvard Apparatus).
6. Modified chest retractors to spread the wound 4–5 mm in width.
7. Betadine® solution, sterile gloves, sutures (Ethicon, Neuchatel, Switzerland), and needle holders (Medicon, Tuttlingen, Germany).
8. Analgesic drugs (buprenorphine).
9. Body temperature controller (Harvard Apparatus, homeothermic blanket system).
10. 1.4-F Millar pressure catheter (Millar Instruments, Houston, TX, USA).
11. Image analysis system (morphometric, Universal Imaging, Downingtown, PA USA).

3. Methods

3.1. Tracheal Intubation

1. Place the mouse into a clear Plexiglas anesthesia box that allows unimpeded visual monitoring of the animals, for example, one can easily determine whether breathing disturbances or apnea are occurring. The box is connected to the anesthesia machine and has two tubes attached for gas delivery and exhaust (*see* **Fig. 1**). The gas delivery carries oxygen mixed with isoflurane (3–4% in oxygen), a halogenated volatile anesthetic, which produces a rapid induction and recovery of anesthesia.

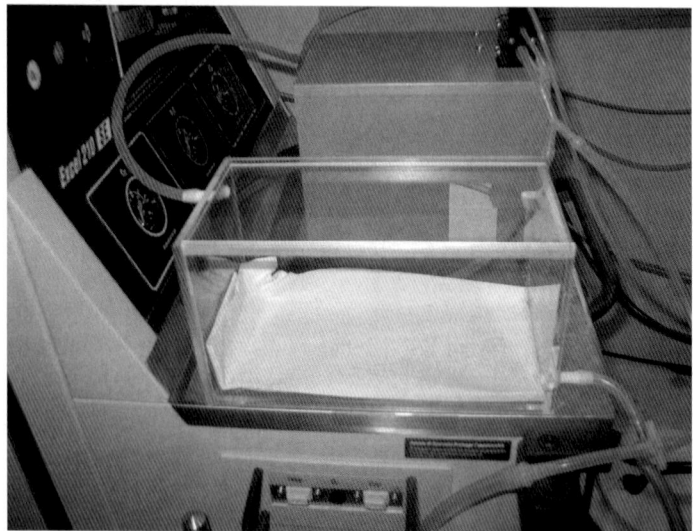

Fig. 1. The Plexiglas anesthesia box connected to the anesthesia machine. Mouse pulmonary ventilator and tubes for gas delivery and exhaust are also shown.

2. After the mice are fully anesthetized (about after 5 min), transfer them from the box to a heated rodent operating table. Place the mouse supine on a heated rodent operating table to avoid a fall in body temperature, with the head and the neck extended.

3. Insert a straight, small bore arthroscope, connected to a light source and a video camera to the left of the incisors (*see* **Note 1**).

4. Move it at the base of the tongue to allow visualization of the glottis and the rhythmic opening of the vocal cords (*see* **Fig. 2**).

5. Remove secretions by a PE cannula connected to a syringe to prevent aspiration during intubation procedure.

6. Thereafter, insert a soft-tip, stiff guidewire in the trachea across the vocal cords during inspiration to facilitate the introduction of the intubation tube, a 3-cm-long PE catheter, over the guide. Apply a drop of 1% lidocaine on the tip of the tube to numb the throat and reduce the gag reflex.

7. Advance the cannula into the trachea for 6 mm starting from the vocal cords, remove the guide, and connect the mouse to the Y-shaped connector of the pulmonary ventilator. Insert the connector of the intubation tube into the Y-shaped connector to minimize the dead space of the breathing circuit. Restraint and intubation of the mouse takes about 3 min.

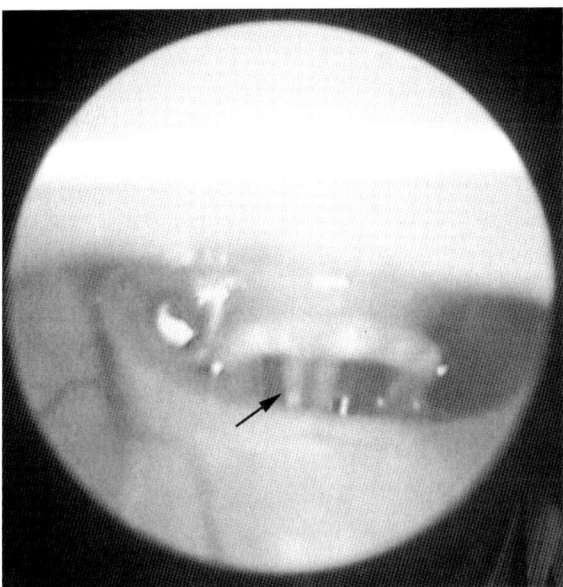

Fig. 2. In this image, the laryngis aditus including glottis plus vocal cords (arrow) as it appears on the Hantarex monitor connected to the video camera of the arthroscope.

8. Evaluate the success of the intubation by the visual observation of the rhythmic expansion of the thorax synchronous with the rate of the mechanical pulmonary ventilation.

3.2. Anesthesia and Pulmonary Ventilation

1. Induce surgical anesthesia by the administration of isoflurane (1.5–2.5% in 100% of oxygen). The most useful indicator of depth of anesthesia is the toe-pinch reflex. Pinch the toe of the hind limb between the fingernails. If the mouse attempts to withdraw its limb, the inhaled anesthetic concentration should be increased. When there is no response to the toe pinch, medium-deep anesthesia has been attained. Thereafter, fix the hind legs and tail to the operating table with strands of tape.

2. Connect the mouse to a rodent volume controlled ventilator to ensure adequate pulmonary ventilation. The tidal volume and ventilation rate can be calculated from formulas provided by the company: $Vt = 0.0062 \times WB^{1.01}$ and $Vr = 53.5 \times WB^{-0.26}$, where Vt is tidal volume in liters, Vr is ventilation rate in strokes/minute, and WB is animal mass in kilogram. For a 30-g mouse, we set the ventilator rate at 133 strokes/min and the tidal volume at 0.3 ml, which accounts for dead space of our breathing circuit. Surgical procedures are done with the aid of a surgical microscope at $\times 1.6$ magnification level for better visualization of the operative field. Before surgery, disinfect the operating field with Betadine solution.

3.3. Thoracic Aortic Banding (see Note 2)

1. Shave the animal's chest with a small animal clipper.
2. Incise the thorax at the level of the second intercostal space, with paw extended at $90°$. Before the incision, inject subcutaneously 0.1 ml of 0.1% lidocaine, a local anesthetic.
3. Perform a transverse incision of 5 mm in length of the skin with scissors 2 mm away from the left sternal border.
4. Cut both layers of thoracic muscles, taking care to avoid the superficial thoracic vein that runs under the skin at the lateral corner of the incision.
5. Open the chest cavity with scissors, taking care not to damage the underlying lung and the internal thoracic artery that runs along the sternal border.
6. Insert the chest retractor to spread the wound 4–5 mm in width.
7. Pull away the thymus and fat with forceps to visualize the great vessels (*see* **Fig. 3**).
8. Place the curved forceps under the transverse aorta between the innominate and the left common carotid arteries. Then, grasp a 7-0 silk string by forceps and move it underneath the aorta.
9. Perform aortic constriction by tying the 7-0 silk suture ligature against an L-shaped 25-G to 27-G needle depending on the amount of stenosis and degree of hypertrophy desired.

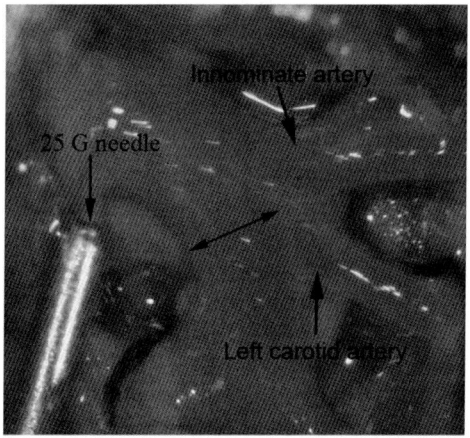

0.5mm

Fig. 3. Transverse aortic banding. Position of a 7-0 silk suture under the transverse aorta. The ligature will be placed between the innominate and the left common carotid arteries. The needle shown in the figure is a 25-G needle, with outer diameter of 0.5 mm. The aortic constriction is performed by tying the 7-0 silk suture ligature against the 25-G needle.

10. Remove rapidly the needle to minimize ischemia and a buildup of systemic arterial pressure. No more of two knots are necessary to secure the tie.
11. Remove the chest retractor and move back the thymus and fat to their normal position.
12. Close the chest cavity by bringing together the ribs with 6-0 nylon, all layers of muscle and skin with 6-0 absorbable suture and 6-0 nylon, respectively.
13. Treat the wound with Betadine. The whole procedure does not take more than 15 min.
14. For the sham operation, put the mice on a similar procedure but stop the intervention when the suture is moved underneath the transverse aorta without placing a ligature.
15. When the mouse begins spontaneous breathing, extubate the mouse taking care of removing the intubation tube slowly to avoid aspiration of oral cavity secretions.
16. Immediately after the operation, give a dose of analgesic (buprenorphine, 0.1 mg/kg) subcutaneously to alleviate pain and place the mouse in a clean individual cage to prevent possible contamination of the surgical site or injury caused by other animals.
17. Remove nylon sutures from the skin 8 days after surgery.

3.4. Evaluation of Hemodynamic Responses

Assessment of left ventricular pressure (LVP) is important for ensuring that the aortic banding operation has been successful and that the LVP is similarly affected within an experimental group.

1. Anesthetize mice through inhalation of 1.5–2% isoflurane mixed with oxygen.
2. Maintain the temperature between 36 and 37 °C using a temperature controller connected to a rectal probe.
3. Expose the right carotid artery for a length of 5 mm.
4. Insert a 1.4-F Millar pressure catheter into the artery and retrogradely advance it into the left ventricle.
5. Connect the catheter to a computerized data-acquisition system to record heart rate, LVP, maximal rate of LVP rise and fall ($+LV$ dP/dt and $-LV$ dP/dt, respectively), ejection time, and isovolumetric relaxation time constant (τ) at a sampling rate of 1 KHz.
6. Confirm the correct position of the catheter tip in the left ventricle by visualizing the waveform of the LVP on a monitor.
7. Insert a venous catheter (PE-10/silicone catheter), made of 4 mm of PE-10 tubing joined to 2 cm of silicone tubing and filled with saline, into the left jugular vein for saline and drug administrations.
8. After a stabilization period of 20 min, record hemodynamic signals for 10–15 min and save them for later analysis.

3.5. Assessment of Cardiac Hypertrophy

The fundamental characteristics of a hypertrophied heart are the increased mass, typically measured as whole heart or isolated chamber wet mass, and the increased size of cardiomyocyte with respect to the normal cell.

1. Fix the left ventricle with 10% buffered formalin solution and embed it in paraffin.
2. Cut sections of $4 - \mu$m thick and stain with hematoxylin and eosin for the measurement of myocyte cross-sectional area or with the Sirius red/picric acid to determine interstitial fibrosis by quantitative morphometry.
3. Accept the visual fields for quantitative analysis if cross-sections of myocytes are present, nuclei are visible, and their cellular membranes are intact.
4. Calculate myocyte mean cross-sectional area by measuring the dimensions of no less than 50 cells/section (four or five sections of each left ventricle). Assess cardiomyocyte width by marking the borders of the cells.
5. Determine the expression of the "fetal gene program," which classically includes the genes of ANP, the β-MHC, and the skeletal muscle isoform of actin, by real-time polymerase chain reaction to subclassify hypertrophy phenotypes.

6. Perform echocardiographic examinations under isoflurane anesthesia, 1 in 100% oxygen. Obtain good quality short-axis images of the left ventricle, print M-mode freeze frames on echocardiographic paper, and digitize them.
7. Measure LV end-systolic diameter, LV end-diastolic diameter, and wall thickness using an image-analysis system from three consecutive beats.

4. Notes

1. At this time, the mouse is ready for the intubation. Orotracheal intubation in mice is a complicated technique because of their oropharyngeal anatomy and the difficulty in visualizing the laryngis aditus. In particular, the relatively large incisors and the restricted mandible opening make this practice difficult to perform, and there is the added risk of bleeding and laryngeal perforation. There have been some reports in the literature regarding the procedure of intubation in mice, but these methods require a direct, surgical exposure of the larynx or an elaborate animal position, specially constructed equipment, and highly trained personnel *(14,15)*. The method described here is simple, reliable, and requires a small bore, straight fiberoptic arthroscope for the visualization of the glottis and the vocal cords on a monitor, a PE for tracheal intubation, and a soft-tip, stiff guidewire to facilitate the introduction of the PE cannula across the vocal cords without damaging the larynx *(16)*.
2. The method has been described by Rockman and colleagues *(17)* in 1991.

References

1. Engelhardt, S., Hein, L., Wiesmann, F., and Lohse, M.J. (1999) Progressive hypertrophy and heart failure in β 1-adrenergic receptor transgenic mice. *Proc. Natl. Acad. Sci. U. S. A.* **96**, 7059–7064.
2. Liggett, S.B., Tepe, N.M., Lorenz, J.N., Canning, A.M., Jantz, T.D., Mitarai, S., et al. (2000) Early and delayed consequences of β2-adrenergic receptor overexpression in mouse hearts: critical role for expression level. *Circulation* **101**, 1707–1714.
3. Kohout, T.A., Takaoka, H., McDonald, P.H., Perry, S.J., Mao, L., Lefkowitz, R.J., et al. (2001) Augmentation of cardiac contractility mediated by the human β3-adrenergic receptor overexpressed in the hearts of transgenic mice. *Circulation* **104**, 2485–2491.
4. Iwase, M., Bishop, S.P., Uechi, M., Vatner, D.E., Shannon, R.P., Kudej, R.K., et al. (1996) Adverse effects of chronic endogenous sympathetic drive induced by cardiac Gsα overexpression. *Circ. Res.* **78**, 517–524.
5. Antos, C.L., Frey, N., Marx, S.O., Reiken, S., Gaburjakova, M., Richardson, J.A., et al. (2001) Dilated cardiomyopathy and sudden death resulting from constitutive activation of protein kinase a. *Circ. Res.* **89**, 997–1004.

6. Du, X.J., Autelitano, D.J., Dilley, R.J., Wang, B., Dart, A.M., and Woodcock, E.A. (2000) β2-adrenergic receptor overexpression exacerbates development of heart failure after aortic stenosis. *Circulation* **101**, 71–77.

7. Rapacciuolo, A., Esposito, G., Caron, K., Mao, L., Thomas, S.A., and Rockman, H.A. (2001) Important role of endogenous norepinephrine and epinephrine in the development of in vivo pressure-overload cardiac hypertrophy. *J. Am. Coll. Cardiol.* **38**, 876–882.

8. Palazzesi, S., Musumeci, M., Catalano, L., Patrizio, M., Stati, T., Michienzi, S., et al. (2006) Pressure overload causes cardiac hypertrophy in beta1-adrenergic and beta2-adrenergic receptor double knockout mice. *J. Hypertens.* **24**, 563–571.

9. Milano, C.A., Dolber, P.C., Rockman, H.A., Bond, R.A., Venable, M.E., Allen, L.F., et al. (1994) Myocardial expression of a constitutively active alpha 1B-adrenergic receptor in transgenic mice induces cardiac hypertrophy. *Proc. Natl. Acad. Sci. U. S. A.* **91**, 10109–10113.

10. Brede, M., Wiesmann, F., Jahns, R., Hadamek, K., Arnolt, C., Neubauer, S., et al. (2002) Feedback inhibition of catecholamine release by two different alpha2-adrenoceptor subtypes prevents progression of heart failure. *Circulation* **106**, 2491–2496.

11. O'Connell, T.D., Swigart, P.M., Rodrigo, M.C., Ishizaka, S., Joho, S., Turnbull, L., et al. (2006) Alpha1-adrenergic receptors prevent a maladaptive cardiac response to pressure overload. *J. Clin. Invest.* **116**, 1005–1015.

12. Perlini, S., Palladini, G., Ferrero, I., Tozzi, R., Fallarini, S., Facoetti, A., et al. (2005) Sympathectomy or doxazosin, but not propranolol, blunt myocardial interstitial fibrosis in pressure-overload hypertrophy. *Hypertension* **46**, 1213–1218.

13. Perlini, S., Ferrero, I., Palladini, G., Tozzi, R., Gatti, C., Vezzoli, M., et al. (2006) Survival benefits of different antiadrenergic interventions in pressure overload left ventricular hypertrophy/failure. *Hypertension* **48**, 93–97.

14. Costa, D.L., Lehmann, J.R., Harold, W.M., and Drew, R.T. (1986) Transoral tracheal intubation of rodents using a fiberoptic laryngoscope. *Lab. Anim. Sci.* **36**, 256–261.

15. Brown, R.H., Walters, D.M., Greenberg, R.S., and Mitzner, W. (1999) A method of endotracheal intubation and pulmonary functional assessment for repeated studies in mice. *J. Appl. Physiol.* **87**, 2362–2365.

16. Vergari, A., Polito, A., Musumeci, M., Palazzesi, S., and Marano, G. (2003) Video-assisted orotracheal intubation in mice. *Lab. Anim.* **37**, 204–206.

17. Rockman, H.A., Ross, R.S., Harris, A.N., Knowlton, K.U., Steinhelper, M.E., Field, L.J., et al. (1991) Segregation of atrial-specific and inducible expression of an atrial natriuretic factor transgene in an in vivo murine model of cardiac hypertrophy. *Proc. Natl. Acad. Sci. U. S. A.* **88**, 8277–8281.

6

Animal Models of Hypertension

Brett M. Mitchell, Thomas Wallerath, and Ulrich Förstermann

Summary

Hypertension affects approximately 25% of adults and is a major risk factor for cardiovascular disease. Although there are currently adequate therapeutic options for humans with hypertension, the molecular mechanisms underlying hypertension are still relatively unknown. The generation of hypertensive animal models provides an excellent modality to not only study the pathophysiology but also test innovative therapeutics. This chapter describes the detailed methods that utilize the drinking water of rats to develop models of nitric oxide synthase (NOS) inhibition-induced, guanosine triphosphate cyclohydrolase (GTPCH) inhibition-induced, and glucocorticoid-induced hypertension.

Key Words: Hypertension; Experimental; Nitric oxide synthase; GTP cyclohydrolase 1; Tetrahydrobiopterin; Glucocorticoids.

1. Introduction

Animal models allow researchers to study the signaling pathways and molecular mechanisms involved in the pathophysiology of hypertension. Additionally, scientists can rapidly test various therapeutic agents and analyze the effects on vascular function and blood pressure regulation. Several genetic models of hypertension exist including the spontaneously hypertensive rat and the Dahl salt-sensitive rat. These are currently the best models of human essential and salt-dependent hypertension, respectively. These genetically altered strains contribute greatly to the study of hypertension pathophysiology; however, these animals have abnormal vascular function prior to the development of elevated blood pressures thus making it difficult to precisely

From: *Methods in Molecular Medicine, Vol. 139: Vascular Biology Protocols*
Edited by: N. Sreejayan and J. Ren © Humana Press Inc., Totowa, NJ

determine the contributing mechanisms. The ability to induce hypertension in normotensive, control animals bypasses the genetic predisposition and enables researchers to elucidate the mechanisms involved in the onset and development of hypertension.

A simple, and more importantly, non-invasive method to induce elevations in blood pressure is by introducing various inhibitors, agonists, or synthetic compounds into the drinking water of rodents. This method can be used to examine the role of a protein in vascular function and blood pressure regulation as well as to mimic human genetic or idiopathic conditions in which the patient presents with hypertension. Here we describe three models of hypertension that we have used to achieve these goals. Two of the models are related to nitric oxide synthase (NOS) and are used to study the role of this enzyme and one of its cofactors in blood pressure regulation, and the last model uses the synthetic glucocorticoid dexamethasone (DEX) to mimic the hypertensive conditions in humans with Cushing's syndrome and those treated with excess synthetic glucocorticoids (i.e., organ transplant recipients).

NO plays a key role in the regulation of vascular tone, and alterations in NO production modulate endothelium-dependent vasodilation and blood pressure. Inhibition of NOS, which produces NO from the conversion of arginine and oxygen to citrulline, with N^{ω}-nitro-L-arginine (L-NNA) in the drinking water of rats increases blood pressure dose and time dependently *(1–4)*. In vivo inhibition of NOS has helped elucidate the numerous roles that NO plays in blood pressure regulation and vascular signaling in the intact animal.

NOS requires the cofactor tetrahydrobiopterin (BH4) for catalytic activity. In the absence of BH4, the nicotinamide adenine dinucleotide phosphate (NADPH) oxidase activity of NOS uncouples from NO synthesis, resulting in an increased production of superoxide at the expense of NO *(5–8)*. BH4 is produced by a de novo pathway from guanosine triphosphate (GTP) by GTP cyclohydrolase (GTPCH), the first and rate-limiting enzyme of a three-step pathway. Previous in vitro studies reported that inhibition of GTPCH with 2,4-diamino-6-hydroxypyrimidine (DAHP) decreased BH4 levels and subsequently NO production *(9–11)*. By pharmacologically reducing intracellular BH4 through in vivo GTPCH inhibition, we found that uncoupled NOS leads to decreased vasodilation and increased blood pressure *(12)*. This model is useful in studying the role of the NOS cofactor BH4 in blood pressure regulation and the cardiovascular effects of uncoupled NOS.

Lastly, excess glucocorticoids in humans and animals can cause Cushing's syndrome that is associated with an elevation in blood pressure. Although several mechanisms contribute to the elevation in blood pressure, collectively we have shown that glucocorticoid-induced down-regulation of eNOS and GTPCH/BH4 contributes greatly to the increased blood pressure *(13–15)*. The elucidation of the molecular mechanisms involved in the detrimental vascular effects of excess glucocorticoids will aid in the development of synthetic gluco-corticoids that do not affect the vasculature and therapeutics to reduce the incidence of Cushing's syndrome.

2. Materials

1. Glass or plastic water bottles (1 L).
2. Tap water.
3. L-NNA (Sigma, St. Louis, MO, USA).
4. DAHP (Sigma, St. Louis, MO, USA).
5. Water-soluble DEX (Sigma, St. Louis, MO, USA).
6. Stir bars.
7. Stirrer.

3. Methods

3.1. NOS Inhibition-Induced Hypertension

1. Add 1 L of tap water to the bottle.
2. Add 0.25–0.5 g of L-NNA to bottle and stir for approximately 4 h until completely dissolved.
3. Fill animal water bottles with solution noting the amount added.
4. Store remaining solution at 4 °C until needed. When more solution needs to be added to the animal's water bottle, remove the stock solution from the refrigerator and let it warm to room temperature.
5. Measure the amount of water the animal ingested daily (*see* **Note 1**).
6. Measure blood pressure through tail-cuff or telemetry (*see* **Note 2**).
7. Determine the effective concentration of the solution (*see* **Note 3**).

3.2. GTPCH Inhibition-Induced Hypertension

1. Add 1 L of tap water to the bottle.
2. Add 1.261 g of DAHP (10 mM) to the bottle and stir overnight until completely dissolved.
3. Fill animal water bottles with the DAHP solution noting the amount added.
4. Store remaining solution at 4 °C until needed. When more solution needs to be added to the animal's water bottle, remove the stock solution from the refrigerator and let it warm to room temperature.
5. Measure the amount of water the animal ingested daily (*see* **Notes 1** and **4**).
6. Measure blood pressure through tail-cuff or telemetry (*see* **Notes 2** and **5**).
7. Determine the effective concentration of the solution (*see* **Note 3**).

3.3. Glucocorticoid-Induced Hypertension

1. Add 1 L of tap water to the bottle.
2. Add 46 mg DEX (this corresponds to ∼3 mg pure DEX, *see* **Note 6**) to bottle and stir for 5–10 min until completely dissolved.
3. Fill animal water bottles with the DEX noting the amount added.
4. Store remaining solution at 4 °C until needed. When more solution needs to be added to the animal's water bottle, remove the stock solution from the refrigerator and let it warm to room temperature.
5. Measure the amount of water the animal ingested daily (*see* **Notes 1** and **7**).
6. Measure blood pressure through tail-cuff or telemetry (*see* **Notes 2** and **8**).
7. Determine the effective concentration of the solution (*see* **Note 3**).

4. Results

4.1. NOS Inhibition-Induced Hypertension

We gave 0.25 g/L of L-NNA in the drinking water of 300-g rats, and after 4 days, systolic blood pressure was 163 ± 9 mmHg (mean $\pm$ SEM). Kanagy's group routinely administers 0.5 g/L of L-NNA in the drinking water of 250-g to 300-g rats, and after 14 days, systolic blood pressures average between 195 and 212 mmHg.

4.2. GTPCH Inhibition-Induced Hypertension

To examine the role of BH4 in blood pressure regulation and NO signaling in the intact animal, we administered 10 mmol/L DAHP in the drinking water of 300-g rats for 3 days and achieved systolic blood pressures that averaged 153 ± 5 mmHg after 1 day, 158 ± 5 mmHg after 2 days, and 155 ± 9 mmHg after 3 days of treatment.

4.3. Glucocorticoid-Induced Hypertension

Based on the average water intake of rats, 3 mg/L of DEX was added to the drinking water to reach a target dose of 0.3 mg/kg/day. After 1 week of treatment, rats were hypertensive with 24-h average blood pressures of $143 \pm 3 / 107 \pm 2$ mmHg (systolic/diastolic) compared with controls that had average blood pressures of $121 \pm 1 / 88 \pm 1$ mmHg.

5. Conclusions

The drinking water of rodents presents a simple, non-invasive route to study the roles of various proteins in vascular function and blood pressure regulation. It is important to remember that the cell permeability and specificity of the drug being studied may have profound effects on experimental results; however, through thorough testing and the inclusion of well-designed control studies, this method can be a powerful experimental tool for your biological questions. Additionally, studies in genetically altered rodents have increased markedly in recent years; however, numerous compensatory mechanisms may take place in the intact animal. Therefore, the inclusion of in vivo pharmacological studies to support genetic models of hypertension is warranted.

6. Notes

1. To determine how much water the animal is drinking daily, we have either measured the remaining amount of water in the animal's water bottle and subtracted this from the initial amount or continuously measured fluid input using a computer-based monitoring system (Feeding/Drinking Monitor, Technical & Scientific Equipment, Bad Homburg, Germany) that records the hydrostatic pressure in the water bottle.

2. Detailed methods regarding the measurement of blood pressures are beyond the scope of this chapter. A recent review paper discusses the advantages and disadvantages of several methods of measuring blood pressure in rodents *(16)*.

3. To estimate the effective concentration of the compound, determine the concentration of the solution in mg/mL (i.e., $0.5\,g/L = 0.5\,mg/mL$). Next, determine the daily intake of water (averages range from $\sim$75 to $\sim$200 mL/kg/day in rodents) and calculate the effective concentration in mg/kg/day.

4. Unpublished reports suggest that mice may be more sensitive to the drinking water containing DAHP; therefore, it may be necessary to supplement the drinking water containing DAHP so that the animals will not reduce their daily fluid intake. We suggest adding a small amount of sugar or something sweet, although not enough to make the animals hyperglycemic, to the DAHP water as well as the water of the control animals. We did not encounter this problem in rats when the DAHP was completely dissolved.

5. When characterizing the blood pressure of rats given DAHP in their drinking water, we observed that the elevated blood pressures from days 1 to 3 started to return to baseline on the 4th day. Additionally, rats tended to lose body weight while being given DAHP in their drinking water.

6. DEX is poorly soluble in water (solubility $<0.01\,mg/mL$); therefore, we recommend using DEX-2-hydroxypropyl-beta-cyclodextrin complex ("water-soluble DEX") from Sigma (solubility $>500\,mg/mL$). This compound contains approximately 65 mg pure DEX per gram.

7. When measuring the daily fluid intake of rats treated with DEX in the drinking water, we observed that the water consumption declined initially; however, after 5 days, the normal average water intake ($\sim$100 ml/kg/day; determined before performing our experiments) was reached.

8. The blood pressure of rats treated with DEX in the drinking water increased continuously over 5 days and reached a plateau after 7 days. Additionally, we observed that the rats lost body weight during DEX treatment.

Acknowledgments

L-NNA-induced and DAHP-induced hypertension studies were supported by an American Heart Association Pre-Doctoral Fellowship. DEX-induced hypertension studies were supported by Collaborative Research Center SFB 553, Project A1 from the Deutsche Forschungsgemeinschaft (Bonn).

References

1. Kanagy, N. L. (1997) Increased vascular responsiveness to alpha2-adrenergic stimulation during NOS inhibition-induced hypertension. *Am. J. Physiol. Heart Circ. Physiol.* **273**, H2756–H2764.
2. Bratz, I. N., Falcon, R., Partridge, L. D., and Kanagy, N. L. (2002) Vascular smooth muscle cell membrane depolarization after NOS inhibition hypertension. *Am. J. Physiol. Heart Circ. Physiol.* **282**, H1648–H1655.
3. Carter, R. W., Begaye, M., and Kanagy, N. L. (2002) Acute and chronic NOS inhibition enhances α_2-adrenoceptor-stimulated rhoA and rho kinase in rat aorta. *Am. J. Physiol. Heart Circ. Physiol.* **283**, H1361–H1369.
4. Mitchell, B. M., Dorrance, A. M., Ergul, A., and Webb, R. C. (2004) Sepiapterin decreases vasorelaxation in nitric oxide synthase inhibition-induced hypertension. *J. Cardiovasc. Pharmacol.* **43**, 93–98.
5. Wever, R. M. F., van Dam, T., van Rijn, H. J. M., de Groot, P. F., and Rabelink, T. J. (1997) Tetrahydrobiopterin regulates superoxide and nitric oxide generation by recombinant endothelial nitric oxide synthase. *Biochem. Biophys. Res. Commun.* **237**, 340–344.
6. Vasquez-Vivar, J., Kalyanaraman, B., Martasek, P., Hogg, N., Masters, B. S., Karoui, H., Tordo, P., and Pritchard, K. A. Jr. (1998) Superoxide generation by endothelial nitric oxide synthase: the influence of cofactors. *Proc. Natl. Acad. Sci. U. S. A.* **95**, 9220–9225.
7. Xia, Y., Tsai, A. L., Berka, V., and Zweier, J. L. (1998) Superoxide generation from endothelial nitric-oxide synthase. A calcium/calmodulin-dependent and tetrahydrobiopterin regulatory process. *J. Biol. Chem.* **273**, 25804–25808.
8. Werner-Felmayer, G., Werner, E. R., Fuchs, D., Hausen, A., Reibnegger, G., Schmidt, K., Weiss, G., and Wachter, H. (1993) Pteridine biosynthesis in human endothelial cells: impact on nitric oxide-mediated formation of cyclic GMP. *J. Biol. Chem.* **268**, 1842–1846.
9. Cosentino, F. and Katusic, Z. S. (1995) Tetrahydrobiopterin and dysfunction of endothelial nitric oxide synthase in coronary arteries. *Circulation* **91**, 139–144.

10. Kinoshita, H., Milstein, S., Wambi, C., and Katusic, Z. S. (1997) Inhibition of tetrahydrobiopterin biosynthesis impairs endothelium-dependent relaxations in canine basilar artery. *Am. J. Physiol. Heart Circ. Physiol.* **273**, H718–H724.
11. Xie, L., Smith, J. A., and Gross, S. S. (1998) GTP cyclohydrolase I inhibition by the prototypic inhibitor 2,4-diamino-6-hydroxypyrimidine. *J. Biol. Chem.* **273**, 21091–21098.
12. Mitchell, B. M., Dorrance, A. M., and Webb, R. C. (2003) GTP cyclohydrolase 1 inhibition attenuates vasodilation and increases blood pressure in rats. *Am. J. Physiol. Heart Circ. Physiol.* **285**, H2165–H2170.
13. Wallerath, T., Witte, K., Schafer, S. C., Schwarz, P. M., Prellwitz, W., Wohlfart, P., Kleinert, H., Lehr, H. A., Lemmer, B., and Forstermann, U. (1999) Downregulation of the expression of endothelial NO synthase is likely to contribute to glucocorticoid-mediated hypertension. *Proc. Natl. Acad. Sci. U. S. A.* **96**, 13357–13362.
14. Mitchell, B. M., Dorrance, A. M., and Webb, R. C. (2003) GTP cyclohydrolase 1 downregulation contributes to glucocorticoid hypertension in rats. *Hypertension* **41**, 669–674.
15. Wallerath, T., Godecke, A., Molojavyi, A., Li, H., Schrader, J., and Forstermann, U. (2004) Dexamethasone lacks effect on blood pressure in mice with a disrupted endothelial NO synthase gene. *Nitric Oxide* **10**, 36–41.
16. Kurtz, T. W., Griffin, K. A., Bidani, A. K., Davisson, R. L., and Hall, J. E. (2005) Recommendations for blood pressure measurement in humans and experimental animals. Part 2: blood pressure measurement in experimental animals. *Hypertension* **45**, 299–310.

7

Rat Models of Cardiac Insulin Resistance

Sanjoy Ghosh, Brian Rodrigues, and Jun Ren

Summary

Cardiovascular disease is the leading cause of death in the industrialized world. Diabetes is a major risk factor for cardiovascular disease. Unchecked diabetes can also lead to renal failure, blindness, heart attack, stroke, and amputation. The focus of this chapter will be to review the different mechanisms of insulin resistance (IR)/type 2 diabetes and various animal models used to study cardiac changes during these conditions.

Key Words: Diabetes; Mechanisms; Animal models; Genetic model; Insulin resistance; type 2 diabetes.

1. Introduction

Diabetes is one of the most commonly encountered chronic illnesses that is fast taking the form of a pandemic. In 1995, an estimated 135 million people worldwide had diabetes. Within 10 years, the number rose to 150 million, with a projected increase to 300 million cases by 2025 *(1)*. Diabetes poses a serious problem as poor treatment can lead to serious consequences such as renal failure, blindness, heart attack, stroke, and amputations. Owing to the seminal discovery of Banting, Best, Collip, and McLeod (1921; University of Toronto), it was revealed that diabetes was a consequence of a lack of insulin (type 1 diabetes) due to autoimmune pancreatic β-cell destruction. Although it has a genetic component, environmental factors (such as viruses-congenital rubella, coxsackievirus B, cow milk, and chemical toxins) have also been suggested to initiate the autoimmune response *(2)*. Another major type can occur because the body is unable to use insulin effectively [type 2 diabetes (T2D)]. Among

From: *Methods in Molecular Medicine, Vol. 139: Vascular Biology Protocols*
Edited by: N. Sreejayan and J. Ren © Humana Press Inc., Totowa, NJ

the susceptible population, obesity and insulin resistance (IR) are believed to be the most common causative factors responsible for the development of T2D. In diabetic patients, cardiovascular disease is the leading cause of death *(3–5)*. Although coronary vessel disease and atherosclerosis have been identified to be the primary reasons for the increased incidence of cardiovascular dysfunction, diabetic patients are also susceptible to cardiomyopathy independent of vascular disease, and likely an outcome of an intrinsic defect in the heart muscle itself *(6)*. Thus, diabetic human and animal hearts demonstrate myocyte hypertrophy, increased collagen and triglyceride (TG) deposition, cell death, and perivascular fibrosis *(6)*.

1.1. IR

In a normal individual, a balance between insulin-dependent glucose utilization and hepatic glucose production maintains blood glucose levels within a relatively narrow range (between 4–6 mM) *(7)*. The term "IR" implies an experimental or clinical condition where insulin is unable to evoke such optimal glucose homeostasis. At the onset of this phenomenon, there is compromised glucose uptake into the peripheral adipose and skeletal muscle leading to excessive circulating glucose levels. At the same time, because of the resistance of insulin action on the liver, hepatic gluconeogenesis (glucose production) increases, which adds to the glucose load in the circulation. To maintain normoglycemia, a greater concentration of insulin is required in insulin-resistant humans or animals, and at least in the initial stages, the pancreas hypersecretes insulin (hyperinsulinemia) and normalizes blood glucose. In the latter stage of IR, pancreatic β-cell dysfunction may occur resulting in a loss of insulin release. At this stage, because of a mismatch between circulating glucose and insulin levels, full-blown diabetes develops. As an initiating factor, it is currently unclear as to whether IR precedes hyperinsulinemia or vice versa *(8–10)*. It is important to remember that although the term "IR" is often used exclusively to glucose metabolism, this phenomenon can affect other insulin-sensitive metabolic pathways such as fatty acid metabolism.

1.2. Insulin Action

At the cellular level, the main action of insulin is to increase cellular glucose uptake by stimulating the translocation of glucose transporters (GLUTs) from intracellular sites to the cell membrane. Hyperglycemic glucose-clamp technique has shown that following an intravenous glucose infusion, there is an immediate release of insulin (within 3–5 min after the start of infusion) followed by a second phase of release that usually appears within 10–20 min and persists for several hours *(11)*. During IR, it is believed that the early phasic release of insulin is compromised, whereas the late phase compensates by increasing insulin secretion (hyperinsulinemia) *(11)*. At the molecular level, insulin binds to an insulin receptor and initiates a cascade of phosphorylation events that links the insulin receptor tyrosine kinase to downstream effector molecules. The insulin receptor subfamily includes tetrameric proteins with two α-subunits and β-subunits each. Under basal conditions, α-subunits inhibit the tyrosine kinase activity of the β-subunits. However, when insulin binds to the α-subunit, its action is lost leading to an increased kinase activity of the β-subunits *(12)*. Rarely, extreme IR in the form of leprechaunism or type A syndrome of IR and acanthosis nigrans can arise because of mutations in the insulin receptor in humans *(13)*. Downstream to the insulin receptor, different subtypes of insulin receptor kinase substrates have been identified, the majority of which belong to a class of proteins known as insulin receptor substrate (IRS-1–6) proteins *(14)*. IRS consists of phosphorylated tyrosines that act as docking sites for Src-homology 2 (SH2) domain containing proteins *(15)*. Multiple natural polymorphisms of the IRS-1 [G972R (glycine 972 → arginine), S892G, G819R, etc.] and IRS-2 (G1057D and G879S) have been established in many T2D patients *(16)*. The most studied one is probably the G972R mutation, prevalence of which is 5.8% in normal and 10.7% in Caucasoid T2D patients.

The major pathway for insulin signaling is through IRS proteins and PI3 kinase pathways that are primarily involved in the modulation of enzyme activities and glucose transport *(7)*. PI3 kinase consists of a p110 catalytic and p85 regulatory subunits. The SH2 domains on these subunits bind to tyrosine-phosphorylated motifs on IRS proteins. This pathway leads to the activation of kinases such as phosphoinositide-dependent kinase-1 that phosphorylates and activates other serine/threonine kinases such as Akt. IR and overt hyperglycemia in Akt2-deficient mice demonstrate the important role of Akt in insulin signaling *(17)*. In addition to its role in glucose transport, Akt phosphorylates glycogen synthase kinase-3 (GSK-3) and increases glycogen synthesis *(18,19)*. Activation of phosphodiesterase 3b and Foxo transcription factors *(20)* by Akt results in inhibition of lipolysis and gene transcription of several gluconeogenic enzymes such as phosphoenolpyruvate carboxykinase and the glucose-6-phosphatase catalytic subunit *(21)*. It has been demonstrated that impaired Foxo1 function is associated with improved fasting glycemia in diabetic mice *(22)* and increase in β-cell proliferation in insulin-resistant mice *(23)*.

Another major pathway of insulin signaling is through the Ras protein and the mitogen-activated protein kinase (MAPK) pathway *(7)*. After IRS proteins have been phosphorylated, it leads to the formation of complexes between the exchange factors Sos and Grb2. Grb2 leads to the activation of Ras signaling through MAPK pathway and affects cellular proliferation and growth *(24)*. A third PI3 kinase-independent model of glucose transport has been proposed that involves activation of the protooncogene c-Cbl that may induce GLUT translocation through remodeling of cortical actin filaments within the cell *(25,26)*.

1.3. Mechanisms Involved in IR and T2D

In view of the vast literature and ongoing controversies behind the etiology of IR and T2D, it is almost impossible to review comprehensively all mechanisms and pathways pertaining to these conditions. Thus, in this section, we will only discuss some of the well-established common mechanisms that characterize IR and/or T2D.

1.3.1. Heredity

A positive family history of T2D increases the probability of acquiring this disease by 2.4-fold *(27)*. The lifetime risk for development of T2D is around 38% if one of the parents is affected with the disease *(28)*, but it jumps to around 60% if both the parents are affected *(29)*. Approximately 15–25% of first-degree relatives of patients develop IR, impaired glucose tolerance or diabetes *(28)*. Further confirmation of the role of heredity comes from studying monozygotic twins (with 100% concordant genes) and dizygotic twins (with 50% concordant genes) where the incidence of both having T2D are 35–58% and only 17–20%, respectively *(30,31)*.

1.3.2. Obesity

Inactive lifestyle and overeating has been positively correlated with both obesity and IR/T2D. Obesity leads to an increase in TGs and fatty acids. Accumulation of TGs in adipose tissues leads to larger adipocytes that are resistant to anti-lipolytic action of insulin *(32)*. Increased release of fatty acids can decrease insulin action in insulin-responsive tissues at multiple sites (IRS-1/IRS-2 and PI3 kinase activation and Akt and GSK signaling) *(33)*. In humans, fatty acid induces a decrease in IRS-1-induced PI3 kinase activity *(34)*. In INS-1 (a pancreatic β-cell line), fatty acids inhibit the compensatory hyperplastic response that is normally observed following IR, through activation of several protein kinase C (PKC) pathways, which could be a factor during transition from IR to T2D *(35)*. Fatty acids can also induce IR through the pro-inflammatory nuclear factor-κB (NF-κB) signaling *(36,37)*. NF-κB is kept in check by its association with inhibitory-κB (I-κB). Fatty acids may inhibit the activity of I-κB, promote activation of NF-κB, and transcription of pro-inflammatory genes *(36)*. Lastly, fatty acids may also induce apoptosis (regulated cell death) of pancreatic β-cells and lead to β-cell failure during IR *(38,39)*.

1.3.3. Mitochondrial Dysfunction

Increase in fatty deposits in the liver of diabetic patients has led to the belief that mitochondrial fatty acid oxidation may be impaired in patients with T2D. This is also supported by the fact that T2D patients have smaller mitochondria in the skeletal muscle *(40)*. Decreased expression of mitochondrial oxidative phosphorylation genes has been identified in normal glucose-tolerant, non-diabetic individuals with a family history of diabetes *(41)*. One of the probable players in such faulty expression of genes is low level of peroxisome proliferator-activated receptor gamma coactivator-1 (PGC-1) in such individuals *(41,42)*. PGC is a co-activator of the nuclear respiratory factor, which in turn is responsible for mitochondrial biogenesis, respiratory capacity, and GLUT-4 synthesis *(43)*.

1.3.4. Inflammatory Cytokines

Tumor necrosis factor-α (TNF-α), interleukin-1 (IL-1), IL-6, and IL-1β are components of the normal immune system. According to the recent theory behind obesity and T2D being "an inflammatory condition," inflammatory cytokines are predisposing factors behind the development of these conditions *(44)*. TNF-α released from the adipose tissue decreases insulin-mediated phosphorylation of IRS proteins *(45,46)*. TNF-α and IL-6 also induces activation of suppressors of cytokine signaling proteins that provide a negative feedback loop on insulin signaling *(45,47)*. Cellular stress induced by TNF-α can also lead to serine phosphorylation of IRS proteins and thereby IR. Anti-TNF-α therapy in insulin-resistant humans leads to improved insulin sensitivity *(48)*. Following hyperglycemia, monocytes start producing cytokines such as monocyte chemoattractant protein-1, TNF-α, and IL-1β *(49)*. There are cytokines like IL-10 which have an opposing effect on inflammation. Interestingly, low levels of IL-10 have been demonstrated with the incidence of T2D *(50)*.

1.3.5. Hyperinsulinemia and Hyperglycemia

IR is often accompanied by hyperinsulinemia. In fact, many T2D patients have higher than average circulating insulin levels. It has been shown that chronically, in L6 muscle cells, high exposure to insulin in the absence of hyperglycemia leads to desensitization of insulin-stimulated PI3 kinase, protein kinase B (PKB), and MAPK pathways *(51)*. Insulin exposure also leads to decreased levels of IRS-1 and IRS-2 by stimulating their degradation *(52)*.

Hyperglycemia in most tissues leads to the generation of free radicals or reactive oxygen species (ROS) leading to oxidative stress *(53)*. Beta cells, normally equipped with low levels of antioxidants such as catalase and super-oxide dismutase, are particularly vulnerable toward oxidative stress *(54)*. Thus, hyperglycemia-induced ROS can initiate NFκB activity leading to apoptosis (cell death) in pancreatic β-cells, which can decrease insulin secretion over time *(27)*. Indeed, in T2D, many patients develop insulin insufficiency only at a later stage of the disease (around 10 years). Glucose also reacts non-enzymatically with proteins to form early glycation (Schiff bases) and intermediate glycation products (Amadori bases) that are then degraded to advanced glycation end products (AGEs) such as glyoxal, methylglyoxal, carboxymethylysine, and so on *(55)*. Although AGEs may be simply a consequence of diabetes, increased AGEs can worsen IR. Interestingly, increased glycation of insulin in the pancreatic β-cells and circulation have been reported in T2D *(56,57)*. Glycated albumin have been implicated in the down-regulation of the metabolic effects of insulin in L6 skeletal muscle cell lines through decreased activation of PI3 kinase, PKB, and GSK *(58)*.

1.3.6. Hexosamine Pathway

The majority of intracellular glucose is utilized in the glycolytic pathway. However, 1–3% of glucose is shunted toward the hexosamine pathway, giving rise to uridine diphosphate (UDP)-N acetyl glucosamine, which acts as a major substrate for glycosylation reactions *(59)*. Increased exposure to glucosamine leads to glycosylation of proteins in insulin signaling (e.g., IRS-1) causing diminished phosphorylation and activation of GSK, glycogen synthase, and Akt and decreased GLUT-4 translocation in adipocytes and skeletal muscle *(60–62)*. Mice overexpressing genes for the hexosamine pathway develop obesity and IR and demonstrate diminished GLUT-4 activation *(63,64)*. Other than its role in insulin-responsive tissues, this pathway can also prevent insulin secretion from the β-cells of the pancreas *(65)*.

1.3.7. Adipokines

Recent research has implicated adipocytes to be an active secretory tissue, in addition to its conventional role as a fat-storage depot. The adipocytes secrete a large number of metabolic hormones that are collectively known as adipokines. Leptin is such an adipokine that acts primarily on the central nervous system to inhibit food intake and promote energy expenditure *(66)*. In animal models of leptin deficiency (low leptin levels) or "resistance" (high leptin levels), hyperphagia, obesity, and IR are noted *(66)*. In some of these models, administration of exogenous leptin improves IR, probably by affecting neuroendocrine system that controls insulin actions on the liver. Resistin is an adipokine that is increased in obese mice and may cause IR *(67)*. On the other hand, adiponectin (Acrp 30 or adipoQ) is an adipokine that is decreased in states of IR and T2D. Adiponectin is believed to act through AMP kinase that stimulates energy-generating metabolic pathways such as suppression of gluconeogenesis, glucose uptake in skeletal muscle, and fatty acid oxidation *(68)*. Adenosine monophosphate (AMP) kinase is also believed to be a mediator of the anti-diabetic actions of drugs such as metformin.

1.4. Cardiac IR

In general, whole-body IR is usually attributed to insulin-sensitive tissues such as liver, skeletal muscle, and adipose, with little emphasis focused on the heart. A recent study that measured the arterial-coronary sinus glucose balance to local intra-coronary insulin levels in diabetic patients reported that type 2 diabetics demonstrate both whole-body and cardiac-specific IR *(69)*. In the heart, insulin promotes glucose uptake by two major pathways. Insulin can directly stimulate glucose uptake by GLUT-4 translocation to the cell surface. In animal models of IR/T2D, GLUT-4 translocation and protein levels decrease, thus hindering insulin-stimulated glucose uptake *(70,71)*. Insulin is also known to inhibit free fatty acid (FFA) release from adipose tissue. During IR/T2D, as insulin's effect on the adipose is diminished, adipose tissue lipolysis is enhanced, resulting in elevated circulating plasma FFA. FFAs are known to inhibit glucose utilization (glucose uptake, glycolysis, and pyruvate oxidation) in the heart *(72)*. Increasing FFA levels within the heart can also impair cardiac insulin signaling. In this regard, increased FA within the myocytes may give rise to long chain fatty acyl CoA that either directly or through the formation of diacylglycerol increases PKCs *(73)*. PKCs are serine/threonine kinases that can decrease tyrosine phosphorylation of both the insulin receptor and the IRS, leading to a decreased insulin signaling *(74)*. Fatty acids can also block Akt activation, another important mediator of insulin signaling through ceramide formation in the heart *(75)*.

Ordinarily the heart utilizes both glucose (25–30%) and fatty acids (65–70%) for its energy requirements *(76,77)*. Under conditions of decreased effect of insulin and elevated circulating FFA levels, the heart switches almost exclusively toward FFA oxidation as its major energy source *(76)*. Increased fatty acid oxidation is often hailed as a contributing factor to cardiac dysfunction. This is because fatty acid oxidation is more "oxygen wasting"; that is, it consumes more oxygen for one molecule of ATP produced than glucose *(78)*. Additionally, it has been proposed that increased fatty acid oxidation may lead to increased cellular acidosis making the diabetic heart more prone to injury *(79)*. Other than fatty acid oxidation per se, non-oxidative metabolism of FA may be a key player in precipitating cardiac dysfunction during IR/T2D *(80)*. Under normal conditions, when the heart is exposed to a high level of FFA, a compensatory rise in β-oxidation ensues through nuclear transcription factors like peroxisome proliferator activator (PPAR)-α-regulated gene expression. Binding of fatty acids to PPAR-α induces the gene expression of various enzymes involved in fatty acid oxidation (e.g., malonyl CoA decarboxylase and carnitine palmitoyl transferase 1). PPAR-α expression is partially regulated by leptin, an adipokine *(81)*. Unfortunately, as leptin action is often compromised during obesity and/or IR, excessive FFA is channeled toward storage in the form of TGs that can precipitate cardiac contractile impairment *(80)*. Increase in glucose or FFA can also generate free radicals that can cause cell death and cardiac dysfunction in diabetes *(82–84)*.

2. Animal Models of IR and T2D

The use of animal models in a complex disease process such as IR/T2D is bound to have its benefits and pitfalls *(85)*. The most compelling reason for the use of animal models in the study of this disease is that diseased animals may be more desirable than human patients with respect to ease of sample collection, genetic homogeneity, and faster development of complications like cardiovascular disease in the smaller species *(86,87)*. However, genetically modified animals often develop alternative pathways to circumvent the specific abnormalities of deletion ("knock out") or overexpression. Closely related isoforms or splice variants of the gene may compensate under such conditions. Thus, the phenotype under such conditions need not be due to either absence or overexpression of the gene in question but may be due to compensatory changes in gene expression *(14)*. The animal models with IR or T2D have been obtained either by chance discovery of compromised glycemic control, followed by inbreeding to preserve the unique characteristics, or by outbreeding the rats with non-diabetic populations and selectively isolating rats with impaired glucose tolerance. The success of the animal models lies in the fact that several gene defects in rodent models of obesity/T2D have also been found in humans *(88)*. In recent years, several mice models, with monogenic and polygenic genetic defects, have been developed, leading to either IR and/or

T2D. They have been reviewed in some excellent recent articles *(89–92)* and will not be repeated in this chapter. This article is focused on the rat models of IR with an emphasis on its implications on the heart.

2.1. Genetically Induced Rat Models

2.1.1. The Zucker Rats

In 1961, a spontaneous mutation in a group of outbred rats gave rise to the Zucker fatty model of IR *(93)*. The mutation is due to a single nucleotide (A–C) substitution at position 880 of the leptin receptor gene that leads to an amino acid substituition at position 269 from Gln to Pro that decreases leptin's affinity for the receptor *(94)*. The mutation at the leptin receptor is often referred to as *fa* or Ob-R^{fa}. As central leptin is responsible for satiety and eating behavior, the rats with homozygous mutations (Zucker Fatty or *fa/fa*) are hyperphagic, leading to obesity and IR *(95)*. Additionally, mechanisms that release insulin following a glucose load are impaired. Thus, insulin is continuously released even at basal glucose levels leading to hyperinsulinemia *(96)*. The hyperinsulinemia is maintained throughout the lifetime of the *fa/fa* rat and prevents it from turning overtly diabetic. Therefore *fa/fa* rats are models of IR and not T2D. The heterozygotes (Zucker lean, $fa/+$) do not develop either obesity or IR.

In 1977, the Zucker Diabetic Fatty (ZDF) rats were established by inbreeding obese male *fa/fa* rats with diabetic traits *(97)*. Unlike *fa/fa* rats, ZDF rats display progressive glucose intolerance and hyperinsulinemia followed by overt hyperglycemia (exceeding 22 mM) between 7 and 10 weeks of age *(98)*. At this time, the β-cells of ZDF rats become insensitive to circulating glucose and display diminished GLUT-2 expression, insulin content, and β-cell number *(99)*. Mimicking humans, other diabetic complications in ZDF rats include hyperlipidemia, nephropathy, impaired wound healing, and cardiovascular disorders. The female ZDF rats do not develop spontaneous diabetes on normal chow diet but develop diabetes on a diabetogenic diet *(98)*.

In the hearts of both *fa/fa* and ZDF rats, decreased GLUT-4 expression and impaired translocation of GLUT-4 to the plasma membrane have been reported and could be responsible for the faulty cardiac glucose uptake *(100,101)*. This impairment in glucose uptake can be corrected by treating the rats with rosiglitazone, a PPAR-γ agonist *(69)*. Additionally, fatty acid uptake increases in *fa/fa* hearts *(102)*. Normally, if fatty acid availability exceeds the rate of fatty acid oxidation, the cell overcomes this problem by increasing fatty acid oxidation, in part by inducing PPAR-α and PPAR-α-regulated genes *(103)*. Although fatty acid oxidation is enhanced, *fa/fa* and ZDF rat hearts have a low reserve for further increasing PPAR-α activity. This was shown when both $fa/+$ and *fa/fa* rats were subjected to fasting that further increased their circulating FFA levels. Unlike $fa/+$, the *fa/fa* rat was unable to increase its fatty acid oxidation in response to increased fatty acid availability *(104)*. This can lead to non-oxidative modifications of the excess FA-like deposition of intramyocardial TGs and increased production of ceramide, leading to myocardial apoptosis *(105)*. Thus, myocardial contractile impairment in ZDF rats may be a function of the extent of TG deposition *(106)*. Indeed, decreasing cardiac TG by troglitazone restores cardiac function *(107)*. Conversely, some studies have suggested that during lipid overload, TG synthesis is actually protective as TGs are relatively inert compared to large amounts of FFAs in causing lipotoxicity *(108)*.

Even though both the above models have a common genetic background, substantial differences are observed between the *fa/fa* and ZDF rats. *fa/fa* rats do not develop diabetes because of adequate β-cell function, whereas ZDF rats develop β-cell insufficiency and overt diabetes. It has been hypothesized that lipotoxicity induces greater damage in the ZDF pancreas *(109,110)*. Similarly, in the heart, the contractile dysfunction in the *fa/fa* rat is modest compared to the ZDF *(104)*. Finally, in the fed ZDF rat heart, there is decreased expression of both PPAR-α and PPAR-α-regulated genes compared to the *fa/fa*, which may eventually be responsible for the prevention of cardiac lipotoxicity in the latter group *(104)*.

2.1.2. JCR: LA-Corpulent Rats

The *fa/fa* and ZDF rats are characterized by a mutated leptin receptor. In contrast, the *cp* (corpulent) mutation in rats is characterized by a stop codon in the extracellular domain of the leptin receptor, leading to a complete loss of leptin receptor in the cell membrane of homozygous *cp/cp* rats *(111)*. The most well known and unique among the various *cp/cp* rats available is the JCR: LA-cp rats *(112)*. Like the *fa/fa* rat, homozygous *cp/cp* rats develop hyperphagia, obesity (3 weeks), hyperinsulinemia (4 weeks), IR, and dyslipidemia characterized by high circulating TG levels, whereas heterozygotes do not develop the above characteristics *(113)*. Unlike the Zucker fatty rats or other *cp/cp* rats, JCR: LA-cp rats develop spontaneous atherosclerosis and myocardial ischemia *(112)*. In the absence of a leptin receptor, circulating leptin levels increase approximately 30-fold as a compensatory response *(112)*. At 12 weeks, insulin-stimulated glucose uptake is impaired *(113)*. However, they do not turn hyperglycemic as a result of extreme hyperinsulinemia that is much higher than that seen in obese Zucker rats *(114)*. Although the females exhibit severe hypertriglyceridemia, they do not develop IR or cardiovascular diseases *(115)*.

Like ZDF rats, a drop in cardiac glycolytic rates and an increased fatty acid uptake and intracardiac TG levels characterize JCR: LA-cp rats *(106)*. Lowering plasma lipids, which in turn should reduce cardiac TG overload, is associated with the improvement of the cardiovascular defect *(113,116)*. The hearts of JCR: LA-cp rats are characterized by atherosclerotic lesions by 9 months of age in the aorta *(112,117)*. Occlusive coronary thrombi with myocardial ischemic lesions in various stages of damage (from early necrotic phase to scarring) are evident. It is suggested that the cardiovascular disorder in JCR: LA-cp rats is mainly a vascular one *(112)*.

2.1.3. Otsuka Long–Evans Tokushima Fatty Rats

In 1992, Kawano et al. developed the Otsuka Long–Evans Tokushima Fatty (OLETF) rat strain by selectively inbreeding Long–Evans rats that developed polyuria, polydypsia, hyperinsulinemia, hypertriglyceridemia, and overt diabetes *(118,119)*. Mild to moderate obesity is seen relatively early (after 4 weeks) with hyperglycemia after 18 weeks. As with most other genetic models, diabetes is predominantly seen with males (by 25 weeks, 80–100% of all males develop diabetes), with females less affected. Insulin-mediated glucose uptake is impaired in these animals by 16 weeks of age. Plasma cholesterol and TGs rise around 21 weeks *(85)*.

Like JCR: LA-cp rats, cardiac dysfunction is absent in OLETF rats until around 62 weeks, when it is characterized by a mild diastolic dysfunction, that can be explained by decreased SERCA2A protein expression and impaired Ca^{2+} uptake *(120)*. One report has suggested that there is increased cardiac interstitial collagen and diastolic filling abnormalities in OLETF rats *(121)*. Perivascular fibrosis has been observed in the left ventricle during this period, and the gene expression of extracellular matrix components such as collagen I, II, III, and laminin was found to be increased *(122)*. Furthermore, heart rate of these rats was decreased during diabetes mellitus, which is believed to be due to altered activity of the sino-atrial node and/or decreased activation of the cardiac sympathetic tone *(123)*. Aortic atherosclerosis and hardening of the coronary arteries are found at the later stage of the disease (62 weeks) *(123)*.

2.1.4. Goto–Kakizaki Rats

Goto and Kakizaki developed this model through inbreeding of normal Wistar rats that demonstrated impaired glucose tolerance over 30 generations *(124)*. This rat model exhibits common features of T2D including fasting hyperglycemia, peripheral and hepatic IR, decreased insulin-stimulated glucose uptake, and complications such as neuropathy and glomerulopathy *(86)*. They are unique from other models in that they do not exhibit obesity, hyperlipidemia, or hypertension and are a preferred choice for investigating the role of hyperglycemia per se without the confounding effects of hyperlipidemia or hypertension *(125)*. In as early as 8 days, glucose-induced insulin release is impaired in these rats *(126)*.

The heart function in hyperglycemic Goto-kakizaki (GK) rats is unchanged up to 8 months compared to Wistar rats. However, unlike normal Wistars, brief hypoxia (10 min) impairs the systolic function and diastolic function in GK rats that are reversible in the presence of increased levels of insulin in the reperfusion buffer *(125)*. This may imply that the cardiac dysfunction during hypoxia is mainly mediated by glucose metabolism defects like impaired glycolysis or by decreased GLUTs in the heart. Unlike most other models of genetic IR, the female GK rats demonstrate more hypertrophy, IR, and cardiac contractile impairment and are believed to be more prone to ischemic injury than males *(127)*. At the molecular level, decreased levels of antioxidants, increased ROS production, and mitochondrial damage have also been shown in 12-month-old Goto–Kakizaki rat hearts *(128)*.

2.2. Carbohydrate-Induced Rat Models

2.2.1. Fructose-Fed IR

Fructose is a monosaccharide obtained from fruits and is also a component of sucrose. In 1987, Hwang et al. *(129)* first reported that fructose feeding in rats causes IR accompanied by hypertension. This high carbohydrate (fructose, 66%) diet also resulted in elevated TG levels and hyperinsulinemia in the absence of obesity or overt hyperglycemia. The exact time frame required for the development of IR depends on the rat strain, concentration of fructose, and the duration of exposure to fructose *(130)*. Development of IR and hypertension in Sprague–Dawley rats takes 3 weeks when fed with 66% fructose *(131)*. However, in Wistar rats, the same amount of fructose required 4–7 weeks prior to detection of hypertension *(132)*. Longer feeding schedules can cause obesity and hyperglycemia in fructose-fed animals. The exact cause of IR in fructose-fed rats is still controversial. Studies have shown that fructose feeding generates metabolites that block phosphofructokinase, the rate-limiting enzyme in glucose metabolism, and increases hepatic glucose production *(133)*. High fructose feeding can also increase the rate of de novo lipogenesis that may explain elevated TGs and partly explain IR in these animals *(134)*. There is also evidence that unlike glucose, a high-fructose meal evokes less insulin release from the pancreas and low leptin release from the adipose *(135)*. To compensate for the deficiency of these hormones, there is a positive feedback, as a result of which hyperleptinemia and hyperinsulinemia are seen in these animals.

IR induced by high-fructose diet is involved in the development of systolic hypertension (pressure increase of around 15–30 mm) *(131)*. In this regard, normalization of hyperinsulinemia and insulin sensitivity, by treating with metformin, ameliorates high blood pressure *(131,136)*. Using other hypoglycemic agents such as thiazolidinediones and vanadium, the development of hypertension in fructose-fed rats can be prevented *(137,138)*. At the molecular level, the increase in blood pressure is believed to be due to an impaired endothelium secondary to IR *(139)*. The dysfunction could be due either to the loss in endothelium-dependent relaxation or increased synthesis or responses to endothelial vasoconstrictors *(136,140,141)*. Reports have been published supporting changes in the functions of both endogenous vasodilators and vasoconstrictors. Interestingly, a role for sex hormones in mediating IR and/or hypertension in fructose-fed rats has recently come to light. The sex hormones estrogen and testosterone have been separately shown to decrease and increase, respectively, the development of hypertension following IR *(142–144)*. Female patients as well as fructose-fed female rats are less susceptible to developing hypertension compared with males. The importance of sex hormones (specifically estrogen) in the development of hypertension has been shown in ovariectomized rats that developed IR and hypertension subsequent to feeding with fructose.

2.2.2. Sucrose-Fed IR

In clinical studies, a dose-dependent increase of serum TGs is observed with sucrose feeding *(145–147)*. In one of the earlier studies, 18% sucrose diet had no effect, whereas 36 and 52% were associated with a sustained increase in serum TGs in healthy young normolipidemic men *(148)*. Sucrose feeding decreases hepatic fatty acid oxidation, with the glucose carbons being shunted toward glycerol-3-phosphate, leading to increased TG synthesis *(149)*. In rats, sucrose feeding can cause whole-body IR that can be prevented with the oral hypoglycemic agent, metformin *(150)*. In animals, sucrose-fed IR can be achieved through two procedures: dietary sucrose feeding for 7–10 weeks (68% of total energy) and 10–30% sucrose in the drinking water *(150–153)*. Interestingly, both these techniques lead to the development of whole-body IR in the absence of overt hyperglycemia. In addition to changes in TG, this model also demonstrates elevated circulating levels of insulin and leptin *(154)*.

Unlike most genetic models of IR/T2D, this model does not develop obesity or hypertension. However, with longer periods of sucrose feeding (around 4 months), there is a risk of developing hypertension and even hyperglycemia.

Cardiac dysfunction has been reported repeatedly with sucrose feeding. Mechanical properties of isolated ventricular myocytes (measured by high-speed video edge detection) from sucrose-fed rats demonstrated significantly longer periods of myocyte shortening and relengthening compared to starch-fed rats *(151,154,155)*. Sucrose-fed rats also exhibited a slower rate of intracellular Ca^{2+} decay. Recent work indicates that the impaired Ca^{2+} decay may be a result of impaired sarcoplasmic reticulum Ca^{2+}-ATPase (SERCA) activity (with normal protein content) in sucrose-fed rats *(155)*. It should be noted that at this stage of the disease, there is no change in Na^+/Ca^{2+} exchanger function, nor are there changes in protein content of phospholamban, factors that could affect cardiac function *(155)*. Clinically relevant therapies with bezafibrate or metformin or even exercise ameliorated these cardiac defects in sucrose-fed rats. With longer sucrose feeding (15 months), Ca^{2+}-induced myofibrillar ATPase activity may be impaired in rat hearts *(156)*. It should be noted that sucrose-fed rat hearts also demonstrate decreased cardiac antioxidant levels that may predispose the heart to increased oxidative damage *(157)*.

2.3. Drug-Induced Rat Models for Diabetes

2.3.1. Glucocorticoid-Induced IR

Glucocorticoids have widespread use as anti-inflammatory and immuno-suppressive agents *(158)*. However, chronic glucocorticoid therapy is often associated with adverse and serious side effects, including Cushing's syndrome, osteoporosis, gastrointestinal bleeding, and dyslipidemia *(158)*. More importantly, both excess endogenous and exogenous glucocorticoids impair insulin sensitivity, contributing to the generation of the metabolic syndrome, including IR, obesity, and hypertension *(159–161)*. Glucocorticoids may exert their action on insulin signaling through either central mechanisms or peripheral effects *(162)*. Infusion of normal rats with dexamethasone (DEX) intracerebroventricularly for 3 days results in hyperphagia, probably through neuropeptide y up-regulation that regulates feeding behavior *(162,163)*. Interestingly, these rats also showed hyperinsulinemia and marked IR. At the same time, central DEX infusion also reduces peripheral glucose uptake in skeletal muscle and adipose tissue *(164)*. The link between glucocorticoid and obesity/IR is confounded by the fact that human obese patients do not consistently demonstrate high plasma cortisol (endogenous glucocorticoid) levels *(162)*. One explanation is that peripheral glucocorticoid receptor number is high among these subjects that may increase their glucocorticoid sensitivity *(165,166)*. In rodent skeletal muscle, glucocorticoids are known to decrease GLUT-4 translocation (but not its expression) and impair glucose uptake *(152)*. Recently, glucocorticoid action was found to be regulated by enzymes such as 11β-hydroxysteroid dehydrogenase (11β-HSD), which converts inactive glucocorticoids into its active forms *(167)*. Mice overexpressing 11β-HSD-1 in adipose tissue were shown to be obese and insulin resistant *(168)*. Adipose tissue 11β-HSD-1 mRNA expression is increased in high fat (HF)-fed animal models and in obese humans and is positively correlated with hyperglycemia. Alternately, 11β-HSD-1 knockout mice demonstrates improved glucose tolerance and impaired hepatic gluco-neogenic enzyme expressions *(169)*.

Although much research has been performed on skeletal muscle and the adipose following DEX, the heart remains perhaps the most understudied among insulin-sensitive tissues. In a recent study, we studied the acute effects of DEX on the heart *(76)*. In this study, adult male Wistar rats (270–290 g) were injected with a single bolus dose of DEX (1 mg/kg, intraperitoneally). As the plasma half-life of DEX is approximately 279 min, the animals were killed 4 h later. Such short treatment had no influence on either plasma insulin or glucose after 4 h. Nevertheless, exploiting the euglycemic-hyperinsulinemic clamp, we demonstrated that DEX induced whole-body and cardiac-specific IR. Assessment of cardiac glycolytic rates revealed no change following DEX, whereas glucose oxidation decreased significantly. Moreover, cardiac glycogen content increased almost 2-fold after DEX. These data suggest that in the short term, DEX is capable of inducing IR and switching cardiac glucose disposal from oxidation to storage in the form of glycogen, likely compromising energy production in the heart. Regarding the molecular mechanisms behind cardiac IR induced by DEX, we initially looked at pyruvate dehydrogenase kinase (PDK). Following glucose uptake and glycolysis, the pyruvate dehydrogenase complex (PDC) facilitates entry of pyruvate into the mitochondria *(170)*. By phosphorylating PDC, PDK can decrease the rate of glucose oxidation. In our study, we demonstrated that in the absence of hyperglycemia, a single bolus dose of DEX was able to decrease PDK-4 gene expression in the heart, impairing cardiac glucose oxidation. As cardiac glucose oxidation is impaired following DEX, compensatory increase in fatty acid oxidation (unpublished data) was also noticed. In this regard, hearts from insulin-resistant DEX animals also demonstrated enlargement of the coronary lipoprotein lipase (LPL) pool within 1 h of DEX infusion. Given the observation that cardiac LPL is a major determinant of plasma TG *(171)*, the increase in cardiac luminal LPL could be associated with the observed decline in circulating TG. However, as no apparent change was noted in plasma fatty acid levels, we hypothesized that following LPL-mediated TG hydrolysis, fatty acid can be taken up rapidly and directly into tissues. In support of this suggestion, visualization using electron microscopy also revealed high TG storage in DEX-treated hearts *(76)*. It has not yet been determined whether these acute effects of DEX on cardiac metabolism can be translated into increased cardiovascular risk.

2.4. Fat-Induced Rat Models for Diabetes

Dietary fatty acids could lead to IR/T2D independent of obesity. However, in clinical studies, factors that preclude confirmation of this direct link are high body mass index (obesity) and/or inactivity that are often associated with increased dietary fat intake *(172)*. One of the mechanisms that could mediate the effects of fatty acids is the phospholipid composition of skeletal muscle membranes *(173)*. As insulin receptors are embedded within the cell membrane, unfavorable alterations of the fatty acid profile of membrane phospholipids could well alter the insulin receptor binding affinity and capacity. In animal studies, both the amount and type of fatty acids have been deemed important in modulation of insulin sensitivity *(174)*. In this regard, it was demonstrated that saturated fatty acid (SFA) feeding leads to IR, whereas polyunsaturated fatty acids (PUFAs) or monounsaturated fatty acids limit such a development *(174)*. In parallel, it has also been shown that by simply increasing fat content to beyond 37% of total energy consumption leads to IR irrespective of the type of dietary fatty acids *(175)*.

FA accumulation and TG accumulation have been shown to cause cardiac IR both in vivo and in vitro, an effect that may be duplicated with dietary fatty acids. HF feeding (50% of total energy) leads to increased postprandial blood glucose, myocardial TG accumulation, and gross structural changes compared with low fat-fed rats. Ca^{2+}-mediated cardiac force generation, PI3 kinase activity, and phosphorylation of Akt and GSK were also decreased *(176)*. A recent observation *(177)* suggested that the detrimental cardiovascular effects of dietary fatty acids on IR could be demonstrated in adult offspring of rats fed with a diet rich in animal fat during pregnancy and suckling. This may imply that changes in membrane phospholipid composition even under gestation and suckling phase may have a profound effect on postnatal insulin signaling. HF feeding in some genetic rodent models of IR can worsen/precipitate diabetes *(178)*. However, there are conflicting reports regarding the effect of HF feeding on non-genetic rat models such as the Wistar rats. It was previously demonstrated that in non-genetic rat models (e.g., in Wistar rats), continued PUFA feeding (60% of total energy) for up to 10 months leads to IR but not diabetes *(179)*. It was postulated that a genetic or congenital susceptibility to β-cell impairment is required for overt hyperglycemia to develop in the presence of severe IR *(179,180)*. One additional reason behind such protection may be the reduced 11β-HSD-1 activity in the Wistar adipose tissue in response to HF diet *(180)* compared to genetic models of IR. It has also been reported that unlike high PUFA diet, feeding 50% HF diet with SFAs, such as palmitic acid as the main component, induces overt diabetes in Wistar rats within 7 weeks *(176)*. In this regard, it has been shown that saturated FFA-like palmitic acid can give rise to excess palmitoyl CoA that can lead to de novo synthesis of ceramide, a naturally occurring sphingolipid and ultimately β-cell apoptosis *(110,181)*. Beta-cell apoptosis can negate the compensatory β-cell hyperplasia following IR, leading to diabetes.

To examine the effect of different fatty acids on the heart, we recently examined the effect of 20% w/w (40% of energy) palm oil (PO) feeding on the rat heart keeping 20% (w/w) sunflower oil (SO), a diet rich in ω-6 PUFAs as an isocaloric control *(84)*. Following 4 weeks of HF feeding, both the groups demonstrated hyperinsulinemia, marked obesity, and marginally elevated glucose levels compared with normal chow-fed animals (10% of energy from fats). As hyperglycemia was absent after 4 weeks of HF feeding, to induce diabetes, rats were injected with streptozotocin (STZ; 55 mg/kg, intravenously). Such a model of fat-fed IR with moderate dose of STZ was used to (i) mimic the human condition, where IR is often followed by overt diabetes, (ii) avoid genetic models of diabetes (like the ZDF) that are accompanied by leptin dysregulation, and (iii) establish less expensive and easily available non-genetic models *(84,182)*. Feeding PO and SO magnified palmitic and linoleic acid contents within lipoproteins and hearts, respectively. Compared with SO, PO diabetic hearts demonstrated significantly higher levels of apoptosis, with an altered Bax : Bcl-2 ratio, and augmented oxidative stress (as measured by lipid peroxidation and nitrotyrosine accumulation). In contrast to PO-fed diabetic animals, SO-fed diabetic animals demonstrated an increase in serum lactate dehydrogenase and myocardial necrotic changes. In a later study, the increase in cardiac necrosis, with a decline in cardiac function in SO-fed animals, was attributed to pro-inflammatory linoleic acid and was associated with mitochondrial abnormalities, impaired substrate utilization, and enhanced TG accumulation *(83)*. An interesting aspect of myocardial necrosis in SO-fed animals was the drop in myocardial glutathione (GSH). It has been demonstrated that GSH, besides being a crucial cardiac antioxidant, can also regulate the mode of cell death *(183–185)*. A major functional impact of a decreased GSH is oxidation of regulatory proteins at cysteine residues *(184)*. For caspase activation and propagation of the apoptotic pathway, an intact cysteine moiety is necessary, which can easily be oxidized under GSH-depleted conditions *(186)*. Such a drastic increase in oxidative environment can switch cell death from a caspase-3-dependent (apoptotic) to a caspase-3-independent mode (necrosis). Although these effects of n-6 PUFA in the diabetic animal would seem contrary to accepted belief as being beneficial, in countries such as Israel, with high dietary n-6 PUFA, there is an excessive incidence of obesity, IR, hypertension, and T2D *(187)*. Thus, the dilemma with recommending vegetable oils such as SO as a source of PUFA is that they contain high levels of n-6 PUFA but negligible amounts of n-3 PUFA, the primary candidate for conferring cardio-vascular benefit *(83)*. Given the concurrent scarcity of fish oil (the main source of n-3 PUFA) in the Western diet, the n-6 PUFA : n-3 PUFA ratio has been amplified from a traditional balance of 1:1 to an astounding 15:1 to 20:1 in the modern diet *(187–190)*. In contrast to other models of diabetic cardiomy-opathy that exhibit cardiac dysfunction only after chronic hyperglycemia, n-6 PUFA feeding coupled with only 4 days of diabetes precipitated metabolic and contractile abnormalities in the heart *(83)*.

References

1. King, H., Aubert, R.E., and Herman, W.H. (1998) Global burden of diabetes, 1995–2025: prevalence, numerical estimates, and projections. *Diabetes Care* **21**, 1414–1431.
2. Batstra, M.R., Aanstoot, H.J., and Herbrink, P. (2001) Prediction and diagnosis of type 1 diabetes using beta-cell autoantibodies. *Clin. Lab.* **47**, 497–507.
3. Kannel, W.B., and McGee, D.L. (1979) Diabetes and cardiovascular disease. The Framingham study. *JAMA* **241**, 2035–2038.
4. Stamler, J., Vaccaro, O., Neaton, J.D., and Wentworth, D. (1993) Diabetes, other risk factors, and 12-yr cardiovascular mortality for men screened in the Multiple Risk Factor Intervention Trial. *Diabetes Care* **16**, 434–444.
5. Schernthaner, G. (1996) Cardiovascular mortality and morbidity in type-2 diabetes mellitus. *Diabetes Res. Clin. Pract.* **31**(Suppl), S3–S13.
6. Rodrigues, B., Cam, M.C., and McNeill, J.H. (1995) Myocardial substrate metabolism: implications for diabetic cardiomyopathy. *J. Mol. Cell Cardiol.* **27**, 169–179.
7. Saltiel, A.R., and Kahn, C.R. (2001) Insulin signalling and the regulation of glucose and lipid metabolism. *Nature* **414**, 799–806.
8. Hall, J.E., Brands, M.W., Hildebrandt, D.A., and Mizelle, H.L. (1992) Obesity-associated hypertension. Hyperinsulinemia and renal mechanisms. *Hypertension* **19**, I45–I55.

9. Del Prato, S., Riccio, A., Vigili de Kreutzenberg, S., et al. (1993) Mechanisms of fasting hypoglycemia and concomitant insulin resistance in insulinoma patients. *Metabolism* **42**, 24–29.

10. Vigili de Kreutzenberg, S., Riccio, A., Dorella, M., et al. (1995) Surgical removal of insulinoma restores glucose recovery from hypoglycaemia but does not normalize insulin action. *Eur. J. Clin. Invest.* **25**, 360–367.

11. Pratley, R.E., and Weyer, C. (2001) The role of impaired early insulin secretion in the pathogenesis of type II diabetes mellitus. *Diabetologia* **44**, 929–945.

12. Patti, M.E., and Kahn, C.R. (1998) The insulin receptor–a critical link in glucose homeostasis and insulin action. *J. Basic Clin. Physiol. Pharmacol.* **9**, 89–109.

13. Krook, A., Kumar, S., Laing, I., Boulton, A.J., Wass, J.A., and O'Rahilly, S. (1994) Molecular scanning of the insulin receptor gene in syndromes of insulin resistance. *Diabetes* **43**, 357–368.

14. Krook, A., Wallberg-Henriksson, H., and Zierath, J.R. (2004) Sending the signal: molecular mechanisms regulating glucose uptake. *Med. Sci. Sports Exerc.* **36**, 1212–1217.

15. White, M.F. (2002) IRS proteins and the common path to diabetes. *Am. J. Physiol. Endocrinol. Metab.* **283**, E413–E422.

16. Sesti, G., Federici, M., Hribal, M.L., Lauro, D., Sbraccia, P., and Lauro, R. (2001) Defects of the insulin receptor substrate (IRS) system in human metabolic disorders. *FASEB J.* **15**, 2099–2111.

17. Garofalo, R.S., Orena, S.J., Rafidi, K., et al. (2003) Severe diabetes, age-dependent loss of adipose tissue, and mild growth deficiency in mice lacking Akt2/PKB beta. *J. Clin. Invest.* **112**, 197–208.

18. Cross, D.A., Alessi, D.R., Cohen, P., Andjelkovich, M., and Hemmings, B.A. (1995) Inhibition of glycogen synthase kinase-3 by insulin mediated by protein kinase B. *Nature* **378**, 785–789.

19. Lajoie, C., Calderone, A., Trudeau, F., et al. (2004) Exercise training attenuated the PKB and GSK-3 dephosphorylation in the myocardium of ZDF rats. *J. Appl. Physiol.* **96**, 1606–1612.

20. Nakae, J., Park, B.C., and Accili, D. (1999) Insulin stimulates phosphorylation of the forkhead transcription factor FKHR on serine 253 through a Wortmannin-sensitive pathway. *J. Biol. Chem.* **274**, 15982–15985.

21. Nakae, J., Kitamura, T., Silver, D.L., and Accili, D. (2001) The forkhead transcription factor Foxo1 (Fkhr) confers insulin sensitivity onto glucose-6-phosphatase expression. *J. Clin. Invest.* **108**, 1359–1367.

22. Altomonte, J., Richter, A., Harbaran, S., et al. (2003) Inhibition of Foxo1 function is associated with improved fasting glycemia in diabetic mice. *Am. J. Physiol. Endocrinol. Metab.* **285**, E718–E728.

23. Kitamura, T., Nakae, J., Kitamura, Y., et al. (2002) The forkhead transcription factor Foxo1 links insulin signaling to Pdx1 regulation of pancreatic beta cell growth. *J. Clin. Invest.* **110**, 1839–1847.

24. Ogawa, W., Matozaki, T., and Kasuga, M. (1998) Role of binding proteins to IRS-1 in insulin signalling. *Mol. Cell. Biochem.* **182**, 13–22.
25. Khan, A.H., and Pessin, J.E. (2002) Insulin regulation of glucose uptake: a complex interplay of intracellular signalling pathways. *Diabetologia* **45**, 1475–1483.
26. Molero, J.C., Jensen, T.E., Withers, P.C., et al. (2004) c-Cbl-deficient mice have reduced adiposity, higher energy expenditure, and improved peripheral insulin action. *J. Clin. Invest.* **114**, 1326–1333.
27. Stumvoll, M., Goldstein, B.J., and van Haeften, T.W. (2005) Type 2 diabetes: principles of pathogenesis and therapy. *Lancet* **365**, 1333–1346.
28. Pierce, M., Keen, H., and Bradley, C. (1995) Risk of diabetes in offspring of parents with non-insulin-dependent diabetes. *Diabet. Med.* **12**, 6–13.
29. Tattersal, R.B., and Fajans, S.S. (1975) Prevalence of diabetes and glucose intolerance in 199 offspring of thirty-seven conjugal diabetic parents. *Diabetes* **24**, 452–462.
30. Kaprio, J., Tuomilehto, J., Koskenvuo, M., et al. (1992) Concordance for type 1 (insulin-dependent) and type 2 (non-insulin-dependent) diabetes mellitus in a population-based cohort of twins in Finland. *Diabetologia* **35**, 1060–1067.
31. Newman, B., Selby, J.V., King, M.C., Slemenda, C., Fabsitz, R., and Friedman, G.D. (1987) Concordance for type 2 (non-insulin-dependent) diabetes mellitus in male twins. *Diabetologia* **30**, 763–768.
32. Boden, G. (1997) Role of fatty acids in the pathogenesis of insulin resistance and NIDDM. *Diabetes* **46**, 3–10.
33. Boden, G. (2002) Interaction between free fatty acids and glucose metabolism. *Curr. Opin. Clin. Nutr. Metab. Care* **5**, 545–549.
34. Dresner, A., Laurent, D., Marcucci, M., et al. (1999) Effects of free fatty acids on glucose transport and IRS-1-associated phosphatidylinositol 3-kinase activity. *J. Clin. Invest.* **103**, 253–259.
35. Wrede, C.E., Dickson, L.M., Lingohr, M.K., Briaud, I., and Rhodes, C.J. (2003) Fatty acid and phorbol ester-mediated interference of mitogenic signaling via novel protein kinase C isoforms in pancreatic beta-cells (INS-1). *J. Mol. Endocrinol.* **30**, 271–286.
36. Sinha, S., Perdomo, G., Brown, N.F., and O'Doherty, R.M. (2004) Fatty acid-induced insulin resistance in L6 myotubes is prevented by inhibition of activation and nuclear localization of nuclear factor kappa B. *J. Biol. Chem.* **279**, 41294–41301.
37. Evans, J.L., Goldfine, I.D., Maddux, B.A., and Grodsky, G.M. (2002) Oxidative stress and stress-activated signaling pathways: a unifying hypothesis of type 2 diabetes. *Endocr. Rev.* **23**, 599–622.
38. Lupi, R., Dotta, F., Marselli, L., et al. (2002) Prolonged exposure to free fatty acids has cytostatic and pro-apoptotic effects on human pancreatic islets: evidence that beta-cell death is caspase mediated, partially dependent on ceramide pathway, and Bcl-2 regulated. *Diabetes* **51**, 1437–1442.

39. El-Assaad, W., Buteau, J., Peyot, M.L., et al. (2003) Saturated fatty acids synergize with elevated glucose to cause pancreatic beta-cell death. *Endocrinology* **144**, 4154–4163.
40. Kelley, D.E., He, J., Menshikova, E.V., and Ritov, V.B. (2002) Dysfunction of mitochondria in human skeletal muscle in type 2 diabetes. *Diabetes* **51**, 2944–2950.
41. Short, K.R., Nair, K.S., and Stump, C.S. (2004) Impaired mitochondrial activity and insulin-resistant offspring of patients with type 2 diabetes. *N. Engl. J. Med.* **350**, 2419–2421; author reply 2419–2421.
42. Patti, M.E., Butte, A.J., Crunkhorn, S., et al. (2003) Coordinated reduction of genes of oxidative metabolism in humans with insulin resistance and diabetes: potential role of PGC1 and NRF1. *Proc. Natl. Acad. Sci. U. S. A.* **100**, 8466–8471.
43. Mootha, V.K., Lindgren, C.M., Eriksson, K.F., et al. (2003) PGC-1alpha-responsive genes involved in oxidative phosphorylation are coordinately downregulated in human diabetes. *Nat. Genet.* **34**, 267–273.
44. Ferroni, P., Basili, S., Falco, A., and Davi, G. (2004) Inflammation, insulin resistance, and obesity. *Curr. Atheroscler. Rep.* **6**, 424–431.
45. Pirola, L., Johnston, A.M., and Van Obberghen, E. (2004) Modulation of insulin action. *Diabetologia* **47**, 170–184.
46. Hotamisligil, G.S. (2000) Molecular mechanisms of insulin resistance and the role of the adipocyte. *Int. J. Obes. Relat. Metab. Disord.* **24**(Suppl 4), S23–S27.
47. Senn, J.J., Klover, P.J., Nowak, I.A., et al. (2003) Suppressor of cytokine signaling-3 (SOCS-3), a potential mediator of interleukin-6-dependent insulin resistance in hepatocytes. *J. Biol. Chem.* **278**, 13740–13746.
48. Yazdani-Biuki, B., Stelzl, H., Brezinschek, H.P., et al. (2004) Improvement of insulin sensitivity in insulin resistant subjects during prolonged treatment with the anti-TNF-alpha antibody infliximab. *Eur. J. Clin. Invest.* **34**, 641–642.
49. Shanmugam, N., Reddy, M.A., Guha, M., and Natarajan, R. (2003) High glucose-induced expression of proinflammatory cytokine and chemokine genes in monocytic cells. *Diabetes* **52**, 1256–1264.
50. van Exel, E., Gussekloo, J., de Craen, A.J., Frolich, M., Bootsma-Van Der Wiel, A., and Westendorp, R.G. (2002) Low production capacity of interleukin-10 associates with the metabolic syndrome and type 2 diabetes: the Leiden 85-Plus Study. *Diabetes* **51**, 1088–1092.
51. Pirola, L., Bonnafous, S., Johnston, A.M., Chaussade, C., Portis, F., and Van Obberghen, E. (2003) Phosphoinositide 3-kinase-mediated reduction of insulin receptor substrate-1/2 protein expression via different mechanisms contributes to the insulin-induced desensitization of its signaling pathways in L6 muscle cells. *J. Biol. Chem.* **278**, 15641–15651.
52. Rui, L., Fisher, T.L., Thomas, J., and White, M.F. (2001) Regulation of insulin/insulin-like growth factor-1 signaling by proteasome-mediated degradation of insulin receptor substrate-2. *J. Biol. Chem.* **276**, 40362–40367.

53. Brownlee, M. (2001) Biochemistry and molecular cell biology of diabetic compli-
    cations. *Nature* **414**, 813–820.
54. Robertson, R.P., Harmon, J., Tran, P.O., Tanaka, Y., and Takahashi, H. (2003)
    Glucose toxicity in beta-cells: type 2 diabetes, good radicals gone bad, and the
    glutathione connection. *Diabetes* **52**, 581–587.
55. Singh, R., Barden, A., Mori, T., and Beilin, L. (2001) Advanced glycation end-
    products: a review. *Diabetologia* **44**, 129–146.
56. Abdel-Wahab, Y.H., O'Harte, F.P., Ratcliff, H., McClenaghan, N.H., Barnett, C.R.,
    and Flatt, P.R. (1996) Glycation of insulin in the islets of Langerhans of normal
    and diabetic animals. *Diabetes* **45**, 1489–1496.
57. Lindsay, J.R., McKillop, A.M., Mooney, M.H., O'Harte, F.P., Bell, P.M., and
    Flatt, P.R. (2003) Demonstration of increased concentrations of circulating glycated
    insulin in human type 2 diabetes using a novel and specific radioimmunoassay.
    Diabetologia **46**, 475–478.
58. Miele, C., Riboulet, A., Maitan, M.A., et al. (2003) Human glycated albumin
    affects glucose metabolism in L6 skeletal muscle cells by impairing insulin-induced
    insulin receptor substrate (IRS) signaling through a protein kinase C alpha-mediated
    mechanism. *J. Biol. Chem.* **278**, 47376–47387.
59. Rossetti, L. (2000) Perspective: hexosamines and nutrient sensing. *Endocrinology*
    141, 1922–1925.
60. Vosseller, K., Wells, L., Lane, M.D., and Hart, G.W. (2002) Elevated nucleocy-
    toplasmic glycosylation by O-GlcNAc results in insulin resistance associated with
    defects in Akt activation in 3T3-L1 adipocytes. *Proc. Natl. Acad. Sci. U. S. A.* **99**,
    5313–5318.
61. Chen, G., Liu, P., Thurmond, D.C., and Elmendorf, J.S. (2003) Glucosamine-
    induced insulin resistance is coupled to O-linked glycosylation of Munc18c. *FEBS
    Lett.* **534**, 54–60.
62. Parker, G.J., Lund, K.C., Taylor, R.P., and McClain, D.A. (2003) Insulin resistance
    of glycogen synthase mediated by o-linked N-acetylglucosamine. *J. Biol. Chem.*
    278, 10022–10027.
63. Cooksey, R.C., Hebert, L.F., Jr., Zhu, J.H., Wofford, P., Garvey, W.T., and
    McClain, D.A. (1999) Mechanism of hexosamine-induced insulin resistance in
    transgenic mice overexpressing glutamine:fructose-6-phosphate amidotransferase:
    decreased glucose transporter GLUT4 translocation and reversal by treatment with
    thiazolidinedione. *Endocrinology* **140**, 1151–1157.
64. Veerababu, G., Tang, J., Hoffman, R.T., et al. (2000) Overexpression of glutamine:
    fructose-6-phosphate amidotransferase in the liver of transgenic mice results
    in enhanced glycogen storage, hyperlipidemia, obesity, and impaired glucose
    tolerance. *Diabetes* **49**, 2070–2078.
65. Monauni, T., Zenti, M.G., Cretti, A., et al. (2000) Effects of glucosamine infusion
    on insulin secretion and insulin action in humans. *Diabetes* **49**, 926–935.

66. Munzberg, H., Bjornholm, M., Bates, S.H., and Myers, M.G., Jr. (2005) Leptin receptor action and mechanisms of leptin resistance. *Cell Mol. Life Sci.* **62**, 642–652.

67. Adeghate, E. (2004) An update on the biology and physiology of resistin. *Cell Mol. Life Sci.* **61**, 2485–2496.

68. Fasshauer, M., Paschke, R., and Stumvoll, M. (2004) Adiponectin, obesity, and cardiovascular disease. *Biochimie* **86**, 779–784.

69. Yue, T.L., Bao, W., Gu, J.L., et al. (2005) Rosiglitazone treatment in Zucker diabetic fatty rats is associated with ameliorated cardiac insulin resistance and protection from ischemia/reperfusion-induced myocardial injury. *Diabetes* **54**, 554–562.

70. Kolter, T., Uphues, I., and Eckel, J. (1997) Molecular analysis of insulin resistance in isolated ventricular cardiomyocytes of obese Zucker rats. *Am. J. Physiol.* **273**, E59–E67.

71. Gibbs, E.M., Stock, J.L., McCoid, S.C., et al. (1995) Glycemic improvement in diabetic db/db mice by overexpression of the human insulin-regulatable glucose transporter (GLUT4). *J. Clin. Invest.* **95**, 1512–1518.

72. Stanley, W.C., Lopaschuk, G.D., and McCormack, J.G. (1997) Regulation of energy substrate metabolism in the diabetic heart. *Cardiovasc. Res.* **34**, 25–33.

73. Bronfman, M., Morales, M.N., and Orellana, A. (1988) Diacylglycerol activation of protein kinase C is modulated by long-chain acyl-CoA. *Biochem. Biophys. Res. Commun.* **152**, 987–992.

74. Russ, M., and Eckel, J. (1995) Insulin action on cardiac glucose transport: studies on the role of protein kinase C. *Biochem. Biophys. Acta.* **1265**, 73–78.

75. Schmitz-Peiffer, C., Craig, D.L., and Biden, T.J. (1999) Ceramide generation is sufficient to account for the inhibition of the insulin-stimulated PKB pathway in C2C12 skeletal muscle cells pretreated with palmitate. *J. Biol. Chem.* **274**, 24202–24210.

76. Qi, D., Pulinilkunnil, T., An, D., et al. (2004) Single-dose dexamethasone induces whole-body insulin resistance and alters both cardiac fatty acid and carbohydrate metabolism. *Diabetes* **53**, 1790–1797.

77. Randle, P.J., Garland, P.B., Hales, C.N., and Newsholme, E.A. (1963) The glucose fatty-acid cycle. Its role in insulin sensitivity and the metabolic disturbances of diabetes mellitus. *Lancet* **1**, 785–789.

78. Wang, P., Lloyd, S.G., Zeng, H., Bonen, A., and Chatham, J.C. (2005) Impact of altered substrate utilization on cardiac function in isolated hearts from Zucker diabetic fatty rats. *Am. J. Physiol. Heart Circ. Physiol.* **288**, H2102–H2110.

79. Liu, Q., Docherty, J.C., Rendell, J.C., Clanachan, A.S., and Lopaschuk, G.D. (2002) High levels of fatty acids delay the recovery of intracellular pH and cardiac efficiency in post-ischemic hearts by inhibiting glucose oxidation. *J. Am. Coll. Cardiol.* **39**, 718–725.

80. Unger, R.H. (2003) The physiology of cellular liporegulation. *Annu. Rev. Physiol.* **65**, 333–347.

81. Wang, M.Y., Lee, Y., and Unger, R.H. (1999) Novel form of lipolysis induced by leptin. *J. Biol. Chem.* **274**, 17541–17544.

82. Sparagna, G.C., Hickson-Bick, D.L., Buja, L.M., and McMillin, J.B. (2000) A metabolic role for mitochondria in palmitate-induced cardiac myocyte apoptosis. *Am. J. Physiol. Heart Circ. Physiol.* **279**, H2124–2132.

83. Ghosh, S., Qi, D., An, D., et al. (2004) Brief episode of STZ-induced hyperglycemia produces cardiac abnormalities in rats fed a diet rich in n-6 PUFA. *Am. J. Physiol. Heart Circ. Physiol.* **287**, H2518–H2527.

84. Ghosh, S., An, D., Pulinilkunnil, T., et al. (2004) Role of dietary fatty acids and acute hyperglycemia in modulating cardiac cell death. *Nutrition* **20**, 916–923.

85. McIntosh, C.H.S., and Pederson, R.A. (1999) Noninsulin-dependent animal models of diabetes mellitus. In: McNeill, J.H. (ed.) *Experimental Models of Diabetes.* CRC Press LLC, Boca Raton, FL, pp. 337–398.

86. Ktorza, A., Bernard, C., Parent, V., et al. (1997) Are animal models of diabetes relevant to the study of the genetics of non-insulin-dependent diabetes in humans? *Diabetes Metab.* **23**(Suppl 2), 38–46.

87. York, D.A. (1996) Lessons from animal models of obesity. *Endocrinol. Metab. Clin. North Am.* **25**, 781–800.

88. Chagnon, Y.C., and Bouchard, C. (1996) Genetics of obesity: advances from rodent studies. *Trends Genet.* **12**, 441–444.

89. Postic, C., Mauvais-Jarvis, F., and Girard, J. (2004) Mouse models of insulin resistance and type 2 diabetes. *Ann. Endocrinol. (Paris).* **65**, 51–59.

90. Nandi, A., Kitamura, Y., Kahn, C.R., and Accili, D. (2004) Mouse models of insulin resistance. *Physiol. Rev.* **84**, 623–647.

91. Wood, P.A. (2004) Genetically modified mouse models for disorders of fatty acid metabolism: pursuing the nutrigenomics of insulin resistance and type 2 diabetes. *Nutrition* **20**, 121–126.

92. Kahn, C.R. (2003) Knockout mice challenge our concepts of glucose homeostasis and the pathogenesis of diabetes. *Exp. Diabetes Res.* **4**, 169–182.

93. Zuzker, L.M., and Zucker, T.F. (1961) Fatty, a new mutation in the rats. *J. Hered.* **52**, 275–287.

94. White, D.W., Wang, D.W., Chua, S.C., Jr., et al. (1997) Constitutive and impaired signaling of leptin receptors containing the Gln –> Pro extracellular domain fatty mutation. *Proc. Natl. Acad. Sci. U. S. A.* **94**, 10657–10662.

95. Bray, G.A., and York, D.A. (1972) Studies on food intake of genetically obese rats. *Am. J. Physiol.* **223**, 176–179.

96. Chan, C.B., Pederson, R.A., Buchan, A.M., Tubesing, K.B., and Brown, J.C. (1985) Gastric inhibitory polypeptide and hyperinsulinemia in the Zucker (fa/fa) rat: a developmental study. *Int. J. Obes.* **9**, 137–146.

97. Peterson, R.G., Shaw, W.N., Neel, M.A., Little, L.A., and Eichberg, J. (1990) Zucker diabetic fatty rats as a model for non-insulin dependent diabetes mellitus. *ILAR News* **32**.

98. Peterson, R.G. (2000) The Zucker diabetic fatty rat. In: Sima, A.F., and Shafrir, S. (eds.) *Animal Models of Diabetes: A Primer.* Harwood Academic Publisher, Newark, NJ, pp. 109–128.

99. Tokuyama, Y., Sturis, J., DePaoli, A.M., et al. (1995) Evolution of beta-cell dysfunction in the male Zucker diabetic fatty rat. *Diabetes* **44**, 1447–1457.

100. Uphues, I., Kolter, T., Goud, B., and Eckel, J. (1995) Failure of insulin-regulated recruitment of the glucose transporter GLUT4 in cardiac muscle of obese Zucker rats is associated with alterations of small-molecular-mass GTP-binding proteins. *Biochem. J.* **311**, 161–166.

101. Slieker, L.J., Sundell, K.L., Heath, W.F., et al. (1992) Glucose transporter levels in tissues of spontaneously diabetic Zucker fa/fa rat (ZDF/drt) and viable yellow mouse (Avy/a). *Diabetes* **41**, 187–193.

102. Luiken, J.J., Arumugam, Y., Dyck, D.J., et al. (2001) Increased rates of fatty acid uptake and plasmalemmal fatty acid transporters in obese Zucker rats. *J. Biol. Chem.* **276**, 40567–40573.

103. Unger, R.H., and Orci, L. (2000) Lipotoxic diseases of nonadipose tissues in obesity. *Int. J. Obes. Relat. Metab. Disord.* **24**, S28–32.

104. Young, M.E., Guthrie, P.H., Razeghi, P., et al. (2002) Impaired long-chain fatty acid oxidation and contractile dysfunction in the obese Zucker rat heart. *Diabetes* **51**, 2587–2595.

105. Zhou, Y.T., Grayburn, P., Karim, A., et al. (2000) Lipotoxic heart disease in obese rats: implications for human obesity. *Proc. Natl. Acad. Sci. U. S. A.* **97**, 1784–1789.

106. Atkinson, L.L., Kozak, R., Kelly, S.E., Onay Besikci, A., Russell, J.C., and Lopaschuk, G.D. (2003) Potential mechanisms and consequences of cardiac triacylglycerol accumulation in insulin-resistant rats. *Am. J. Physiol. Endocrinol. Metab.* **284**, E923–E930.

107. Wang, M.Y., and Unger, R.H. (2005) Role of PP2C in cardiac lipid accumulation in obese rodents and its prevention by troglitazone. *Am. J. Physiol. Endocrinol. Metab.* **288**, E216–E221.

108. Listenberger, L.L., Han, X., Lewis, S.E., et al. (2003) Triglyceride accumulation protects against fatty acid-induced lipotoxicity. *Proc. Natl. Acad. Sci. U. S. A.* **100**, 3077–3082.

109. Zhou, Y.T., Shimabukuro, M., Lee, Y., et al. (1998) Enhanced de novo lipogenesis in the leptin-unresponsive pancreatic islets of prediabetic Zucker diabetic fatty rats: role in the pathogenesis of lipotoxic diabetes. *Diabetes* **47**, 1904–1908.

110. Shimabukuro, M., Zhou, Y.T., Levi, M., and Unger, R.H. (1998) Fatty acid-induced beta cell apoptosis: a link between obesity and diabetes. *Proc. Natl. Acad. Sci. U. S. A.* **95**, 2498–2502.

111. Wu-Peng, X.S., Chua, S.C., Jr., Okada, N., Liu, S.M., Nicolson, M., and Leibel, R.L. (1997) Phenotype of the obese Koletsky(f) rat due to Tyr763Stop mutation in the extracellular domain of the leptin receptor (Lepr): evidence for deficient plasma-to-CSF transport of leptin in both the Zucker and Koletsky obese rat. *Diabetes* **46**, 513–518.

112. Brindley, D.N., and Russell, J.C. (2002) Animal models of insulin resistance and cardiovascular disease: some therapeutic approaches using JCR:LA-cp rat. *Diabetes Obes. Metab.* **4**, 1–10.

113. Russell, J.C., Shillabeer, G., Bar-Tana, J., et al. (1998) Development of insulin resistance in the JCR:LA-cp rat: role of triacylglycerols and effects of MEDICA 16. *Diabetes* **47**, 770–778.

114. Pederson, R.A., Campos, R.V., Buchan, A.M., Chisholm, C.B., Russell, J.C., and Brown, J.C. (1991) Comparison of the enteroinsular axis in two strains of obese rat, the fatty Zucker and the JCR:LA-corpulent. *Int. J. Obes.* **15**, 461–470.

115. Russell, J.C., Graham, S., and Hameed, M. (1994) Abnormal insulin and glucose metabolism in the JCR:LA-corpulent rat. *Metabolism* **43**, 538–543.

116. Russell, J.C., Koeslag, D.G., Dolphin, P.J., and Amy, R.M. (1990) Prevention of myocardial lesions in JCR:LA-corpulent rats by nifedipine. *Arteriosclerosis* **10**, 658–664.

117. Richardson, M., Schmidt, A.M., Graham, S.E., Achen, B., DeReske, M., and Russell, J.C. (1998) Vasculopathy and insulin resistance in the JCR:LA-cp rat. *Atherosclerosis* **138**, 135–146.

118. Kawano, K., Hirashima, T., Mori, S., and Natori, T. (1996) Spontaneously diabetic rat "OLETF" as a model for NIDDM in humans. In: Shafrir, E. (ed.) *Lessons from Animal Diabetes*. Birkhauser, Boston, MA, p. 225.

119. Kawano, K., Hirashima, T., Mori, S., Saitoh, Y., Kurosumi, M., and Natori, T. (1992) Spontaneous long-term hyperglycemic rat with diabetic complications. Otsuka Long-Evans Tokushima Fatty (OLETF) strain. *Diabetes* **41**, 1422–1428.

120. Abe, T., Ohga, Y., Tabayashi, N., et al. (2002) Left ventricular diastolic dysfunction in type 2 diabetes mellitus model rats. *Am. J. Physiol. Heart Circ. Physiol.* **282**, H138–H148.

121. Mizushige, K., Yao, L., Noma, T., et al. (2000) Alteration in left ventricular diastolic filling and accumulation of myocardial collagen at insulin-resistant prediabetic stage of a type II diabetic rat model. *Circulation* **101**, 899–907.

122. Yagi, K., Kim, S., Wanibuchi, H., Yamashita, T., Yamamura, Y., and Iwao, H. (1997) Characteristics of diabetes, blood pressure, and cardiac and renal complications in Otsuka Long-Evans Tokushima fatty rats. *Hypertension* **29**, 728–735.

123. Saito, F., Kawaguchi, M., Izumida, J., Asakura, T., Maehara, K., and Maruyama, Y. (2003) Alteration in haemodynamics and pathological changes in the cardiovascular system during the development of Type 2 diabetes mellitus in OLETF rats. *Diabetologia* **46**, 1161–1169.

124. Goto, Y., Suzuki, K., Sasaki, M., Ono, T., and Abe, S. (1988) GK rat as a model of nonobese, non-insulin dependent diabetes. Selective breeding over 35 generations. In: Shafrir, E., and Reynold, A.E. (eds.) *Frontiers in Diabetes Research. Lessons from Animal Diabetes II.* John Libbey and Co, London, pp. 301–303.

125. El-Omar, M.M., Yang, Z.K., Phillips, A.O., and Shah, A.M. (2004) Cardiac dysfunction in the Goto-Kakizaki rat. A model of type II diabetes mellitus. *Basic Res. Cardiol.* **99**, 133–141.

126. Guenifi, A., Abdel-Halim, S.M., Hoog, A., Falkmer, S., and Ostenson, C.G. (1995) Preserved beta-cell density in the endocrine pancreas of young, spontaneously diabetic Goto-Kakizaki (GK) rats. *Pancreas* **10**, 148–153.

127. Desrois, M., Sidell, R.J., Gauguier, D., Davey, C.L., Radda, G.K., and Clarke. K. (2004) Gender differences in hypertrophy, insulin resistance and ischemic injury in the aging type 2 diabetic rat heart. *J. Mol. Cell Cardiol.* **37**, 547–555.

128. Santos, D.L., Palmeira, C.M., Seica, R., et al. (2003) Diabetes and mitochondrial oxidative stress: a study using heart mitochondria from the diabetic Goto-Kakizaki rat. *Mol. Cell Biochem.* **246**, 163–170.

129. Hwang, I.S., Ho, H., Hoffman, B.B., and Reaven, G.M. (1987) Fructose-induced insulin resistance and hypertension in rats. *Hypertension* **10**, 512–516.

130. Dai, S., and McNeill, J.H. (1995) Fructose-induced hypertension in rats is concentration- and duration-dependent. *J. Pharmacol. Toxicol. Methods* **33**, 101–107.

131. Verma, S., Bhanot, S., and McNeill, J.H. (1994) Antihypertensive effects of metformin in fructose-fed hyperinsulinemic, hypertensive rats. *J. Pharmacol. Exp. Ther.* **271**, 1334–1337.

132. Galipeau, D., Arikawa, E., Sekirov, I., and McNeill, J.H. (2001) Chronic thromboxane synthase inhibition prevents fructose-induced hypertension. *Hypertension* **38**, 872–876.

133. Vasdev, S., Longerich, L., and Gill, V. (2004) Prevention of fructose-induced hypertension by dietary vitamins. *Clin. Biochem.* **37**, 1–9.

134. Elliott, S.S., Keim, N.L., Stern, J.S., Teff, K., and Havel, P.J. (2002) Fructose, weight gain, and the insulin resistance syndrome. *Am. J. Clin. Nutr.* **76**, 911–922.

135. Havel, P.J. (2005) Dietary fructose: implications for dysregulation of energy homeostasis and lipid/carbohydrate metabolism. *Nutr. Rev.* **63**, 133–157.

136. Verma, S., Yao, L., Dumont, A.S., and McNeill, J.H. (2000) Metformin treatment corrects vascular insulin resistance in hypertension. *J. Hypertens.* **18**, 1445–1450.

137. Bhanot, S., Micholaus, A., and McNeill, J.H. (1995) Antihypertensive effects of vanadium compounds in hyperinsulinemic, hypertensive rats. *Mol. Cell Biochem.* **153**, 205–209.

138. Lee, M.K., Miles, P.D., Khoursheed, M., Gao, K.M., Moossa, A.R., and Olefsky, J.M. (1994) Metabolic effects of troglitazone on fructose-induced insulin resistance in the rat. *Diabetes* **43**, 1435–1439.

139. Katakam, P.V., Ujhelyi, M.R., Hoenig, M.E., and Miller, A.W. (1998) Endothelial dysfunction precedes hypertension in diet-induced insulin resistance. *Am. J. Physiol.* **275**, R788–R792.

140. Verma, S., Bhanot, S., Yao, L., and McNeill, J.H. (1996) Defective endothelium-dependent relaxation in fructose-hypertensive rats. *Am. J. Hypertens.* **9**, 370–376.

141. Verma, S., Skarsgard, P., Bhanot, S., Yao, L., Laher, I., and McNeill, J.H. (1997) Reactivity of mesenteric arteries from fructose hypertensive rats to endothelin-1. *Am. J. Hypertens.* **10**, 1010–1019.

142. Galipeau, D., Verma, S., and McNeill, J.H. (2002) Female rats are protected against fructose-induced changes in metabolism and blood pressure. *Am. J. Physiol. Heart Circ. Physiol.* **283**, H2478–H2484.

143. Song, D., Arikawa, E., Galipeau, D., Battell, M., and McNeill, J.H. (2004) Androgens are necessary for the development of fructose-induced hypertension. *Hypertension* **43**, 667–672.

144. Vasudevan, H., Xiang, H., and McNeill, J.H. (2005) Differential regulation of insulin resistance and hypertension by sex hormones in fructose-fed male rats. *Am. J. Physiol. Heart Circ. Physiol.* **289**(4): H1335–42.

145. Fried, S.K., and Rao, S.P. (2003) Sugars, hypertriglyceridemia, and cardiovascular disease. *Am. J. Clin. Nutr.* **78**, 873S–880S.

146. Marckmann, P., Raben, A., and Astrup, A. (2000) Ad libitum intake of low-fat diets rich in either starchy foods or sucrose: effects on blood lipids, factor VII coagulant activity, and fibrinogen. *Metabolism* **49**, 731–735.

147. Raben, A., Holst, J.J., Madsen, J., and Astrup, A. (2001) Diurnal metabolic profiles after 14 d of an ad libitum high-starch, high-sucrose, or high-fat diet in normal-weight never-obese and postobese women. *Am. J. Clin. Nutr.* **73**, 177–189.

148. Albrink, M.J., and Ullrich, I.H. (1986) Interaction of dietary sucrose and fiber on serum lipids in healthy young men fed high carbohydrate diets. *Am. J. Clin. Nutr.* **43**, 419–428.

149. Frayn, K.N., and Kingman, S.M. (1995) Dietary sugars and lipid metabolism in humans. *Am. J. Clin. Nutr.* **62**, 250S–261S; discussion 261S–263S.

150. Dutta, K., Podolin, D.A., Davidson, M.B., and Davidoff, A.J. (2001) Cardiomyocyte dysfunction in sucrose-fed rats is associated with insulin resistance. *Diabetes* **50**, 1186–1192.

151. Davidoff, A.J., Mason, M.M., Davidson, M.B., et al. (2004) Sucrose-induced cardiomyocyte dysfunction is both preventable and reversible with clinically relevant treatments. *Am. J. Physiol. Endocrinol. Metab.* **286**, E718–E724.

152. Coderre, L., Vallega, G.A., Pilch, P.F., and Chipkin, S.R. (1996) In vivo effects of dexamethasone and sucrose on glucose transport (GLUT-4) protein tissue distribution. *Am. J. Physiol.* **271**, E643–E648.

153. Kawasaki, T., Kashiwabara, A., Sakai, T., et al. (2005) Long-term sucrose-drinking causes increased body weight and glucose intolerance in normal male rats. *Br. J. Nutr.* **93**, 613–618.

154. Hintz, K.K., Aberle, N.S., and Ren, J. (2003) Insulin resistance induces hyperleptinemia, cardiac contractile dysfunction but not cardiac leptin resistance in ventricular myocytes. *Int. J. Obes. Relat. Metab. Disord.* **27**, 1196–1203.

155. Wold, L.E., Dutta, K., Mason, M.M., et al. (2005) Impaired SERCA function contributes to cardiomyocyte dysfunction in insulin resistant rats. *J. Mol. Cell Cardiol.* **39**(2): 297–307.

156. Pierce, G.N., Lockwood, M.K., and Eckhert, C.D. (1989) Cardiac contracile protein ATPase activity in a diet induced model of noninsulin dependent diabetes mellitus. *Can. J. Cardiol.* **5**, 117–120.

157. Busserolles, J., Zimowska, W., Rock, E., Rayssiguier, Y., and Mazur, A. (2002) Rats fed a high sucrose diet have altered heart antioxidant enzyme activity and gene expression. *Life Sci.* **71**, 1303–1312.

158. Schacke, H., Docke, W.D., and Asadullah, K. (2002) Mechanisms involved in the side effects of glucocorticoids. *Pharmacol. Ther.* **96**, 23–43.

159. Phillips, D.I., Barker, D.J., Fall, C.H., et al. (1998) Elevated plasma cortisol concentrations: a link between low birth weight and the insulin resistance syndrome? *J. Clin. Endocrinol. Metab.* **83**, 757–760.

160. Rosmond, R., Dallman, M.F., and Bjorntorp, P. (1998) Stress-related cortisol secretion in men: relationships with abdominal obesity and endocrine, metabolic and hemodynamic abnormalities. *J. Clin. Endocrinol. Metab.* **83**, 1853–1859.

161. Stojanovska, L., Rosella, G., and Proietto, J. (1990) Evolution of dexamethasone-induced insulin resistance in rats. *Am. J. Physiol.* **258**, E748–E756.

162. Asensio, C., Muzzin, P., and Rohner-Jeanrenaud, F. (2004) Role of glucocorticoids in the physiopathology of excessive fat deposition and insulin resistance. *Int. J. Obes. Relat. Metab. Disord.* **28**, S45–S52.

163. Zakrzewska, K.E., Cusin, I., Stricker-Krongrad, A., et al. (1999) Induction of obesity and hyperleptinemia by central glucocorticoid infusion in the rat. *Diabetes* **48**, 365–370.

164. Cusin, I., Rouru, J., and Rohner-Jeanrenaud, F. (2001) Intracerebroventricular glucocorticoid infusion in normal rats: induction of parasympathetic-mediated obesity and insulin resistance. *Obes. Res.* **9**, 401–406.

165. Reynolds, R.M., Chapman, K.E., Seckl, J.R., Walker, B.R., McKeigue, P.M., and Lithell, H.O. (2002) Skeletal muscle glucocorticoid receptor density and insulin resistance. *JAMA* **287**, 2505–2506.

166. Whorwood, C.B., Donovan, S.J., Flanagan, D., Phillips, D.I., and Byrne, C.D. (2002) Increased glucocorticoid receptor expression in human skeletal muscle cells may contribute to the pathogenesis of the metabolic syndrome. *Diabetes* **51**, 1066–1075.

167. Seckl, J.R., Morton, N.M., Chapman, K.E., and Walker, B.R. (2004) Glucocorticoids and 11beta-hydroxysteroid dehydrogenase in adipose tissue. *Recent Prog. Horm. Res.* **59**, 359–393.

168. Masuzaki, H., Paterson, J., Shinyama, H., et al. (2001) A transgenic model of visceral obesity and the metabolic syndrome. *Science* **294**, 2166–2170.

169. Kotelevtsev, Y., Holmes, M.C., Burchell, A., et al. (1997) 11beta-hydroxysteroid dehydrogenase type 1 knockout mice show attenuated glucocorticoid-inducible responses and resist hyperglycemia on obesity or stress. *Proc. Natl. Acad. Sci. U. S. A.* **94**, 14924–14929.

170. Bowker-Kinley, M.M., Davis, W.I., Wu, P., Harris, R.A., and Popov, K.M. (1998) Evidence for existence of tissue-specific regulation of the mammalian pyruvate dehydrogenase complex. *Biochem. J.* **329**, 191–196.

171. Levak-Frank, S., Hofmann, W., Weinstock, P.H., et al. (1999) Induced mutant mouse lines that express lipoprotein lipase in cardiac muscle, but not in skeletal muscle and adipose tissue, have normal plasma triglyceride and high-density lipoprotein-cholesterol levels. *Proc. Natl. Acad. Sci. U. S. A.* **96**, 3165–3170.

172. Riccardi, G., Giacco, R., and Rivellese, A.A. (2004) Dietary fat, insulin sensitivity and the metabolic syndrome. *Clin. Nutr.* **23**, 447–456.

173. Storlien, L.H., Baur, L.A., Kriketos, A.D., et al. (1996) Dietary fats and insulin action. *Diabetologia* **39**, 621–631.

174. Lichtenstein, A.H., and Schwab, U.S. (2000) Relationship of dietary fat to glucose metabolism. *Atherosclerosis* **150**, 227–243.

175. Rivellese, A.A., and Lilli, S. (2003) Quality of dietary fatty acids, insulin sensitivity and type 2 diabetes. *Biomed. Pharmacother.* **57**, 84–87.

176. Ouwens, D.M., Boer, C., Fodor, M., et al. (2005) Cardiac dysfunction induced by high-fat diet is associated with altered myocardial insulin signalling in rats. *Diabetologia* **48**, 1229–1237.

177. Khan, I.Y., Dekou, V., Douglas, G., et al. (2005) A high-fat diet during rat pregnancy or suckling induces cardiovascular dysfunction in adult offspring. *Am. J. Physiol. Regul. Integr. Comp. Physiol.* **288**, R127–R133.

178. Kaneko, T., Wang, P.Y., Wang, Y., and Sato, A.. (2000) The long-term effect of low-carbohydrate/high-fat diet on the development of diabetes mellitus in spontaneously diabetic rats. *Diabetes Metab.* **26**, 459–464.

179. Chalkley, S.M., Hettiarachchi, M., Chisholm, D.J., and Kraegen, E.W. (2002) Long-term high-fat feeding leads to severe insulin resistance but not diabetes in Wistar rats. *Am. J. Physiol. Endocrinol. Metab.* **282**, E1231–E1238.

180. Drake, A.J., Livingstone, D.E., Andrew, R., Seckl, J.R., Morton, N.M., and Walker, B.R. (2005) Reduced adipose glucocorticoid reactivation and increased hepatic glucocorticoid clearance as an early adaptation to high-fat feeding in Wistar rats. *Endocrinology* **146**, 913–919.

181. Shimabukuro, M., Higa, M., Zhou, Y.T., Wang, M.Y., Newgard, C.B., and Unger, R.H. (1998) Lipoapoptosis in beta-cells of obese prediabetic fa/fa rats. Role of serine palmitoyltransferase overexpression. *J. Biol. Chem.* **273**, 32487–32490.

182. Reed, M.J., Meszaros, K., Entes, L.J., et al. (2000) A new rat model of type 2 diabetes: the fat-fed, streptozotocin-treated rat. *Metabolism* **49**, 1390–1394.
183. Ghosh, S., Ting, S., Lau, H., et al. (2004) Increased efflux of glutathione conjugate in acutely diabetic cardiomyocytes. *Can. J. Physiol. Pharmacol.* **82**, 879–887.
184. Das, D.K. (2001) Redox regulation of cardiomyocyte survival and death. *Antioxid. Redox. Signal.* **3**, 23–37.
185. Ghosh, S., Pulinilkunnil, T., Yuen, G., et al. (2005) Cardiomyocyte apoptosis induced by short-term diabetes requires mitochondrial GSH depletion. *Am. J. Physiol. Heart Circ. Physiol.* **289**(2): H768–76.
186. Hampton, M.B., Fadeel, B., and Orrenius, S. (1998) Redox regulation of the caspases during apoptosis. *Ann. N. Y. Acad. Sci.* **854**, 328–335.
187. Berry, E.M. (2001) Who's afraid of n-6 polyunsaturated fatty acids? Methodological considerations for assessing whether they are harmful. *Nutr. Metab. Cardiovasc. Dis.* **11**, 181–188.
188. Simopoulos, A.P. (1999) Evolutionary aspects of omega-3 fatty acids in the food supply. *Prostaglandins Leukot. Essent. Fatty Acids.* **60**, 421–429.
189. Simopoulos, A.P. (1999) Essential fatty acids in health and chronic disease. *Am. J. Clin. Nutr.* **70**, 560S–569S.
190. Simopoulos, A.P., Leaf, A., and Salem, N., Jr. (1999) Essentiality of and recommended dietary intakes for omega-6 and omega-3 fatty acids. *Ann. Nutr. Metab.* **43**, 127–130.

8

The Isolated, Perfused Pseudo-Working Heart Model

Gary J. Grover and Rajni Singh

Summary

Increased interest in cardiac safety and renewed interest in drugs for treating myocardial ischemia and congestive heart failure have led to increased use of cardiovascular models. Unfortunately, many molecular or cell-based screens are not perfect predictors of activity in vivo. One rapid means to test "proof of principle" in hearts is the isolated or Langendorff perfused heart system. This allows direct measurement of cardiac contractile function and coronary flow without interference from changes in the systemic circulation. The setup basically is a heart whose coronary arteries are retrogradely perfused through the aortic root with heated and oxygenated buffer solutions. We will describe the constant pressure technique as this is the most commonly used and allows us to directly measure changes in coronary flow without changes in turgor noted with constant flow methods. Instead of a classical working heart preparation, we describe the use of a balloon properly fitted into the left ventricle to measure end diastolic and peak systolic pressure making this a much simpler and less expensive technique without loss of scientific quality. Such a setup can be used to measure effects of compounds or treatments on coronary flow, contractile function either in the normal or pathological states.

Key Words: Heart; Langendorff; Isolated heart; Coronary flow; Cardiac function.

1. Introduction

In an age when molecular-based drug discovery and research predominate, it is still critical to perform "proof of concept" studies in living systems. The requirements for such in vivo or ex vivo assays are predictability of efficacy or safety, depending on the purpose of the study. In the cardiovascular field, direct effects of compounds on the heart are critical both in terms of efficacy and in terms of safety. There is no better means of determining the direct

From: *Methods in Molecular Medicine, Vol. 139: Vascular Biology Protocols*
Edited by: N. Sreejayan and J. Ren © Humana Press Inc., Totowa, NJ

effects of compounds on cardiac function (inotropy and lusitropy), heart rate, coronary flow, and electrophysiology than isolated perfused hearts *(1,2)*. The outstanding advantage of this preparation is that true measurements of inotropy (force of contraction at a given pre-load) and heart rate can be measured without the confounding influence of changes in peripheral hemodynamic status seen in whole-animal preparations. The other advantage is the speed with which these studies can be done, allowing rapid determination of critical questions, and if need be, focused structure-activity efforts can be successfully done using this relatively high throughput ex vivo screen *(3)*. This is particularly valuable for safety determinations and studies in complicated diseases such as myocardial ischemia in which in vitro assays may not always be predictive. In the **Subheading 3**, we discuss about the technical setup for small animal hearts. We will use what has been called the "pseudo-working" heart that involves measurement of inotropy and lusitropy using a left ventricular balloon. This setup is infinitely easier than and just as predictive as the true working heart preparation. The technique described here can be used for guinea pigs, rats, mice, and rabbits.

2. Materials

2.1. Animal Preparation

1. Sodium pentobarbital.
2. Sodium heparin (1000 U/kg).
3. Small animal ventilator.

2.2. Removal of the Heart and Attachment to Langendorff Apparatus

1. Krebs–Henseleit solution containing 112 mM NaCl, 25 mM $NaHCO_3$, 5 mM KCl, 1.2 mM $MgSO_4$, 1 mM KH_2PO_4, 1.2 mM $CaCl_2$, 11.5 mM glucose, and 2 mM pyruvate for rats. If guinea pigs or rabbits are used, increase $CaCl_2$ to 2–2.5 mM to achieve reproducible contractile function.
2. Metzenbaum (curved, with flat heads) scissors, small hemostats, tissue clamps, and forceps.
3. Suture material, 3-0 silk.
4. Heating exchanger jacket with coiled tubes for perfusate. The perfusate tube will have bubble traps in perfusion line (*see* **Fig. 1**).
5. Incubator and pump for perfusion of heat exchanger at 37 °C.
6. Large buffer container with oxygenator stone connected to 95% O_2 and 5% CO_2 tank (*see* **Note 1**).
7. Small container (0.5 L) with oxygenated Krebs–Henseleit and cannula for retrograde perfusion (cannula attached to a three-way stopcock) for moving heart to main Langendorff setup.

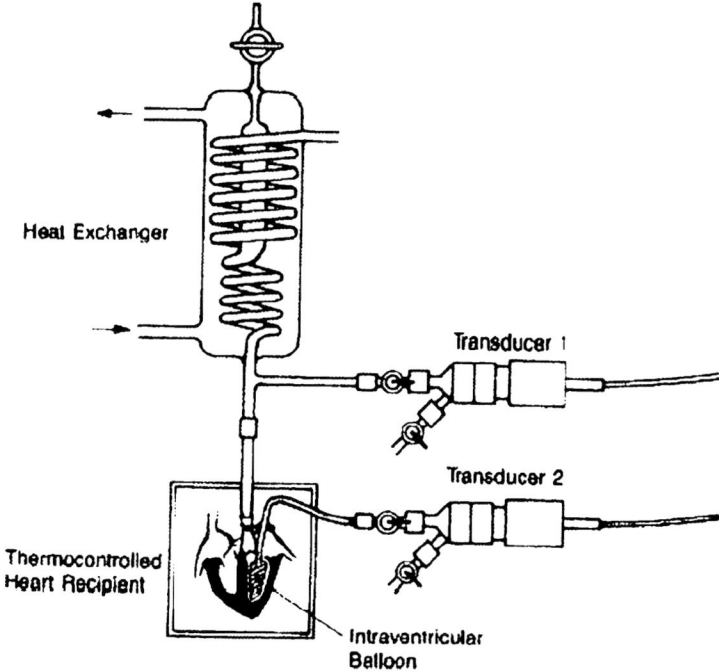

Fig. 1. Diagrammatic setup for isolated "pseudo-working" heart preparation. Such preparations can be set up as constant pressure or constant flow using a pump. For constant pressure, the buffer container height is adjusted to achieve proper perfusion pressure that must be measured at the bottom of the setup, as close to the heart as possible. The heat exchanger is basically a water-jacketed system in which the perfusate tubing is coiled to allow temperature equilibration to 37 °C. Transducers will measure coronary perfusion pressure and balloon pressure.

2.3. The Balloon

1. One pressure transducer at bottom of heat exchanger (measurement of coronary perfusion pressure) and one for the left ventricular balloon.
2. Spindle syringe for transducer for balloon. Attach to three-way stopcock on transducer.
3. Left ventricular balloon (small latex finger cots or balloons purchased through vendor).
4. Three-way stopcocks for transducers.
5. Heated tissue chamber (water jacketed, heated to 37 °C) for placement of heart. Obtain size that best fits the size of the heart to be used.

6. Extracorporeal electromagnetic flow probe placed in-line with perfusion medium flowing at bottom of heat exchanger (if flow probes not available, measurement of flow can be done by simply collecting fluid draining heart at time intervals of interest).
7. Physiological saline (0.9%) for balloon.

3. Methods

3.1. Animal Preparation

1. Withdraw food 24 h prior to the start of the experiment.
2. Inject the animal with 100 mg/kg sodium pentobarbital into the peritoneum (i.p.).
3. Give an i.v. injection of 1000 U/kg sodium heparin to prevent blood clotting (through tail vein or jugular).
4. The animals will have their tracheas intubated and ventilated mechanically.

3.2. Removal of the Heart and Attachment to Langendorff Apparatus

1. Incise the skin from the middle of the abdomen up to the throat by a longitudinal cut and open the abdomen up to the diaphragm.
2. Cut the diaphragm off the ribs following the anterior part of the inferior thoracic aperture.
3. Split the sternum from the xiphoid process along the midline and spread the ribs as far lateral as possible. Then turn the two thorax halves upward using clamps.
4. Remove the pericardium up to its vascular stem attachment.
5. Free the ascending aorta from connective tissue (separate it from the pulmonary artery using only forceps) and position two threads around the aorta.
6. Fill aortic cannula with oxygenated perfusate connected to heated smaller buffer container (*see* **Subheading 2.2., step 7**).
7. Clamp the inferior vena cava above the diaphragm to inhibit bleeding.
8. To prevent the right ventricle from overstretching, incise the pulmonary artery.
9. Use bent (Metzenbaum) scissors to cut out the heart while keeping it elevated. Take care not to injure the posterior part of the right atrium, insert the cannula (toward heart) and open stopcock for retrograde perfusion. While inserting, allow the perfusion fluid to flow through the tube. Secure the cannula to the aorta using suture material so that no flow escapes and the heart is secure. Move the cannula stopcock to the main perfusion system and then disconnect the stopcock from the small perfusion container (*see* **Note 2**).
10. Attach the stopcock along with the heart to the heat exchanger's nozzle and allow flow from the large buffer container to begin. Observe the complete fluid system (from the reservoir to the lower end of the spiral in the heat exchanger) and the cannula to ensure the absence of air bubbles. The heart will turn pale pink due to the removal of blood from the capillaries and start to beat vigorously.
11. Remove any extraneous pieces on the surface of the heart (fat tissues, lung pieces, vessel roots, etc.).

12. Make sure that cannula is not inserted too low so as to impede the closure of the aortic valves and tighten the lower thread.
13. The height of the perfusion container above the heat exchanger should be set to 85 mmHg normally or can be altered depending on the study parameters. The perfusion pressure is measured using a pressure transducer attached to a side port just above the level of the heart to accurately assess the coronary perfusion pressure. This side port can also be used to inject drugs that cannot be exposed to an oxidizing environment (e.g., sulfhydryl-containing drugs such as captopril or zofenopril). Inotropic agents such as epinephrine must be further protected by the addition of ascorbic acid even if given through the side port.
14. Make sure that the height of the column remains constant so that pressure remains constant, so either use a large capacity buffer container or use a servo-controlled system where lowering of fluid level completes a circuit allowing a high torque pump to restore fluid at constant level (this can be made from common household materials or can be purchased). Although labor intensive, one can simply keep adding buffer to the buffer container to maintain levels.
15. Place the heart into the water-jacketed tissue chamber and put rubber stopper on bottom drain and allow heart to be surrounded by 37 °C buffer (*see* **Note 3**) (*see* **Fig. 1**).

3.3. The Balloon Method

1. Attach a small balloon to a bent metal catheter with a bell-shaped tip. Connect the other end of the catheter to a pressure transducer, and using a three-way stopcock, attach a 2-mL, water-filled spindle syringe to the second, angled opening of the transducer dome (*see* **Note 4**).
2. Insert the balloon into the isolated heart's left ventricle through one of the pulmonary veins (which can be widened with a pair of scissors), the left atrium, and the mitral valve.
3. Using the spindle syringe, fill balloon until an end diastolic pressure of 5 mmHg is obtained.
4. Allow 15 min for equilibration time. Maintain this balloon volume for the duration of the experiment unless generation of a Starling curve is desired.
5. After a baseline measurement is taken, the appropriate experiment can be started.
6. Compounds of interest are most often given through the perfusion chamber unless stability of compound in salt solution or in an oxidizing environment is a concern. Obviously, compounds will be administered on a molar basis and great care must be taken to assure proper dosing.

3.4. Summary

With this setup, effects of compounds or treatments on coronary flow and cardiac function can be determined. It is crucial that the hearts be sufficiently oxygenated to allow for a significant vasodilator reserve. This setup is also ideal for ischemia studies as coronary flow can be reduced by shutting off perfusate or ligating a coronary artery. Ischemic injury can be followed during ischemia by the time to onset of contracture (5 mmHg increase in end diastolic pressure above baseline) and reperfusion recovery of contractile function *(2)*. Electrophysiological and arrhythmia studies can also be done with this setup *(4)*.

4. Notes

1. Use a stone allowing a fine spray of bubbles to give as high a PO_2 as possible. If properly done, PO_2 should be in the 550–600 mmHg range, allowing for a coronary vasodilator reserve. We have found that many investigators do not pay enough attention to this detail. It is critical that there is sufficient oxygen so that there is a coronary dilator reserve.

2. Some investigators remove heart and put on ice while transferring hearts. This will increase the time for equilibration, and by transferring the heart while being perfused, the hearts are nearly completely perfused throughout the procedure allowing for rapid equilibration and more consistent results.

3. Experience has taught us that even small convective currents in the room due to door opening or even people walking by can change heart temperature by 1–2 °C. Keeping the hearts submerged in heated buffer solution keeps the hearts at constant temperature and reduces the sample size needed. This is particularly important for ischemia studies where a 0.5 °C temperature difference can cause significantly different degrees of damage (cardioplegic effect).

4. The compliance and size of the balloon are critical as it must exert no pressure when empty, yet exert pressure when filled with physiological saline. Balloons can be purchased or some people use the tips of reservoir condoms. Our best results have come from finger cots.

References

1. Doring H.J. and Dehnert H. (1988) *The Isolated Perfused Warm-Blooded Heart according to Langendorff*. Biomesstechnik-Verlag March GmbH, West Germany.
2. Garlid K.D., Paucek P., Yarov-Yarovoy V., Murray H.N., Darbenzio R.B., D'Alonzo A.J., Lodge N.J., Smith M.A., and Grover G.J. (1997) Cardioprotective effect of diazoxide and its interaction with mitochondrial ATP-sensitive K^+ channels. *Circulation* **81**, 1072–1082.
3. Atwal K.S., Ferrara F.N., Ding C.Z., Grover G.J., Sleph P.G., Dzwonczyk S., Baird A.J., and Normandin D.E. (1996) Cardioselective antiischemic ATP-sensitive potassium channel openers 4: structure activity studies on benzopyranyl cyanoguanidines; replacement of the benzopyran portion. *J. Med. Chem.* **39**, 304–313.
4. D'Alonzo A.J., Grover G.J., Darbenzio R.B., Sewter J.C., Hess T.A., Dzwonczyk S., and Sleph. P.G. (1996) Hemodynamic and cardiac effects of BMS-180448, a novel ATP-sensitive potassium channel opener in anesthetized dogs and isolated perfused rat hearts. *Pharmacology* **52**, 101–112.

9

Altering and Analyzing Glucose Metabolism in Perfused Hearts of Transgenic Mice

Rajakumar V. Donthi and Paul N. Epstein

Summary

Glucose metabolism plays an important role in cardiac bioenergetics that changes under various stress conditions including hypertrophy, diabetic cardiomyopathy, and ischemia-reperfusion injury. To understand the role of glycolysis under these conditions, we have altered several steps of the glycolytic pathway specifically in the heart. In this chapter, we describe methods used to produce cardiac-targeted transgenic mice and procedures for measuring various glucose metabolites including glucose-6-phosphate, fructose-6-phosphate, fructose-1,6-bisphosphate, and glycogen. Also, we describe methods for measuring glucose transport and glycolysis in perfused mouse hearts. Using these methods, we show that mice over-expressing cardiac-specific kinase-deficient 6-phosphofructo-2-kinase/fructose-2,6-bisphosphatase (Mykd-PFK-2) show reduced glucose transport and reduced glycolysis when compared with control. The metabolites glucose-6-phosphate, fructose-6-phosphate, and glycogen were elevated, whereas fructose-1,6-bisphosphate was reduced in the transgenic Mykd-PFK-2 mouse hearts.

Key Words: Transgenic mice; Cardiac-specific expression; Diabetes; Glycolysis; Glycolytic metabolites; Insulin sensitivity.

1. Introduction

Heart depends on various substrates such as carbohydrates, fatty acids, and amino acids for normal function. However, in some pathological conditions, the heart preferentially uses carbohydrates over other substrates. For example, Bishop and Altschuld (*1*) reported that glycolytic metabolism is increased in cardiac hypertrophy and congestive heart failure. Glycolysis is critical to cardiac survival in hypoxia, because glycolysis is the sole source of adenosine

From: *Methods in Molecular Medicine, Vol. 139: Vascular Biology Protocols*
Edited by: N. Sreejayan and J. Ren © Humana Press Inc., Totowa, NJ

triphosphate (ATP) production in the absence of oxygen *(2)*. In diabetes, there is a chronic reduction in cardiac glycolytic capacity *(3,4)*, and this may contribute to diabetic cardiomyopathy. The mechanisms involved in these stress conditions are diverse and complicated. Many studies have been carried out to elucidate the mechanism, but still our understanding of various control mechanisms in cardiac bioenergetics is not complete. The development of transgenic technology has allowed us to modify the various steps in the glycolytic pathway and study its effects on the myocardial glucose metabolism under various stress conditions.

We have developed transgenic mouse models with altered glucose metabolism. Using this technique, we have produced mice over-expressing hexokinase *(5)* and kinase-deficient 6-phosphofructo-2-kinase/fructose-2,6-bisphosphatase *(6)*. Glucose metabolism has been altered specifically in the heart using the cardiac-specific α-myosin heavy chain (α-MHC) promoter. These mice have the advantage that any change in glucose metabolism must be due to changes in cardiomyocytes as metabolism is normal in all other tissues. The hexokinase transgenic mice showed increased glycolysis, whereas the kinase-deficient 6-phosphofructo-2-kinase/fructose-2,6-bisphosphatase (kd-PFK-2) mice decreased glycolysis in the heart. The over-expression of kd-PFK-2 also altered various relevant metabolites, including glucose-6-phosphate, fructose-6-phosphate, glycogen, and fructose-1,6-bisphosphate, and it reduced insulin sensitivity in perfused hearts. Hence, these mouse models are very helpful in elucidating the various mechanisms regulating glucose metabolism under normal and pathological conditions.

2. Materials

2.1. Genetic Models of Altered Glucose Metabolism

In our laboratory, we use MyHX and Mykd-PFK-2 transgenic mice that over-express hexokinase-deficient or kinase-deficient kd-PFK-2, respectively. Each transgene uses the α-MHC promoter to produce cardiomyocyte-specific high-level expression.

2.2. Production of Cardiac-Specific Transgenes

1. cDNA or gene of choice.
2. Plasmid containing mouse α-MHC promoter.
3. Primers for polymerase chain reaction (PCR) amplification of cDNA or gene.
4. Restriction enzymes KpnI, HindIII, SalI, and NotI.
5. Agarose gels in tris/acetate/ethylenediamine tetraacetic acid (EDTA) buffer containing $0.5\,\mu g/ml$ ethidium bromide.
6. StrataPrep DNA gel extraction kit (Stratagene, La Jolla, California, USA).
7. Bio-spin 6 spin chromatography columns (BioRad, Hercules, California, USA).

8. Molecular mass markers (GIBCO-BRL, Carlsbad, California, USA).
9. Mini-gel apparatus.
10. Luria Broth (LB) agar and LB media containing $100\,\mu g/ml$ ampicillin.
11. Epicurian Coli XL1-blue competent cells (Stratagene).
12. DNeasy tissue kit (Qiagen, Valencia, California, USA).

2.3. Analysis of Metabolites: Glucose-6-Phosphate, Fructose-6-Phosphate, Fructose-1,6-Bisphosphate, and Glycogen

1. $KHCO_3$ (2 M) solution in distilled water.
2. Assay buffer for glucose-6-phosphate and fructose-6-phosphate estimation: 25 mM Tris–HCl, pH 8.1. Add DL-dithiothreitol (DTT) $(200\,\mu M)$ and Nicotinamide Adenine Dinucleotide Phosphate (NADP) $(50\,\mu M)$ prior to use.
3. $HClO_4$ (1 M).
4. Assay buffer for fructose-1,6-biphosphate estimation: 25 mM Tris–HCl, pH 7. Add imidazole (30 mM), imidazole HCl (20 mM), DTT $(100\,\mu M)$, and NADP $(10\,\mu M)$ prior to use.
5. Potassium hydroxide solution (30% w/v).
6. Saturated solution of sodium sulfate.
7. Acetate buffer: 0.05 M acetic acid with 0.05 M sodium acetate.
8. Amyloglucosidase solution (0.01 mg/ml).
9. Assay buffer for glycogen estimation: 50 mM Tris–HCl, pH 8.1, containing 5 mM $MgCl_2$ (5 mM), ATP $(150\,\mu M)$, $40\,\mu M$ NADP $(40\,\mu M)$, DTT $(100\,\mu M)$, glucose-6-phosphate dehydrogenase (0.2U/ml), and hexokinase (0.7U/ml).
10. Black 96-well Fluotrac 200 fluorescence plates from Greiner®, Monroe, North Carolina, USA.

2.4. Perfusion of Isolated Heart

1. Perfusion buffer: 120 mM NaCl, 20 mM $NaHCO_3$, 4.63 mM KCl, 1.17 mM KH_2PO_4, 1.2 mM $MgCl_2$, 1.25 mM $CaCl_2$, and 5 mM glucose, pH 7.4, bubbled with 95% O_2 and 5% CO_2.
2. Ketamine HCl/xylazine HCl solution (Sigma, St. Louis, MO, USA).
3. Water bath heating circulator.
4. Water-jacketed heat exchange glass coil.
5. Peristaltic pump.
6. Fine scissors.
7. Two fine forceps.
8. 6-0 silk tread.
9. 95% O_2 and 5% CO_2.

2.5. Measurement of Glycolysis Using 5-³H-Glucose and Glucose Transport/Phosphorylation Using 2-³H-Glucose

1. D-[5-³H]glucose (Amersham, Piscataway, New Jersey, USA).
2. D-[2-³H]Glucose (Amersham, Piscataway, New Jersey, USA).

3. Methods

3.1. Production of Cardiac-Specific Transgenic Mice

3.1.1. Production of Mykd-PFK-2 Transgene

As an example of transgene construction, we will describe production of our methods for developing the Mykd-PFK-2 cardiac-specific transgene regulated by the mouse α-MHC promoter (*see* **Notes 1** and **2**). This method is similar to what we have used successfully for several other transgenes derived from cDNA or genomic sequences of yeast and mammalian origin *(5–11)*. It can be readily applied to any gene the user is interested in.

1. Development of the plasmid-containing kd-PFK-2 mutant has been previously described, and the same sequences were used for preparing the transgene *(6,12,13)*. This plasmid was provided by Dr. Alex Lange.
2. The kd-PFK-2 cDNA was inserted between the α-MHC fragment, containing the promoter and parts of the first three exons, and the polyadenylation region of the rat insulin II gene in the plasmid MY. Construction was accomplished by cutting this plasmid with Sal I, followed by blunt ending the termini with the Klenow fragment of DNA polymerase I. It was further digested with Hind III. One microgram of this fragment was purified on a 0.7% agarose gel.
3. One microgram of the kd-PFK-2 cDNA fragment was obtained by cleavage of BIF-K plasmid with KpnI and blunting with T4 DNA polymerase. It was further digested with Hind III to release the kd-PFK-2-coding fragment. This fragment was purified on a 0.7% agarose gel.
4. Fifty nanograms each of the vector and insert were ligated at 15°C for 16 h in a total volume of 10 µl with 2000 units of T4 DNA ligase.
5. Two microliters of the ligation reaction were used to transform 50 µl of Epicurian Coli XL1-blue competent cells and plated on LB agar plates containing 100 µg/ml ampicillin. Plasmids were isolated from single colonies.
6. Correct construction and orientation of the Mykd-PFK-2 clone and myosin promoter was determined by restriction enzyme digestion and verified by sequencing across the ligated regions.

3.1.2. Production of Mykd-PFK-2 Transgenic Mouse (6)

Before embryo microinjection, 20 µg of Mykd-PFK-2 was digested with NotI, and the plasmid-free transgene was purified on a 0.5% Seakem Gold agarose gel. The fragment was purified with StrataPrep DNA gel extraction kit and then passed through a Bio-spin 6 chromatography column prior to microinjection.

The procedures we use for embryo microinjection and embryo implantation are standard for production of transgenic mice *(14)* and widely performed in commercial facilities or university core facilities.

3.1.3. Breeding of Transgenic Mice (6)

The transgenic mice were produced on the FVB background, and they were mated with normal FVB mice. Each litter will yield approximately equal numbers of transgenic and normal mice. To determine the presence of a cardiac transgene, DNA from a 1–3 mm slice of tail is PCR amplified and analyzed by agarose gel electrophoresis using REDExtract-N-Amp tissue PCR kit from Sigma. Approximately 50 ng of DNA is subject to standard PCR conditions using one oligonucleotide complimentary to the mouse MHC promoter and another oligonucleotide complimentary to the coding strand of the transgene.

3.2. Analysis of Metabolites

3.2.1. Estimation of Glucose-6-Phosphate (15)

1. Homogenize 100 mg tissue in 1 M $HClO_4$ (100 μl) with Biospec tissue tearor. Centrifuge the homogenate and neutralize the supernatant with 2 M $KHCO_3$ (105 μl). Dilute 25 μl supernatant with 125 μl of assay buffer.
2. Place 25 μl standard (0.03–4 nmol)/sample in black 96-well plates. Add 295 μl of the above assay buffer and measure fluorescence readings from the top on Tecan® plate reader using excitation filter, 360 nm, and emission filter, 460 nm.
3. Add 10 μl glucose-6-P dehydrogenase (0.06 U/ml).
4. Read fluorescence after 15 min from the top on Tecan® plate reader using excitation filter, 360 nm, and emission filter, 465 nm.

3.2.2. Estimation of Fructose-6-Phosphate (15)

1. Homogenize 100 mg tissue in 1 M $HClO_4$ (100 μl) with Biospec tissue tearor. Centrifuge the homogenate and neutralize the supernatant with 2 M $KHCO_3$ (105 μl). Dilute 25 μl supernatant with 125 μl of assay buffer.
2. Place 25 μl standard (0.03–3 nmol)/sample in black 96-well plates. Add 285 μl assay buffer and measure fluorescence from the top on Tecan® plate reader using excitation filter, 360 nm, and emission filter, 460 nm.
3. Add 10 μl glucose-6-P dehydrogenase and measure fluorescence after 15 min from the top on Tecan® plate reader using excitation filter, 360 nm, and emission filter, 465 nm.
4. Continue reaction by adding 10 μl phosphoglucose isomerase (0.175 U/ml) for measuring fructose-6-phosphate. Read fluorescence from the top on Tecan® plate reader after 25 min using excitation filter, 360 nm, and emission filter, 465 nm (*see* **Note 3**).
5. Subtract the values of step 4 from values in step 3 to get the concentration of fructose-6-phosphate.

3.2.3. Estimation of Fructose-1,6-Bisphosphate (15)

1. Homogenize 100 mg tissue in 1 M $HClO_4$ (100 μl) with Biospec tissue tearor. Centrifuge the homogenate and neutralize the supernatant with 2 M $KHCO_3$ (105 μl). Dilute 25 μl supernatant with 125 μl assay buffer.
2. Place 25 μl standard (0.016–1.33 nmol)/sample in black 96-well plates. Add 285 μl of assay buffer and measure fluorescence from the top on Tecan® plate reader using excitation filter, 360 nm, and emission filter, 460 nm.
3. Add 10 μl of triosephosphate isomerase (1.2 U/ml) and α-glycerophosphate dehydrogenase (0.04 U/ml), incubate for 10 min, and measure fluorescence from the top on Tecan® plate reader using excitation filter, 360 nm, and emission filter, 460 nm.
4. Continue reaction by adding 10 μl of aldolase (0.004 U/ml) and incubate for 15–20 min. Read fluorescence from the top on Tecan® plate reader using excitation filter, 360 nm, and emission filter, 465 nm.
5. Subtract the values in step 4 from the values in step 3 to get the concentration of fructose-1,6-bisphosphate.

3.2.4. Estimation of Glycogen (15)

1. Powder the heart, weigh 20 mg, and homogenize with Biospec tissue tearor in 0.6 ml of potassium hydroxide (*see* **Note 4**). Heat the mixture at 80 °C for about 20 min to assist in the solubilization. (Also this step destroys any glucose present in the homogenate.)
2. Add 0.15 ml of saturated sodium sulfate and precipitate the glycogen by addition of 1.2 vol of 95% ethanol (1 ml). Vortex and heat the tubes until the mixture begins to boil. Cool and centrifuge at 14,000 × *g* on tabletop for 5 min.
3. Rinse the tubes with 1 ml of 95% ethanol, decant, and dry the tubes at 70 °C to remove the remaining alcohol.
4. Dissolve the precipitated glycogen in 500 μl of the acetate buffer.
5. Place 3.3 μl glycogen standard (0.06–1 mM)/sample in a black 96-well plates. Add 35 μl of acetate buffer and 5 μl containing 50 ng amyloglucosidase (Roche Applied Science, Indianapolis, IN USA) in acetate buffer and incubate at room temperature for 30 min.
6. Add 300 μl of assay buffer and incubate for 30 min at room temperature.
7. Measure fluorescence from the top on Tecan® plate reader using excitation filter, 360 nm, and emission filter, 460 nm.

3.3. Perfusion of Isolated Hearts (5,6,8)

1. Anesthetize the mouse with ketamine HCl/xylazine HCl and coadminister 100 U heparin by i.p.
2. Isolate the heart and place it in ice-cold perfusion buffer. Mount the heart onto the Lagendorff apparatus by cannulating the aorta and perfuse the heart retrogradely. The buffer is pre-filtered with a sterile 0.22 μm filter and is continuously bubbled with 95% O_2 and 5% CO_2, resulting in a pH of 7.4. Maintain the perfusion pressure at approximately 65 mmHg and the temperature of the water jacket and

the buffer at 37 °C. The flow rate is maintained with the help of peristaltic pump at 1.8 ml/min.

3. A stainless steel hook is attached to the heart apex and connected by a silk suture to a force transducer (Grass FT03 or CB sciences ETH 400) to measure the contractility.

4. Adjust the baseline force and pressure to zero when the suture connecting the hook and the transducer is loose. The preload (rest tension or diastolic force) is set at 3 g. The cardiac contractile force is continuously recorded with a Power Lab system.

5. The heart is paced at 5 Hz (6 V, 3 ms) with a built-in stimulator from Power Lab system. After a 20-min baseline recording (starting at the moment the heart is hooked up and connected to the F transducer), the heart if found to be stable is subjected to various interventions. Throughout the whole procedure, the heart will be electrically paced.

6. The perfusate is allowed to drip through inferior vena cava. The buffer is collected frequently to measure glycolysis and glucose transport.

7. After all the procedures are finished, the heart is removed rapidly from the perfusion apparatus, blotted dry, weighed, and frozen in liquid nitrogen and stored at −80 °C until further analysis for metabolites.

3.4. Measurement of Glucose Transport/Phosphorylation and Glycolysis (5,6)

1. After mounting the heart onto a Lagendorf apparatus and stabilization for 20 min with well-oxygenated perfusion buffer, perfuse the heart with oxygenated perfusion buffer containing D-[5-^{3}H]-glucose (15 μCi/100 ml) for 30 min at a flow rate of 1.8 ml/min (*see* **Note 5**). Collect the perfusate at regular time intervals to measure the glycolysis. We can use a similar protocol to measure glucose transport/phosphorylation using D-[2-^{3}H]-glucose (15 μCi/100 ml) as the substrate (*see* **Notes 6** and **7**).

2. Place 400 μl of the collected perfusate in 1.5 ml eppendorf tube with the cap cut-off. Add 25 μl of 0.6 M HCl and place the tube in a scintillation vial containing 2 ml of H_2O. Allow the ^3H_2O to diffuse for 72 h at 37 °C.

3. At the end of 72-h diffusion, discard the eppendorf tube and measure the radioactivity by adding 10 ml of liquid scintillation cocktail.

4. Results

We measured the concentration of metabolites in the hearts of control and Mykd-PFK-2 transgenic mice. As shown in **Table 1**, the levels of glucose-6-phosphate, fructose-6-phosphate, and glycogen were elevated, whereas the levels of fructose-1,6-bisphosphate were reduced in the transgenic mouse hearts. These results are consistent with the function of the transgene in these mice, which is to down-regulate the function of PFK-1. This down-regulation will

Table 1
Metabolites in Normal and Transgenic Mouse Hearts

	FVB	Mykd-PFK-2
Glucose-6-phosphate (nmol/g)	148.67 ± 16.95	$293.84 \pm 45.21^{*}$
Fructose-6-phosphate (nmol/g)	32.64 ± 6.01	$97.88 \pm 14.42^{**}$
Fructose-1,6-bisphosphate (nmol/g)	498.54 ± 62.55	$261.76 \pm 62.56^{*}$
Glycogen (μmol/g)	18.57 ± 4.19	$50.19 \pm 9.23^{*}$

Mykd-PFK-2, cardiac-specific kinase-deficient 6-phosphofructo-2-kinase/fructose-2,6-bisphosphatase. The values shown are means $\pm$ SE for at least four mice in each group. The values were compared by student's t-test.
$^{*}p < 0.05$.
$^{**}p < 0.01$.

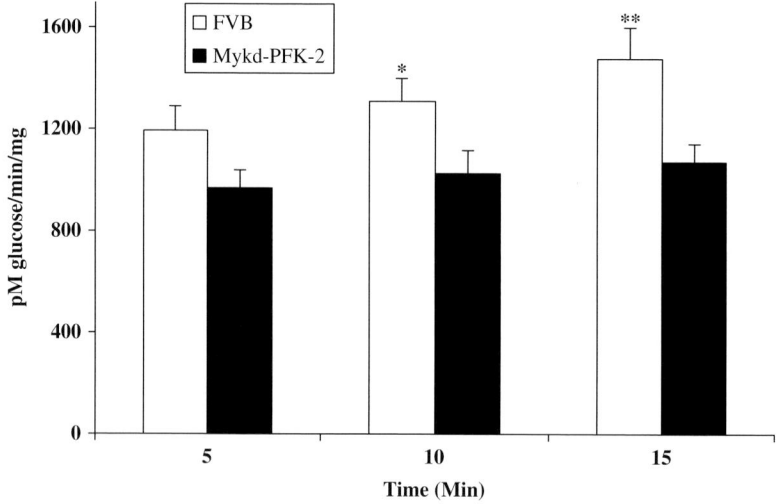

Fig. 1. Glycolysis in the perfused mouse hearts. Over-expression of cardiac-specific kinase-deficient 6-phosphofructo-2-kinase/fructose-2,6-bisphosphatase (Mykd-PFK-2) transgene reduces glycolysis in in vitro perfused hearts. Glycolysis was measured by consumption of [5-^{3}H]glucose. The values shown are means $\pm$ SE for FVB and Mykd-PFK-2 mice *(14)*. The values were analyzed by two-way repeated measures analysis of variance followed by Tukey post hoc test. $^{*}p < 0.05$ and $^{**}p < 0.004$.

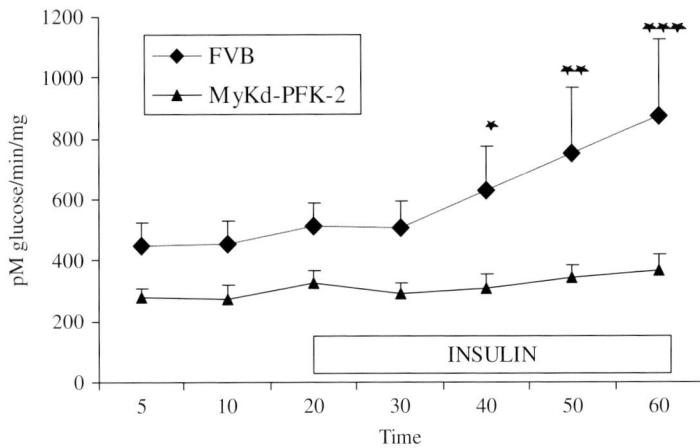

Fig. 2. Glucose transport/phosphorylation in perfused mouse hearts. Over-expression of cardiac-specific kinase-deficient 6-phosphofructo-2-kinase/fructose-2,6-bisphosphatase (Mykd-PFK-2) transgene reduces glucose transport/phosphorylation in in vitro perfused hearts. Glucose transport/phosphorylation was measured by consumption of [2-³H]glucose. The values shown are means ± SE for FVB *(6)* and Mykd-PFK-2 *(7)* mice. The values were analyzed by two-way repeated measures analysis of variance followed by Tukey post hoc test. $^*p < 0.05$, $^{**}p < 0.01$, and $^{***}p < 0.002$.

lead to the accumulation of metabolites upstream of PFK-1 and decreases the levels of metabolites down-stream of PFK-1.

The glycolysis in the kd-PFK-2 over-expressing mice was measured using D-[5-³H]glucose as the substrate. The transgenic mouse showed statistically significant down-regulation of glycolysis as shown in **Fig. 1**. The glucose transport in the FVB and Mykd-PFK-2 over-expressing mice was measured using D-[2-³H]glucose as the substrate. The hearts were perfused with and without 200 µU/ml insulin (Sigma), and the results are shown in **Fig. 2**. The glucose transport study also showed a reduced response in Mykd-PFK-2 over-expressing mice compared with the normal FVB mice.

5. Notes

1. We have produced several lines of transgenic mice using the α -MHC promoter with no examples of transgene not expressing.
2. The mice over-expressing Mykd-PFK-2 had normal life span and breed efficiently without any problems.

3. The use of fluorescence plates is important to prevent the noise from the adjacent wells.
4. The heart samples for glycogen assay were prepared in KOH solution. This helps in solubilizing the tissue and also destroys any glucose contaminating the homogenate.
5. ^3H_2O is released from the metabolism of D-[5-^{3}H]-glucose by triose-phosphate isomerase and enolase steps of the glycolytic pathway *(16)*.
6. D-[2-^{3}H]-glucose is transported into cells and phosphorylated by hexokinase, and the resulting glucose-6-phosphate is converted to fructose-6-phosphate by phospho-glucose isomerase. Cycling between glucose-6-phosphate and fructose-6-phosphate by this enzyme releases ^3H_2O during the isomerization step *(17)*.
7. D-[5-^{3}H]-glucose and D-[2-^{3}H]-glucose were vacuum spin-dried and dissolved in perfusion buffer before use. This will reduce the background counts significantly.

Acknowledgments

The work was supported by NIH grant number HL66778.

References

1. Bishop, S.P., and Altschuld R.A. (1970) Increased glycolytic metabolism in cardiac hypertrophy and congestive failure. *Am. J. Physiol.* **218**,153–159.
2. Neely, J.R., and Rovetto, M.J. (1975) Techniques for perfusing isolated rat hearts. *Methods Enzymol.* **39**, 43–60.
3. Avogaro, A., Nosadini, R., Doria, A., Fioretto, P., Velussi, M., Vigorito, C., Sacca, L., Toffolo, G., Cobelli, C., and Trevisan, R. (1990) Myocardial metabolism in insulin-deficient diabetic humans without coronary artery disease. *Am. J. Physiol.* **258**,E606–E618.
4. Stanley, W.C., Lopaschuk, G.D., and McCormack, J.G. (1997) Regulation of energy substrate metabolism in the diabetic heart. *Cardiovasc. Res.* **34**, 25–33.
5. Liang, Q., Donthi, R.V., Kralik, P.M., and Epstein, P.N. (2002) Elevated hexokinase increases cardiac glycolysis in transgenic mice. *Cardiovasc. Res.* **53**,423–430.
6. Donthi, R.V., Ye, G., Wu, C., McClain, D.A., Lange, A.J., and Epstein, P.N. (2004) Cardiac expression of kinase-deficient 6-phosphofructo-2-kinase/fructose-2,6-bisphosphatase inhibits glycolysis, promotes hypertrophy, impairs myocyte function, and reduces insulin sensitivity. *J. Biol. Chem.* **279**,48085–48090.
7. Liang, Q., Carlson, E.C., Borgerding, A.J., and Epstein, P.N. (1999) A transgenic model of acetaldehyde overproduction accelerates alcohol cardiomyopathy. *J. Pharmacol. Exp. Ther.* **291**,766–772.
8. Liang, Q., Carlson, E.C., Donthi, R.V., Kralik, P.M., Shen, X., and Epstein, P.N. (2002) Overexpression of metallothionein reduces diabetic cardiomyopathy. *Diabetes* **51**,174–181.

9. Ye, G., Metreveli, N.S., Donthi, R.V., Xia, S., Xu, M., Carlson, E.C., and Epstein, P.N. (2004) Catalase protects cardiomyocyte function in models of type 1 and type 2 diabetes. *Diabetes* **53**,1336–1343.

10. Kang, Y.J., Chen, Y., and Epstein, P.N. (1996) Suppression of doxorubicin cardiotoxicity by overexpression of catalase in the heart of transgenic mice. *J. Biol. Chem.* **271**,12610–12616.

11. Shen, X., Zheng, S., Thongboonkerd, V., Xu, M., Pierce, W.M., Jr., Klein, J.B., and Epstein, P.N. (2004) Cardiac mitochondrial damage and biogenesis in a chronic model of type 1 diabetes. *Am. J. Physiol. Endocrinol. Metab.* **287**,E896–E905.

12. Wu, C.D., Okar, D.A., Peng, L., and Lange, A.J. (2002) Decreasing fructose-2,6-bisphosphate leads to diabetic phenotype in normal mice. *Diabetes* **51**,A319

13. Wu, C., Okar, D.A., Stoeckman, A.K., Peng, L.J., Herrera, A.H., Herrera, J.E., Towle, H.C., and Lange, A.J. (2004) A potential role for fructose-2,6-bisphosphate in the stimulation of hepatic glucokinase gene expression. *Endocrinology* **145**, 650–658.

14. Palmiter, R.D., and Brinster, R.L. (1986) Germ-line transformation of mice. *Annu. Rev. Genet.* **20**,465–499.

15. Lowry, O.H., and Passonneau, J.V. (1993) *A Flexible System of Enzymatic Analysis.* New York: Academic Press, 111–228.

16. Barr, R.L., and Lopaschuk, G.D. (2000) Methodology for measuring in vitro/ex vivo cardiac energy metabolism. *J. Pharmacol. Toxicol. Methods* **43**,141–152.

17. Perriott, L.M., Kono, T., Whitesell, R.R., Knobel, S.M., Piston, D.W., Granner, D.K., Powers, A.C., and May, J.M. (2001) Glucose uptake and metabolism by cultured human skeletal muscle cells: rate-limiting steps. *Am. J. Physiol. Endocrinol. Metab.* **281**,E72–E80.

10

Methods in the Evaluation of Cardiovascular Renin Angiotensin Aldosterone Activation and Oxidative Stress

Camila Manrique, Guido Lastra, Javad Habibi, Yongzhong Wei, E. Matthew Morris, Craig S. Stump, and James R. Sowers

Summary

Renin angiotensin aldosterone system (RAAS) activation plays an essential role in the development of cardiovascular disease (CVD). Multiple pathophysiologic processes are able to activate RAAS, among which hypertension, obesity, diabetes mellitus 2, and chronic kidney disease deserve special attention, because they are the main contributors to CVD. Adding to the well-known effects of RAAS overactivity on the vasculature and water and electrolyte balance, current evidence links abnormal activation of the RAAS to increased production of reactive oxygen species (ROS) and oxidative stress. This association is mediated at least partially through interaction of angiotensin II (Ang II) with its receptor angiotensin receptor 1 (AT1R) in cardiovascular tissue, and subsequent activation of the nicotinamide adenine dinucleotide phosphate (NADPH) enzymatic complex, which finally leads to increased ROS production. This resulting state of enhanced oxidative stress contributes largely to generalized atherosclerosis and finally to CVD. The generation of animal models of increased RAAS and Ang II expression, in particular the Ren2 rodent model, provides important opportunities to better characterize the relationship between this system and the production of ROS. This chapter describes methods to evaluate, characterize, and quantify the activity of the RAAS and NADPH oxidase, as well as the production of ROS production in animal model of RAAS.

Key Words: Renin angiotensin aldosterone system; NADPH oxidase; Ren2 transgenic mice.

From: *Methods in Molecular Medicine, Vol. 139: Vascular Biology Protocols*
Edited by: N. Sreejayan and J. Ren © Humana Press Inc., Totowa, NJ

1. Introduction

Since its discovery, the renin angiotensin aldosterone system (RAAS) has proven to be of paramount importance in the regulation of the cardiovascular system function, and disorders in the modulation of its different components play a key role in the pathogenesis of arterial hypertension (HTN), chronic kidney disease (CKD), and congestive heart failure. During the 1990s, the participation of the RAAS in endothelial dysfunction, atherosclerosis, vascular, and myocardial remodeling has gained mounting importance.

Angiotensin II (Ang II), one of the principal effectors of the RAAS, is a multifunctional octapeptide involved in the regulation of blood pressure, water, and sodium balance, and it is also involved in the neurohumoral homeostasis of other systems *(1)*. The effects of Ang II are mediated through interaction with two specific plasma membrane-bound receptors angiotensin receptors 1 and 2 (AT1R and AT2R, respectively) *(2)*. However, most of the known properties of Ang II have been related to the activation of AT1R *(3)*, which are expressed abundantly in vascular smooth muscle cells (VSMCs), heart, lungs, brain, liver, kidney, and adrenal glands *(3)*. AT1R belongs to the seven-membrane-domain superfamily of G-protein-coupled receptors and is able to interact with numerous intracellular signaling molecules and second messenger systems, including phospholipases, adenylate cyclase, and voltage-dependent calcium channels *(4)*. Stimulation of these intracellular pathways leads to vasoconstriction, cellular hypertrophy, proliferation, or apoptosis depending on the specific cell type in which AT1R is stimulated. In addition, during the 1990s, growing evidence implies the AT1R activation in the pathophysiology of reactive oxygen species (ROS) and oxidative stress *(5)*. Despite their physiologic important role in the defense system against infectious agents and in its participation in the growth of VSMCs *(6)*, inappropriately increased oxidative stress plays a key role in the development of endothelial dysfunction, atherogenesis, and also CKD *(7)*. Consequences of excessive oxidative stress include modifications of DNA, lipid peroxidation, oxidative protein modification, and increased expression of adhesion molecules, such as vascular cell adhesion molecules 1 and 2 and monocyte chemoattractant protein-1, which actively participate in inflammation and atherogenesis. On the other hand, ROS stimulate extracellular matrix proliferation and vascular remodeling, at least partially through activation of matrix metalloproteinases *(8)*.

The production of ROS, such as superoxide and hydrogen peroxide, is necessary for different normal biological process, but excess production, when compared with the endogenous antioxidant system, has been implicated in multiple cardiovascular and metabolic pathological conditions *(3)*. The role of the RAAS in the induction of oxidative stress has been intensively studied over the past several years. Ang II and aldosterone promote the production of ROS in adipose tissue, skeletal muscle, and cardiovascular tissue *(9)*. In turn, ROS induce a shift toward proinflammatory and proatherogenic patterns and mitogenic actions in VSMC *(3,9)*.

In mammalian cells, the NADPH oxidase, the nitric oxide synthase, the cytochrome p450, the mitochondrial electron transport system, and the xanthine oxidase systems are capable of ROS production. However, the NADPH oxidase is probably the most important system implicated in excessive oxidative stress leading to vascular dysfunction in cardiovascular tissue *(10)*. The NADPH oxidase enzymatic complex is a multisubunit enzyme composed of cytosolic proteins (small GTPase Rac1, p47^{phox}, and p67^{phox}) and membrane catalytic proteins Nox 2 (gp91) and p22^{phox} *(11)*. Nevertheless, our recent studies showed that Nox 2 and p22^{phox} are also abundant in perinuclear and cytoskeletal localization of Opossum kidney cortex Proximal Tubule Epithelial Cells (OK) and VSMCs (Habibi et al., unpublished data) (*see* **Fig. 1**). Activation of the NADPH

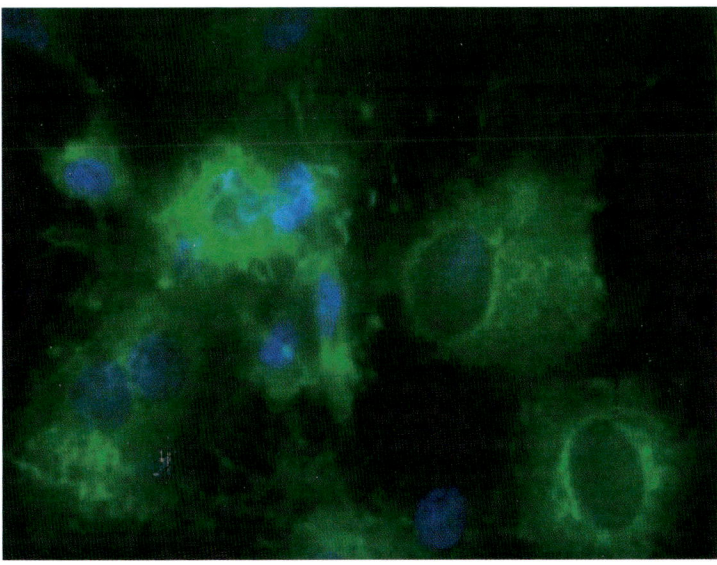

Fig. 1. Live aortic vascular smooth muscle cells stained with CDFDA, a reactive oxygen species indicator, and counterstained with Hoechst, showing strong signal on perinuclear and cytoskeleton regions indicating the presence of all the NADPH subunits in the perinuclear area.

oxidase produces assembly of cytosolic and plasma membrane subunits to generate superoxide (O_2^-) by means of electron transfer from subunit gp91 of NADPH to O_2.

Ang II, through AT1R, stimulates intracellular pathways that result in translocation of all subunits to the plasma membrane, a key step in NADPH oxidase activation *(3)*. Zuo et al. *(12)* recently demonstrated that Ang II cell stimulation promotes Rac1 association with caveolin 1, as well as its migration into caveolin-enriched lipid rafts. It is highly likely that in those lipid rafts, Rac1 is activated (through Guanosine diphosphate (GDP) for Guanosine triphosphate (GTP) exchange) and that this activation involves the trafficking of activated AT1R into the caveolin-enriched lipid rafts *(13)*.

2. Materials

2.1. Ren2 Rat Model

1. Hypertensive transgenic rat (mRen2).

2.2. Measurement of NAD(P)H Consumption From Muscle Tissue

1. Homogenization buffer: 5 ml of 0.25 M sucrose; 20 µl of 0.5 mM EDTA, pH 8; 1 ml of 50 mM 4-(2-hydroxyethyl)-1-piperazineethanesulfonic acid (HEPES), pH 7.7; 40 µl of 2 µg/ml aprotinin; 20 µl of 5 µg/ml leupeptin; 100 µl of 0.2 mM phenylmethanesulphonyl fluoride (PMSF); 60 µl of 3 mM Dithiothreitol (DTT); and 13.7 ml ddH$_2$0.
2. Resuspension buffer: 0.25 M sucrose and protease inhibitors/DTT in ddH$_2$O.

2.3. Tissue Preparation and Immunostaining

1. Paraformaldehye solution: 70 mM NaCl, 30 mM HEPES, 2 mM CaCl$_2$, and 3% paraformaldehyde (Sigma, St. Louis, MO, USA).
2. HEPES wash buffer: 900 ml distilled water, 4.10 g NaCl, 7.14 g HEPES, and 0.29 g CaCl$_2$, pH 7.4.
3. Blocking solution: 5% Bovine serum albumin (BSA), 5% serum of the animal where secondary antibodies were generated, and 0.001% Na$_3$N.
4. Paraplast (Fisher Scientific Pittsburgh, PA, USA).
5. Automated microtome (Micron HM355, Fisher Scientific).

2.4. Lucigenin-Derived Chemiluminescence and Quantification of ROS Production

1. Krebs–Henseleit buffer (KHB): 118 mM NaCl, 4.7 mM KCl, 24 mM NaHCO$_3$, 1.2 mM KH$_2$PO$_4$, 1.2 mM MgSO$_4$, 2.5 mM CaCl$_2$, 5 mM HEPES, and 5.5 mM glucose, pH 7.5.

3. Methods

3.1. Experimental Models of Increased Renin Angiotensin System (RAS) Activity: The Ren2 Rat Model

1. The hypertensive transgenic rat TGR (mRen2), designed by Mullins and coworkers *(14)*, was the first transgenic rat model to be used in HTN research. Transgenic models usually have only one genetic alteration (gain or loss of function), and as only one gene is targeted for modification, the results obtained permit a better understanding of the physiological role of genes under in vivo conditions *(15)*.

2. The hypertensive transgenic rat TGR (mRen2) is constructed by induction of superovulation in female Sprague–Dawley (SD) rats, posterior harvesting of oocytes, and microinjection of a linear DNA fragment containing the entire DBA/2J Ren2 gene, including 5.3 and 9.5 kilobases of 5′ and 3′ flanking sequence, into fertilized oocytes once the pronuclei has been formed *(16)*.

3. The manipulated oocyte is transferred to pseudopregnant female rats (foster mothers).

4. The presence of the transgene in the offspring of the animals can be confirmed by Southern blot analysis or polymerase chain reaction.

5. *Ren2* gene from DBA/2 mice differs from other mammalians renin genes in that its protein is not glycosylated, and it has a special pattern of extrarenal expression previously established in mice model *(17)*.

6. Transgenic rats develop HTN early in their lives (4–5 weeks of age). By week 9, blood pressure reaches a maximum, up to 240 mmHg in heterozygous male rats and 250–290 in homozygous rats compared with 120–130 mmHg seen in normotensive littermates *(16)*. This model exhibits marked sexual dimorphism, as female rats are 30–40 mmHg less hypertensive than male counterparts. This effect appears to be mediated by androgens *(18)*.

7. Although the genetic overexpression on Ren2 is clearly related to the appearance of HTN in this model, the exact mechanism explaining this phenomena is still elusive. Differences in the circulating plasmatic levels of the components of the RAAS do not fully explain the existence of HTN in this transgenic model, and opposite data regarding the levels of the different components of the system have been found in a model traditionally described as low plasma renin *(19)*.

8. The tissue expression of the transgene is highest in the adrenal gland (where expression is low in control animals), followed by the thymus, brain, gastrointestinal and urinary tracts, kidney, and lungs. In some tissues, such as the brain, adrenal gland, and kidney, the expression of the transgene precedes the appearance of HTN. This resulting tissue-specific overexpression of the *Ren2* gene can help to explain this model as a high-tissue renin animal.

In addition to the *Ren2* gene, the whole genetic background of the specific experimental model seems to be involved in the existence of HTN, as more severe HTN is seen in animals generated from SD as compared to models derived from Lewis rats *(20)*.

The cardiovascular and renal complications demonstrated in the TGR (mRen2)27 animals appeared to be secondary in part to the systemic HTN but also in part secondary to the increased activity of Ang II, as is demonstrated by the beneficial effects of RAAS blockade *(21)*. Finally, this animal model also exhibits impaired glucose metabolism and insulin resistance. These abnormalities have been shown to be blocked by A1TR blockade *(22)*.

3.2. Measurement of NAD(P)H Consumption From Muscle Tissue

1. Weigh fresh *soleus* muscles and add 6× (w/v) of ice-cold homogenization buffer.
2. Homogenize samples using a motorized pestle (Duall Addison, IL, USA).
3. Centrifuge homogenate for 10 min at $1000 \times g$ at 4 °C.
4. Transfer to a 1.5-ml tube and centrifuge for 20 min at $9000 \times g$ at 4 °C to remove the lysosomes and peroxisomes (pellet).
5. Centrifuge the resulting supernatant for 20 min at $25,000 \times g$ at 4 °C to remove Golgi and sarcoplasmic reticulum (pellet).
6. Ultracentrifuge the final supernatant at $100,000 \times g$ for 90 min at 4 °C.
7. Resuspend the pellet containing the plasma membrane in 200 μl pellet resuspension buffer and freeze at −80° until analysis.
8. Incubate aliquots of plasma membrane (50 mg protein) with NADPH (100 mM) at 37 °C and monitor the rate of NAD(P)H consumption by measuring the decline in absorbance (340 nm) every 10 min using a spectrophotometer for 2 h.
9. To assess the molecular sources of NAD(P)H consumption, repeat the procedure in the presence of diphenylene iodonium (DPI) (10 μM, flavin-containing enzyme inhibitor) or apocynin (1 mM, specific NAD(P)H oxidase inhibitor).
10. Results are finally statistically analyzed and expressed as exemplified in **Fig. 2**.

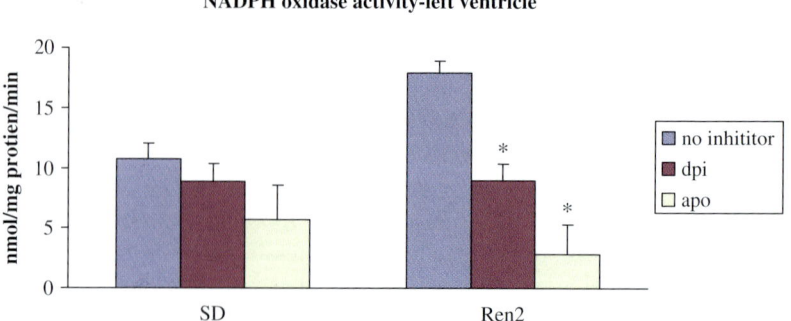

Fig. 2. NADPH oxidase activity in left ventricles. Left ventricles were homogenized, and membrane protein (50 μg) was used for measurement of NADPH consumption (340 nm). The results are mean ± SEM from five to six animals for each group. $^{*}p < 0.05$. apo, apocynin; SD, Sprague–Dawley.

3.3. Tissue Preparation and Immunofluorescent Studies

1. Harvest about 2–3 mm of thoracic aorta and the left ventricle of heart from anesthetized-dissected non-treated SD and Ren2 rats.
2. Immediately immerse the tissues in fixative.
3. Under a dissecting microscope, cut the tissues into small pieces (3 mm in thickness) in the same fixative to allow the fixative to penetrate into the entire tissues.
4. After fixing, place the tissues in histological cassettes and dehydrate with ethanol series, infiltrate with low-melting (50 °C) paraplast, and embed in high-melting (56 °C) paraplast cross-sectioned by 4 μm using an automated microtome. The sections are placed on surface of 60 °C water in a water bath to straighten, are collected from the surface of water, and two or three of them are placed on each slide. Finally, the slides are stored in slide boxes at room temperature until further use.
5. For each experiment, select 10 sections of aorta or heart and deparaffinize in CitriSolv, rehydrated in ethanol series and HEPES wash buffer. Epitopes are retrieved (antigen retrieval) in citrate buffer for 25 min at 95 °C with a steamer.
6. Immediately transfer the slides into a humidity chamber.
7. Block non-specific-binding sites with a blocking agent at room temperature for 4 h.
8. Wash the first section (three times for 15 min) with HEPES wash buffer and mount with Mowiol (first control level). Wash the second section and incubate with 1:100 primary antibodies in 10-fold diluted blocking agent. Wash the third section and keep in the blocker.
9. Over the course of 48 h, on each experiment, incubate the rest of the sections with 1:100 anti-gp91, anti-p22, anti-p40, anti-p47, and anti-p67 that were generated in goat (Santa Cruz, Santa Cruz, CA, USA) and mouse anti-Rac1 (Upstate, Lake Placid, NY, USA) antibodies in 10-fold diluted blocker at room temperature as primary antibodies.
10. After 24 h, wash the slides for 45 min (three times for 15 min) and mount the section number 2 with Mowiol (second control level).
11. Incubate the rest of the sections with 1:300 of Alexa flour rabbit anti-goat 647 in 10-fold diluted blocker at room temperature in a dark light-tight humidity chamber.
12. Stain the last section with 1:300 of Alexa flour goat anti-mouse 647 at room temperature in a dark light-tight humidity chamber.
13. After 4 h, wash the slides for 45 min (three times for 15 min) and mount with Mowiol and store them in light-tight slide boxes at 4 °C.
14. Observe the slides with confocal laser scanning microscope (60% laser, 2.3 Iris, 66 gain, zoom 1 and 00 offset) and capture the images using the LaserSharp software. The signal intensities on the normalized areas of the media and adventitia are measured by MetaVue and analyzed using the SAS system and Excel.

Results obtained in our laboratory using the abovementioned methods are summarized in the **Subheading 3.3.1.**

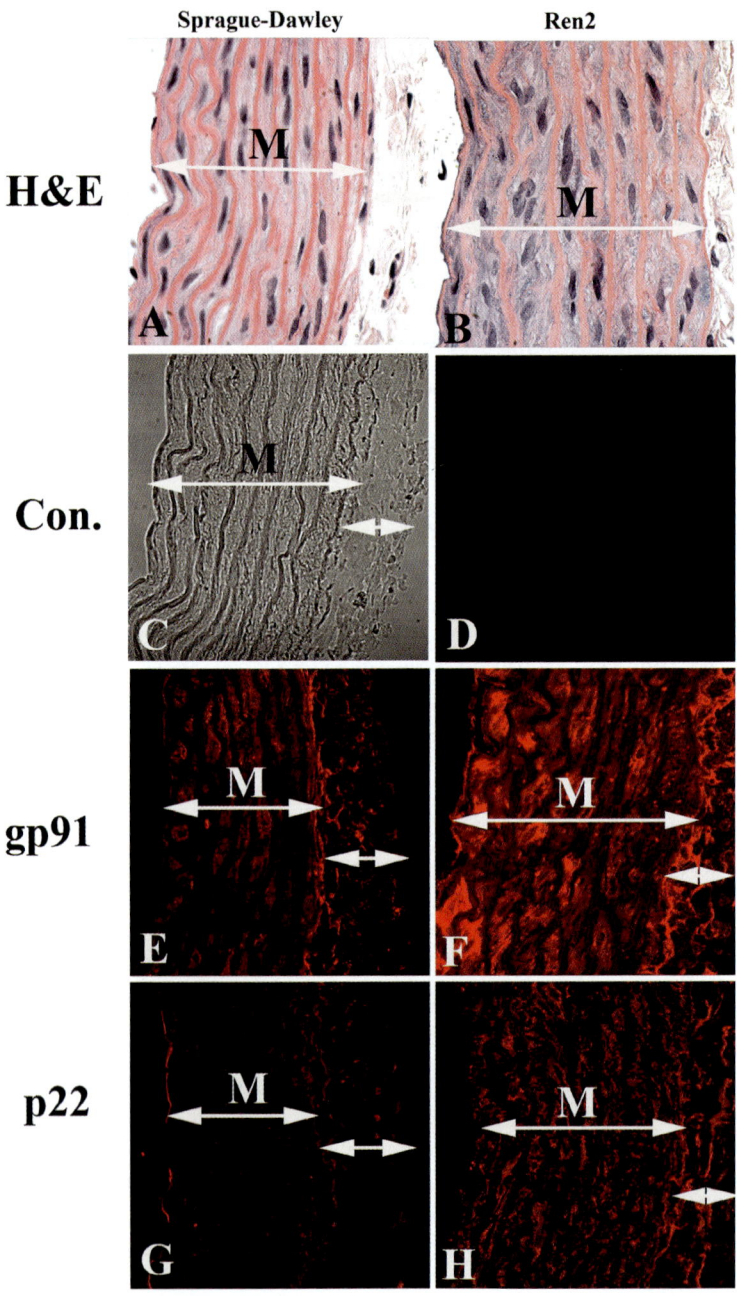

3.3.1. Expression Level of Different Subunits of NADPH Oxidase in the Aorta

Analysis of images from sections 1, 2, and 3 on each experiment showed no signal on the media or adventitial tissue of either SD or Ren2 rats, indicating that the tissues, primary, and secondary antibodies were not auto-fluorescent at 647 nm (*see* **Fig. 3C and D**). In contrast, some signals are observed on the tissues stained with the different subunits of NADPH oxidase (*see* **Figs 3 and 4**). Overall, the expression of gp91, p67, and Rac1 is higher than p22, p40, and p47 in both SD and Ren2 rats (*see* **Figs 3** and **4**). In adventitial tissue, the expression of gp91 ($p = 0.0147$), p22 ($p = 0.0024$), p40 ($p = 0.0083$), p47 ($p = 0.0368$), p67 ($p = 0.0172$), and Rac1 ($p = 0.0101$) are significantly lower in the SD than those in the Ren2 rat especially p40 which is highly expressed in the adventitial tissue of the Ren2 rat (*see* **Fig. 4B**). In the media, only the expression of Rac1 in SD rats appears to be significantly different than Ren2 rat ($p = 0.0365$) (*see* **Figs 3** and **4**). The other subunits do not show any significant difference between the SD and Ren2 rats (*see* **Figs 3** and **4**).

Analysis of bright field and fluorescent images shows significant increase of width of the media of the Ren2 rat in comparison of SD rat (*see* **Figs 3** and **4**). Likewise, the number of nuclei in the Ren2 rat significantly increased in comparison with the SD rat (*see* **Fig. 3A and B**). No significant differences are observed in the areas of the lumen of the aorta of the two rats (data not shown).

Fig. 3. Thick paraffin sections of aorta of Sprague–Dawley and Ren2 rats stained with the hematoxylin and eosin (H&E) and different subunits of NADPH oxidase. (**A**) The Sprague–Dawley aorta that was stained with H&E showing the layers of the media, longer nuclei, and adventitia. (**B**) The Ren2 aorta stained with the H&E showing a thicker media and shorter nuclei than those in the Sprague–Dawley rat. (**C**) Merged images of the transmitted images and the fluorescent images of the abort section showing no signal but the media and adventitia. (**D**) A representative of the different control levels showing no signal indicating that the tissue, primary antibodies, and the secondary antibodies were not auto-fluorescent. (**E** and **F**) The cross-section of the aorta of Sprague–Dawley and Ren2 rats was stained with gp91 showing strong signal on the Ren2 rat indicating higher expression of gp91 in the aorta of Ren2 rat. (**G** and **H**) The cross-section of the aorta of Sprague–Dawley and Ren2 rats was stained with p22 subunit showing strong signal on the Ren2 rat indicating higher expression of p22 in the aorta of Ren2 rat. M: Mascular; Con: Control; Short arrows: adventita.

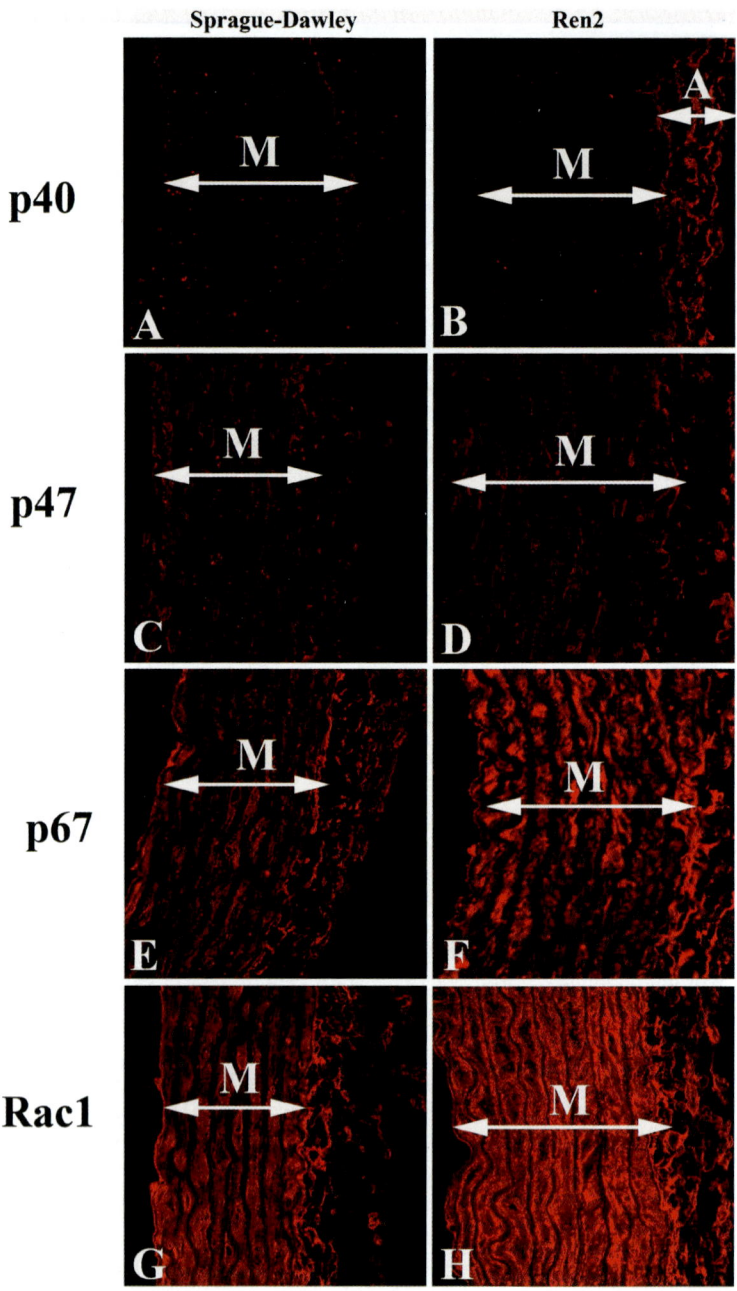

3.3.2. Expression Level of Different Subunits of NADPH Oxidase in the Heart

Like aorta, no signal is observed on the different control level stained with primary antibodies, secondary antibodies, or no staining at all *see* **Fig. 5A and B)**. Overall, the expression of gp91, p67, and Rac1 is higher than the 922, 940, and p47 on both SD and Ren2 rats (*see* **Figs 5** and **6**). Some disruption and deformation are observed in heart muscle fiber of the Ren2 rat (*see* **Figs 5** and **6**). Average gray scale analysis indicate significantly higher expression level of subunits in the media of the Ren2 compared with SD (*see* **Fig. 7**), whereas the average gray scale intensities demonstrate higher expression of Rac1 in the Ren2 animals (*see* **Fig. 8**).

3.4. Lucigenin-Derived Chemiluminescence and Quantification of ROS Production

Lucigenin-derived chemiluminescence (LDCL) has proven to be a useful tool in the quantification of ROS. The mechanism of reaction and application of the assay have been studied thoroughly *(23)*. The assay has been used to detect ROS in cell suspensions, tissue homogenates, and tissue segments. In examining the manifestation of oxidative stress, LDCL has been utilized to evaluate elevated ROS production in left ventricle sections of the mRen2 rat compared with SD as exemplified in **Fig. 9**. The method is as follows:

1. Prepare a working solution of 6.7 μM lucigenin by diluting the stock solution in a modified KHB (mKHB). The working solution is dark adapted at least 1 h prior to use.
2. Collect left ventricle sections and store in 500 μl of mKHB on ice until measurement.
3. Assay samples by adding 1.5 ml of the working lucigenin solution.

◄───

Fig. 4. Thick paraffin sections of aorta of Sprague–Dawley and Ren2 rats stained with the different subunits (p40, p47, p67, and Rac1) of NADPH oxidase. (**A** and **B**) The cross-section of the aorta of Sprague–Dawley and Ren2 rats was stained with p40 showing strong signal on the Ren2 rat especially in adventitia indicating higher expression of p40 in the aorta of Ren2 rat. (**C** and **D**) The cross-section of the aorta of Sprague–Dawley and Ren2 rats was stained with p47 showing almost the same signal intensities on both Sprague–Dawley and Ren2 rats. (**E** and **F**) The cross-section of the aorta of Sprague–Dawley and Ren2 rats was stained with p67 showing strong signal on the Ren2 rat indicating higher expression of p67 in the aorta of Ren2 rat. (**G** and **H**) The cross-section of the aorta of Sprague–Dawley and Ren2 rats was stained with Rac1 showing strong signal on the Ren2 rat indicating higher expression of p40 in the aorta of Ren2 rat; M: muscular.

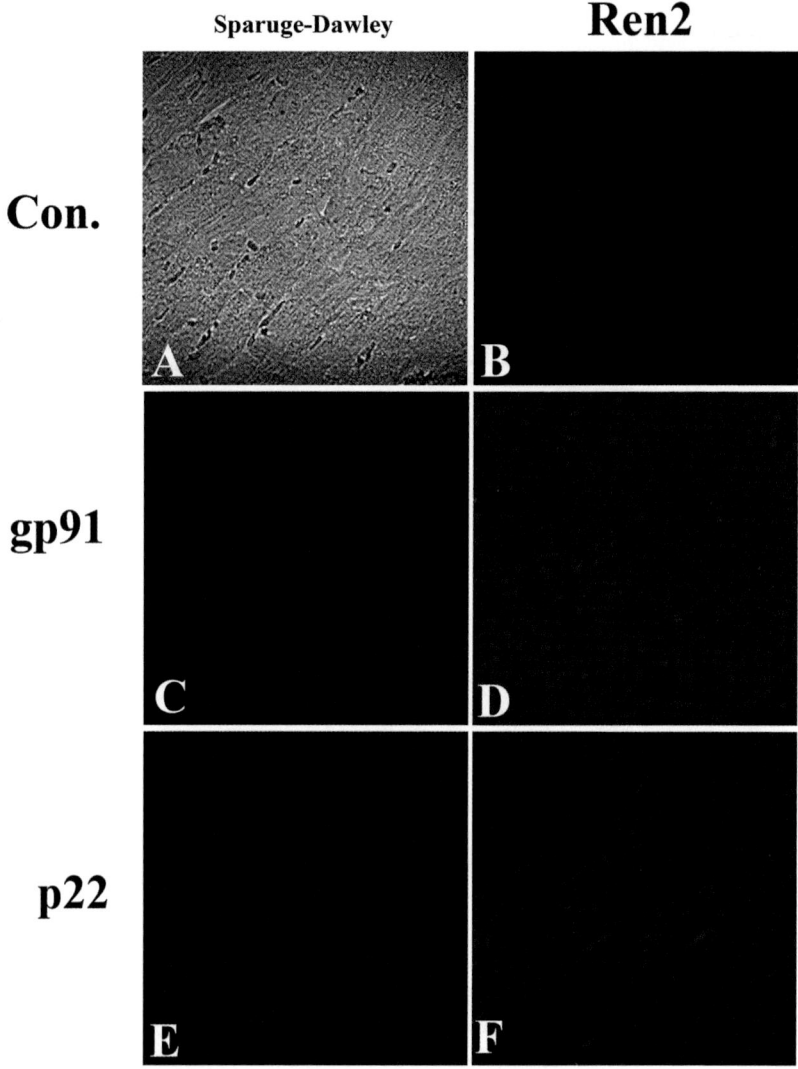

Fig. 5. Thick paraffin sections of left ventricle of Sprague–Dawley and Ren2 rats stained with gp91 and p22 of NADPH oxidase. (**A**) Combination of transmitted and fluorescent images of the left ventricle showing absence of signal on the tissue. (**B**) A representative of the different control levels showing no signal indicating that the tissue, primary antibodies, and the secondary antibodies were not auto-fluorescent. (**C** and **D**) The cross-section of left ventricle immunostained for gp91 showing much stronger signal on Ren2 rat. (**E** and **F**) The cross-section of left ventricle immunostained for p22 showing stronger signal on Ren2 rat. Con: Control.

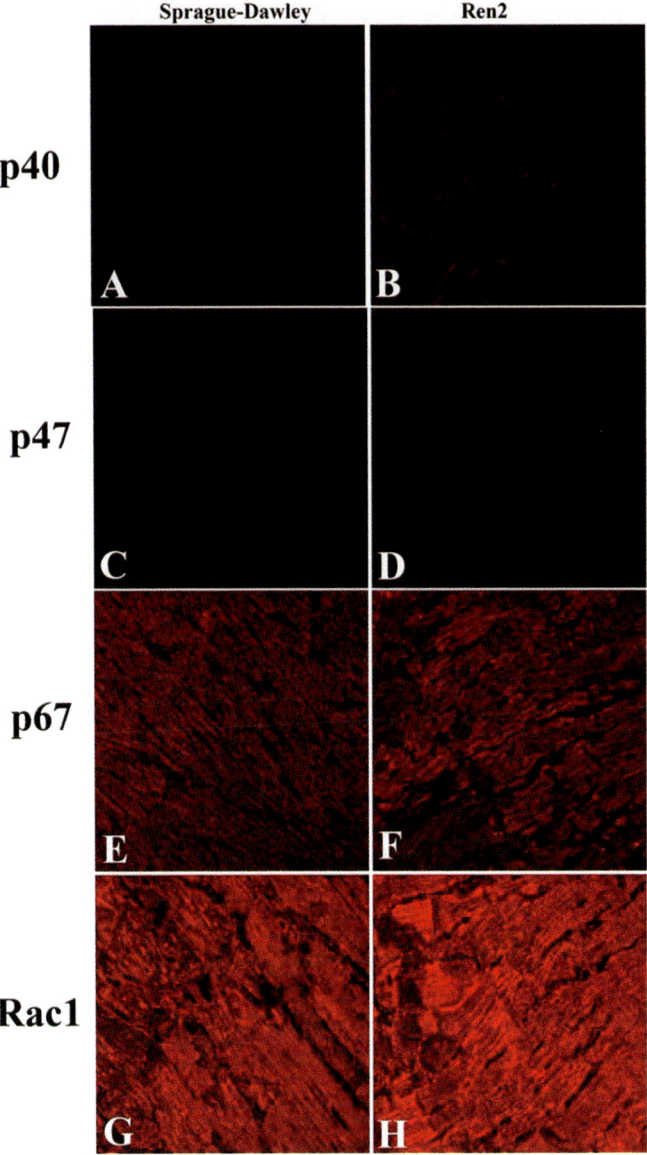

Fig. 6. Thick paraffin sections of left ventricle of Sprague–Dawley and Ren2 rats immunostained with (**A** and **B**) p40, (**C** and **D**) p47, (**E** and **F**) p67, and (**G** and **H**) Rac1 showing stronger signals of NADPH oxidase subunits for Ren2 compared with Sprague–Dawley.

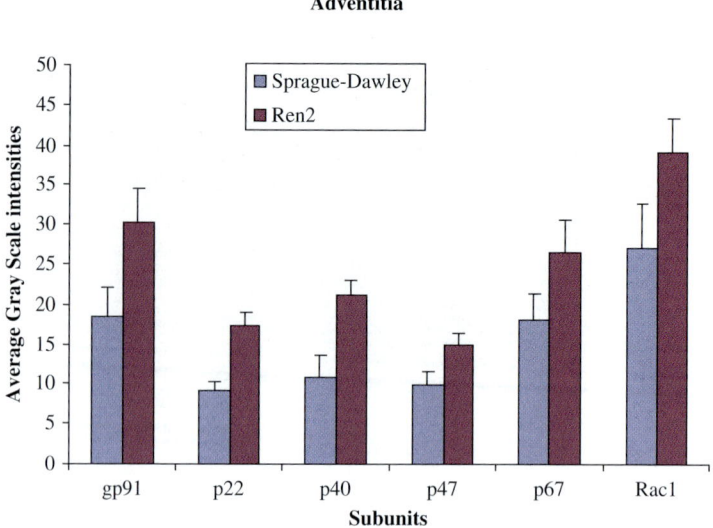

Fig. 7. Average gray scale intensities associated with the media of the Sprague–Dawley and Ren2 rat immunostained with the different subunits of the NADPH oxidase, indicating significantly higher expression level of subunits in the media of the Ren2 compare with Sprague–Dawley.

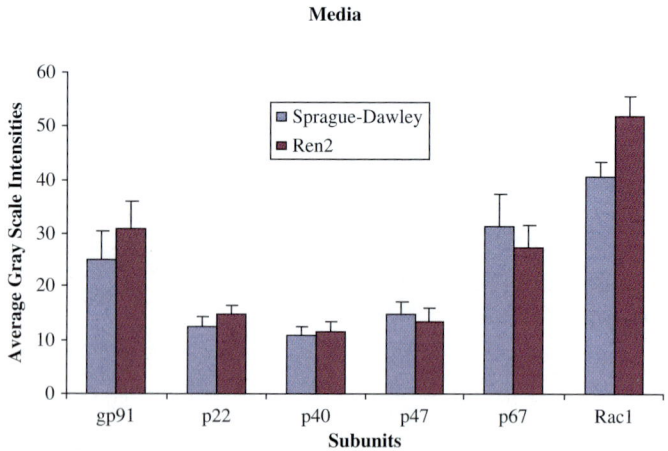

Fig. 8. Average gray scale intensities associated with the adventitia of the Sprague–Dawley and Ren2 rat immunostained with the different subunits of the NADPH oxidase, indicating significantly higher expression level of only Rac1 of the Ren2 compared with Sprague–Dawley.

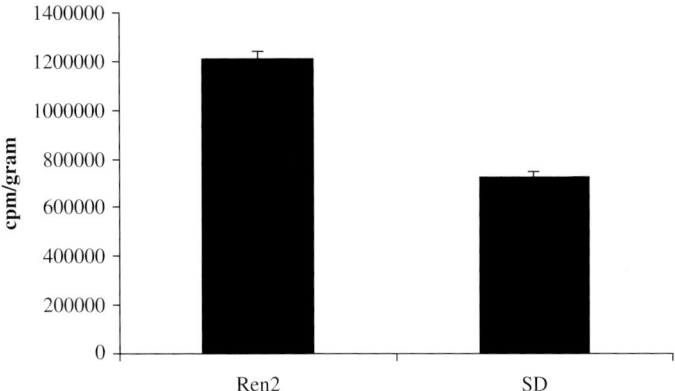

Fig. 9. Lucigenin-derived chemiluminescence of left ventricle sections from Ren2 and Sprague–Dawley (SD) rats. As described, left ventricle sections were assayed in a modified Krebs–Henseleit buffer containing 5 mM lucigenin. Samples were analyzed in a liquid scintillation counter, collecting data every 30 s for 10 min. The data obtained in the last 5 min were averaged, corrected for background, and normalized to tissue section weight. Data are presented as mean ± SEM.

4. Monitor photon emission using a Beckman 6500 in single-photon monitoring mode, as counts per minute (cpm) every 30 s for 10 min.
5. Average the cpm values for the final 5 min, subtract the blank, and normalize to left ventricle weight.
6. Prior to sample analysis and periodically throughout the day assay several blanks (500 μlmKHB) to ensure that the lucigenin solution is dark adapted and background is stable.

References

1. Griending, K.K., Murphy, T.J., and Alexander, R.W. (1993) Molecular biology of the renin-angiotensin system. *Circulation* **87**, 1816–1828.
2. Timmermans, P.B., et al. (1993) Angiotensin II receptors and angiotensin II receptors antagonists. *Pharmacol. Rev.* **45**, 205–251.
3. Nickenig, G., and Harrison, D.G. (2002) The AT1-type angiotensin receptor in oxidative stress and atherogenesis. Part I: oxidative stress and atherogenesis. *Circulation* **105**, 393–396.
4. Griendling, K.K., Lassegue, B., and Alexander, R.W. (1996) Angiotensin receptors and their therapeutic implications. *Annu. Rev. Pharmacol. Toxicol.* **36**, 281–306.
5. Romero, J.C., and Recklehoff, J.F. (1999) State-of-the-art lecture: role of angiotensin and oxidative stress in essential hypertension. *Hypertension* **34**, 943–949.

6. Griendling, K.K., and Harrison, D.G. (1999) Dual role of reactive oxygen species in vascular growth. *Circ. Res.* **85**, 562–563.

7. El-Atat, F.A., Stas, S.N., McFarlane, S.I., et al. (2004) The relationship between hyperinsulinemia, hypertension and progressive renal disease. *J. Am. Soc. Nephrol.* **15**, 2816–2827.

8. Siwik, D.A., Pagano, P.J., and Colucci, W.S. (2001) Oxidative stress regulates collagen synthesis and matrix metalloproteinase activity in cardiac fibroblasts. *Am. J. Physiol.* **280**, C53–C60.

9. Sowers, J.R. (2002) Hypertension, angiotensin II and oxidative stress. *N. Engl. J. Med.* **346**, 1999–2001.

10. Ushio-Fukai, M., et al. (2002) Novel role of gp91phox containing NADPH oxidase in vascular endothelial growth factor induced signaling and angiogenesis. *Circ. Res.* **91**, 1160–1167.

11. Ushio-Fukai, M., and Alexander, W. (2004) Reactive oxygen species as mediators of angiogenesis signaling: role of NADPH oxidase. *Mol. Cell Biochem.* **264**, 85–97.

12. Zuo, L., et al. (2004) Microtubules regulate angiotensin II type 1 receptor and Rac1 localization in caveolae/lipid rafts: role in redox signaling. *Arterioscler. Thromb. Vasc. Biol.* **24**, 1223–1228.

13. Zuo, L., et al. (2005) Caveolin 1 is essential for activation of Rac-1 and NADPH oxidase after angiotensin II type 1 receptor stimulation in vascular smooth muscle cells: role in redox signaling and vascular hyperthophy. *Arterioscler. Thromb. Vasc. Biol.* **25**, 1824–1830.

14. Mullins, J.J., Peters, J., and Ganten, D. (1990) Fulminant hypertension in transgenic rats harboring the mouse Ren-2 gene. *Nature* **344**, 541–544.

15. Lee, M.A., et al. (1996) Physiological characterization of the hypertensive transgenic rat TGR(mREN2)27. *Am. J. Physiol.* **270**, E919–E929.

16. Engler, S., et al. (1998) The TGR(mRen2)27 transgenic rat model of hypertension. *Regul. Pept.* **77**, 3–8.

17. Mullins, J.J., Sigmund C.D., Kane-Haas C., McGowan R.A., and Gross K.W. (1989) Expression of the DBA/2J Ren2 gene in the adrenal gland of transgenic mice. *EMBO J.* **8**, 4065–4072.

18. Baltatu, O., et al. (2003) Abolition of end-organ damage by antiandrogen treatment in female hypertensive transgenic rats. *Hypertension* **41**, 830–833.

19. Pinto, Y.W., et al. The TGR(mRen2)27 transgenic rat model of hypertension. *Regul. Pept.* **77**, 3–8.

20. Pinto, Y.M., Paul, M., and Ganten, D. (1998) Lessons from rat models of hypertension: from Goldblatt to genetic engineering. *Cardiovas. Res.* **39**, 77–88.

21. Brosnan, M.J., et al. (1999) Different effects of antihypertensive agents on cardiac and vascular hypertrophy in the transgenic rat line TGR(mRen2)27. *Am. J. Hypertens.* **12**, 724–731.

22. Sloniger, J.A., Saengsirisuwan, V., Diehl, C.J., Kim, J.S., and Henriksen, E.J. (2005) Selective angiotensin II receptor antagonism enhances whole-body insulin sensitivity and muscle glucose transport in hypertensive TG(mREN2)27 rats. *Metabolism* **54**, 1659–1668.
23. Li, Y., Zhu, H., Kuppusamy, P., Roubaud, V., Zweier, JL., and Trush, MA. (1998) Validation of lucigenin (Bis-*N*-methylacridinium) as a chemilumigenic probe for detecting superoxide anion radical production by enzymatic and cellular systems. *J. Biol. Chem.* **273**, 2015–2023.

11

Detection and Quantification of Apoptosis in the Vasculature

Wendy A. Boivin and David J. Granville

Summary

The integral role of apoptosis in the pathogenesis of cardiovascular diseases has been extensively studied and characterized in recent years. The study of cell death in the vasculature has significantly contributed to our knowledge of vascular disease pathology and has played a role in identifying potential therapeutic strategies for these diseases. This chapter describes a number of standard, widely used protocols for detecting and quantifying apoptosis in vessel wall cells and tissue. These techniques include terminal deoxynucleotidyl transferase dUTP nick-end labeling staining for DNA fragmentation, Hoechst staining for chromatin condensation, Annexin V staining, labeling for phosphatidylserine externalization, Western blot assessment of caspase cleavage, immunofluorescence detection of caspase activation, assessment of mitochondrial membrane depolarization and cytochrome c release, and a splenocyte assay for quantifying susceptibility to immune cell-mediated apoptosis.

Key Words: Annexin V; Aneurysm; Apoptosis; Atherosclerosis; Blood vessel; Caspase; Cytochrome c; Cytotoxic T lymphocyte; DNA fragmentation; Endothelial cells; Heart; Hoechst; Membrane potential; Mitochondria; Protease; Smooth muscle cells; Splenocytes; TUNEL; Vascular injury.

1. Introduction

Apoptosis is an extensively regulated, adenosine triphosphate-dependent form of cell death that is characterized by cell shrinkage, DNA fragmentation, and membrane blebbing. Membrane-enclosed cellular debris is then engulfed by neighboring phagocytes. Many apoptotic pathways are utilized by cells; however, many of these pathways are regulated by Bcl-2 pro-apoptotic and

From: *Methods in Molecular Medicine, Vol. 139: Vascular Biology Protocols*
Edited by: N. Sreejayan and J. Ren © Humana Press Inc., Totowa, NJ

anti-apoptotic proteins as well as by caspases, a family of cysteinyl-aspartate proteases that cleave numerous substrates and can be dependent or independent on mitochondrial depolarization *(1,2)*. Apoptosis plays a key role in ailments such as atherosclerosis, myocardial infarction, hypertrophy, transplant rejection, aneurysms, and numerous other cardiovascular conditions *(1,3–8)*. In the vasculature, apoptosis of endothelial cells, smooth muscle cells, lymphocytes, and macrophages all contribute to the pathogenesis of atherosclerosis and other vascular abnormalities. In atherosclerosis, apoptosis of vascular smooth muscle cells (VSMCs) was detected in the shoulder regions of atherosclerotic plaques, the area that is most prone to rupture. In aortic aneurysms, there is an increase in VSMC apoptosis when compared with the normal aorta *(8)*. However, more research in these areas is required to further define the role of apoptosis in vascular disease.

The standard methods of detecting vascular cell apoptosis are described in this chapter. However, many intrinsic cell-death pathways are redundant, and it is unlikely that all these pathways are activated during apoptosis. The pathways triggered may depend on cell type and apoptotic stimuli; thus, it is strongly recommended that more than one apoptosis detection method is utilized to confirm or rule out apoptosis.

DNA fragmentation is a hallmark of apoptosis and can be detected by terminal deoxynucleotidyl transferase (TdT) dUTP nick-end labeling (TUNEL) staining, a standard, widely utilized methodology. In this technique, labeled dUTP is added onto the 3' ends of DNA strand breaks by TdT, a DNA repair enzyme *(9,10)*. The labeled nucleotides are detected by enzyme-conjugated antibodies and can be quantified by flow cytometry or microscopy.

To detect chromatin condensation, which is a common feature of apoptosis, DNA can be stained using nuclear dyes. Hoechst is a fluorescent cell-permeable dye that binds in the minor groove of DNA and labels condensed chromatin as small, bright, fragmented nuclei.

Externalization of phosphatidylserine (PS) on the outer membrane of the cell is another hallmark of apoptosis. PS normally resides in the inner membrane leaflet of viable cells, but during apoptosis, PS is externalized to the outer leaflet *(11)*. PS localization on the outer leaflet of the membrane is a pro-phagocytotic signal that facilitates the detection and engulfment of apoptotic cells by both professional and non-professional phagocytes such as neighboring myocytes. To detect PS localization on the outer leaflet of the cell membrane, non-permeablized cells can be stained with a fluorescently labeled Annexin V antibody, which has specific affinity for PS. To differentiate between early/late apoptotic and necrotic cells, propidium iodide (PI) can be added to the sample as increased membrane permeability is observed in late apoptotic and necrotic cells resulting in increased uptake of PI.

As there are many apoptotic pathways, there are a wide variety of pathway-specific assays to detect apoptosis that concentrate on modification or localization changes of apoptotic proteins and their substrates. These pathway-specific events can be detected by using sodium dodecyl sulfate–polyacrylamide gel electrophoresis (SDS–PAGE) and Western blotting or by immunofluorescence. Caspase assays are commonly utilized as they are implicated in several pathways. Caspase activation occurs as a result of the proteolytic cleavage of pro-caspases to active forms. This can be easily detected through immunoblotting for the appearance of cleavage fragments or by loss of the pro-form *(12)*. Another procedure for detecting caspase activation is a caspase activity assay. For both these assays, whole-cell lysates can be utilized to detect caspase activity. It is recommended that several caspases (caspase-2, caspase-3, caspase-6, caspase-7, caspase-8, and caspase-9) are examined with this assay to ensure coverage of various caspase-independent apoptotic cascades *(13)*. Furthermore, it is recommended that both assays be used to confirm caspase activation as detection of caspase subunits alone by Western blotting does not demonstrate that the caspases are active and not inhibited by intrinsic caspase inhibitor proteins such as inhibitor of apoptosis proteins.

Mitochondria are organelles crucial to energy production and also play an important role in apoptosis, which occurs as a result of inner and/or outer mitochondrial membrane permeabilization. Mitochondrial membrane depolarization can be detected using tetramethylrhodamine methyl ester (TMRM), a fluorescent lipophilic cation that accumulates in mitochondria as a function of transmembrane potential. Thus, as membrane potential drops, mitochondria will retain proportionally less dye.

Another method for detecting mitochondrial integrity is the detection of cytochrome c leakage from mitochondria into the cytosol. This can be quantified by isolating intact mitochondrial and cytosolic cell fractions. These fractions can be run on an SDS–PAGE and probed for cytochrome c by Western blotting. The presence of cytochrome c in the cytosol indicates a loss of mitochondrial membrane integrity and, ultimately, cell death. The relationship between cytochrome c release and mitochondrial membrane depolarization will vary according to the apoptotic stimulus and cell type, thus analyzing mitochondrial status using procedures for both membrane potential and cytochrome c release can be run in parallel without redundancy.

Vascular cells also undergo apoptosis in response to cytotoxic lymphocyte (CTL)-mediated recognition and killing. Certain stimuli can increase vascular cell susceptibility to CTL-induced apoptosis. The effect of various apoptotic stimuli on cell vulnerability to apoptosis can be quantified in vitro using a splenocyte-based killing assay *(14)*. In this splenocyte assay, purified activated mouse splenocytes can be incubated with labeled target vascular cells to quantify apoptosis. Detection of cell death is based on reduced plasma membrane integrity. This compromised cellular membrane integrity results in target cell release of the fluorescent calcein probe into the surrounding culture media.

2. Materials

2.1. In Situ TUNEL Labeling of Paraffin-Embedded Slides

1. Phosphate-buffered saline (PBS): 2.69 mM KCl, 137 mM NaCl, 1.47 mM KH_2PO_4, 8.1 mM Na_2HPO_4, pH 7.6.
2. Xylene.
3. Ethanol.
4. Three percent H_2O_2: diluted from 30% H_2O_2 in distilled H_2O (dH_2O).
5. Proteinase K (diluted in PBS).
6. Equilibration buffer: 200 mM potassium cacodylate, pH 6.6, 25 mM Tris–HCl, pH 6.6, 0.2 mM dithiothreitol (DTT), 0.25 mg/ml bovine serum albumin (BSA), 2.5 mM cobalt chloride.
7. TdT (Boehringer, Ingelheim, Germany).
8. Reaction mixture: 50 U/ml TdT and 1:50 dilution of digoxigenin (DIG) DNA labeling mixture (Boehringer) in equilibration buffer.
9. Horseradish peroxidase (HRP)-conjugated anti-DIG antibody.
10. 3, 3′-Diaminobenzidine (DAB).
11. Hematoxylin.

2.2. TUNEL on Cell Cultures

1. Ten percent neutral-buffered formalin or 4% paraformaldehyde.
2. PBS: 2.69 mM KCl, 137 mM NaCl, 1.47 mM K_2PO_4, 8.1 mM Na_2HPO_4, pH 7.6.
3. Triton X-100 (0.01%) (diluted in PBS).
4. Proteinase K (diluted in PBS).
5. Equilibration buffer: 200 mM potassium cacodylate, pH 6.6, 25 mM Tris–HCl, pH 6.6, 0.2 mM DTT, 0.25 mg/ml BSA, 2.5 mM cobalt chloride.
6. TdT (Boehringer).
7. Reaction mixture: 50 U/ml TdT and 1:50 dilution of DIG DNA labeling mixture (Boehringer) in equilibration buffer.
8. HRP-conjugated anti-DIG antibody.
9. DAB.
10. Hematoxylin.

2.3. Hoechst Nuclear Staining

1. PBS: 2.69 mM KCl, 137 mM NaCl, 1.47 mM K_2PO_4, 8.1 mM Na_2HPO_4, pH 7.6.
2. Fixative: 10% neutral-buffered formalin or 4% paraformaldehyde.
3. Hoechst 33342 (Molecular Probes, Eugene, OR).

2.4. Annexin V Labeling

1. Annexin V staining buffer: 10 mM 4-(2-Hydroxyethyle) piperazine-1-ethane-sulfonic acid(HEPES), 140 mM NaCl, 2.5 mM $CaCl_2$, pH 7.6.
2. PBS: 2.69 mM KCl, 137 mM NaCl, 1.47 mM K_2PO_4, 8.1 mM Na_2HPO_4, pH 7.6.
3. Alexa-488-conjugated Annexin V (Molecular Probes).
4. PI (Molecular Probes).

2.5. Immunodetection of Caspase Processing

2.5.1. Preparation of Whole-Cell Lysates from Non-Adherent Cells, Adherent Cells, and Tissue Sections

1. PBS: 2.69 mM KCl, 137 mM NaCl, 1.47 mM K_2PO_4, 8.1 mM Na_2HPO_4, pH 7.6.
2. Whole-cell lysate buffer: 1% IGEPAL, also known as NonidetP-40 (NP-40), 20 mM Tris–HCl, pH 8, 137 mM NaCl, 10% glycerol, supplemented with 1 mM phenyl-methylsulfonyl fluoride (PMSF), 0.15 U/ml aprotinin, 1 mM sodium orthovanadate. Prepare immediately prior to use.
3. Cell scrapers.
4. Tissue homogenizer.

2.5.2. Western Blot Detection of Caspase Processing

1. Lysates (prepared as described in **Subheadings 3.4.1., 3.4.2., and 3.4.3.**).
2. 2× SDS/sample buffer: 100 mM Tris–HCl, pH 6.8, 25% glycerol, 2% SDS, 0.01% bromophenol blue, 10% β-mercaptoethanol.
3. Ten percent separating polyacrylamide gel with 4% stacking polyacrylamide gel.
4. Running buffer: 25 mM Tris-base, 192 mM glycine, 1% SDS.
5. Nitrocellulose membrane (Amersham, Piscataway, NJ).
6. Transfer buffer: 20 mM Tris-base, pH 8, 150 mM glycine, 20% methanol.
7. Blocking buffer: 125 mM NaCl, 25 mM Tris-base, pH 8, 5% skim milk, 0.01% Tween-20.
8. Primary antibody against caspase of interest and complementary secondary antibody conjugated to HRP.
9. Enhanced chemiluminescence (ECL) developing solution (Amersham).
10. X-ray film.

2.6. Caspase Proteolytic Activity Assay

1. Lysates (prepared as described in **Subheadings 3.4.1., 3.4.2., and 3.4.3.**).
2. Ninety-six-well flat bottom plate.

3. Reaction buffer: 1% NP-40, 20 mM Tris, pH 7.5, 137 mM NaCl, 10% glycerol.
4. Whole-cell lysate buffer: 1% NP-40, 20 mM Tris-base, pH 8, 137 mM NaCl, 10% glycerol, supplemented with 1 mM PMSF, 0.15 U/ml aprotinin, 1 mM sodium orthovanadate. Prepare immediately prior to use.
5. Caspase substrate (Calbiochem, San Diego, CA) (*see* **Note 1**).
6. Fluorescent plate reader (PerSeptive Biosystems, Framingham, MA).

2.7. Measurement of Mitochondrial Membrane Potential

1. TMRM (Molecular Probes).
2. Dimethylsulfoxide (DMSO).
3. Cell culture media.

2.8. Detection of Cytochrome C Release from Mitochondria

1. PBS: 2.69 mM KCl, 137 mM NaCl, 1.47 mM K_2PO_4, 8.1 mM Na_2HPO_4, pH 7.6.
2. Trypsin (0.05%) and 0.53 mM ethylenediaminetetraacetic acid (EDTA).
3. Potter–Elvehjem homogenizer and Teflon pestle.
4. Homogenization buffer: 250 mM sucrose, 20 mM K^+ HEPES, 10 mM KCl, 1.5 mM $MgCl_2$, 0.1 mM EDTA, 1 mM Ethylene glycol tetraacetic acid (EGTA), 1 mM DTT, 0.1 mM PMSF, pH 7.5.
5. Monoclonal antibody for cytochrome c (Pharmingen, San Diego, CA).

2.9. Splenocyte Killing Assay

1. Mouse spleen.
2. Sterile PBS: 2.69 mM KCl, 137 mM NaCl, 1.47 mM K_2PO_4, 8.1 mM Na_2HPO_4, pH 7.6.
3. Red blood cell (RBC) lysis buffer: 10 mM $KHCO_3$, 150 mM NH_4Cl, 0.1 mM EDTA, pH 8.
4. Concanavalin A (Con A, Sigma, St. Louis, MO).
5. Roswell Park Memorial Institute (RPMI) media.
6. Fetal calf serum (FCS).
7. Target cell-specific media.
8. Calcein acetomethyl (AM) conjugate (Molecular Probes).
9. Triton X-100 (2%).

3. Methods

3.1. TUNEL Staining

3.1.1. In Situ TUNEL Staining on Paraffin-Embedded Slides

1. Fix tissue with 10% neutral buffered formalin overnight and embed in paraffin (*see* **Note 2**).
2. Cut 3-μm tissue sections onto glass slides and incubate slides overnight at 60 °C in oven.

3. Deparaffinize and rehydrate slides in a series of xylene and ethanol washes:

   a. Xylene: 3× with fresh xylene every 5 min.
   b. Ethanol, 100%: 2× for 5 min each using fresh ethanol after each wash.
   c. Ethanol, 90%: once for 5 min.
   d. Ethanol, 70%: once for 5 min.
   e. dH$_2$O: once for 5 min.

4. Block endogenous peroxide by incubation of tissue in 100 μl of 3% H$_2$O$_2$ for 20 min at room temperature (RT) (*see* **Note 3**).
5. Wash 3× with PBS for 2 min.
6. Digest DNA-associated proteins by incubating tissues in 100 μl for 20 μg/ml proteinase K for 15 min (*see* **Note 4**).
7. Wash 3× for 5 min with PBS.
8. Add 75 μl of equilibration buffer to tissue for at least 1 min at RT.
9. Label DNA breaks with 75 μl of TdT diluted in reaction buffer for 1 h at 37 °C (*see* **Note 4**).
10. Stop TdT reaction by washing slides in PBS 3× at 5 min each.
11. Detect DIG-conjugated dUTP by incubating tissue with 100 μl of anti-DIG HRP conjugate for 30 min at RT. Optimal antibody concentration varies, depending on the vendor and should be determined and optimized by the operator.
12. Wash 2× with PBS for 5 min.
13. Wash slide 2× with dH$_2$O for 5 min.
14. Incubate tissue in 100 μl of DAB (HRP substrate) for 5–10 min for color development (*see* **Note 3**).
15. Wash excess DAB from tissue with two dH$_2$O washes at 5 min each.
16. Counterstain nuclei in hematoxylin for 5–10 s.
17. Wash slides in dH$_2$O.
18. Dehydrate tissue in the following sequence of washes:

    a. Ethanol, 70%, for 1 min.
    b. Ethanol, 90%, for 1 min.
    c. Ethanol, 100%, 2× for 1 min.
    d. Xylene 2× for 1 min.

19. Apply coverslip to slides and visualize.

3.1.2. TUNEL on Cell Cultures

This method is similar to the paraffin-embedded section procedure (refer to the previous protocol in **Subheading 3.1.1.**). Deviations from the protocols in **Subheading 3.1.1** are highlighted below:

1. Wash cells cultured in eight chambered slides with PBS.
2. Fix cells in 10% neutral buffered formalin or 4% paraformaldehyde for 15 min.

3. Wash cells 2× for 5 min in PBS.
4. Permeabilize cells in 0.01% Triton X-100 for 15 min at RT.
5. Wash in PBS 2× for 5 min.
6. Follow **steps 8–19** as outlined in **Subheading 3.1.1**.

3.2. Hoechst Nuclear Staining

1. Wash cells with PBS.
2. Fix cells with 10% neutral buffered formalin (or 4% paraformaldehyde) for 15 min at RT (*see* **Note 5**).
3. Wash cells in PBS.
4. Incubate cells in 1 μg/ml Hoechst 33342 for 10 min at RT.
5. Wash cells in PBS once.
6. Mount with aqueous mounting medium and visualize using fluorescence microscopy (excitation 350 nm and emission 460 nm).

3.3. Annexin V Labeling

This staining procedure should only be performed on live cells and should be initiated swiftly without fixation (*see* **Note 6**).

1. Wash cells with PBS.
2. Incubate cells in Alexa-488-conjugated Annexin V (1:50 dilution) and 1 μg/ml PI diluted in Annexin V staining buffer for 15 min at 37 °C (*see* **Note 7**).
3. Wash cells 2× in Annexin V staining buffer.
4. Visualize in phenol-free medium or PBS (emission 495 nm and excitation 519 nm).

3.4. Immunodetection of Caspase Processing

This section describes the preparation of whole-cell lysates from non-adherent cells (*see* **Subheading 3.4.1.**), adherent cells (*see* **Subheading 3.4.2.**), and tissue sections (*see* **Subheading 3.4.3.**) as well as analysis for caspase processing by immunodetection (*see* **Subheading 3.4.4.**) and caspase protease activity (*see* **Subheading 3.4.5.**).

3.4.1. Preparation of Whole-Cell Lysates from Non-Adherent Cultured Cells

1. Pellet cells by centrifugation.
2. Wash cell pellet 2× in ice-cold PBS and resuspend the pellet in 1 ml of whole-cell lysis buffer (*see* **Note 8**) for 15 min on ice.
3. Centrifuge samples for 10 min at 15, 800 *g* at 4 °C.
4. Aliquot supernatant and store at −80 °C.

3.4.2. Preparation of Whole-Cell Lysates from Adherent Cultured Cells

1. Transfer conditioned media from plates of cultured adherent cells to centrifuge tubes and pellet detached cells by centrifugation (*see* **Note 9**).
2. Wash adherent cells and pellet detached cells twice with ice-cold PBS. Be gentle as not to dislodge cells.
3. Treat adherent cell plates with 0.5–1.0 ml of whole-cell lysis buffer for 20 min on ice (*see* **Note 8**).
4. Scrape cells from plate and transfer cell lysates to 1.5-ml microcentrifuge tubes.
5. Centrifuge samples for 10 min at 15, 800 *g* at 4 °C.
6. Resuspend detached cell pellets from conditioned media (*see* **step 2**) in the cell lysate obtained from the plate for the particular treatment (*see* **Note 9**).
7. Lyse cells on ice for 20 min and centrifuge for 10 min at 15, 800 *g* at 4 °C.
8. Aliquot supernatant and store at −80 °C.

3.4.3. Preparation of Whole-Cell Lysates from Tissue Sections

1. Homogenize 100–500 mg of frozen tissue samples in 1 ml lysis buffer and leave on ice for 20 min.
2. Remove insoluble materials by centrifugation for 10 min at 15, 800 *g* at 4 °C and transfer the supernatant to new tubes.
3. Aliquot and store lysates at −80 °C.

3.4.4. Western Blot Detection of Caspase Processing

1. Mix lysates prepared (as described in **Subheading 3.4.3.**) 1:1 in 2× SDS/sample buffer.
2. Boil samples for 5 min and cool samples on ice.
3. Load 20–50 μg of whole-cell lysate per lane of an SDS–PAGE mini-gel.
4. Electrophorese samples at 20 mA/gel until dye front reaches the bottom of the gel in running buffer.
5. Transfer proteins to nitrocellulose blotting membrane at 250 mA in transfer buffer for 2–4 h.
6. Incubate blot in blocking buffer for 1 h at RT.
7. Incubate blot for 1 h with primary antibody diluted in blocking buffer at RT. Check the supplier's recommendation for antibody concentration.
8. Wash blot 4× for 5 min in washing buffer with shaking at RT.
9. Incubate blot with appropriate secondary antibody conjugated to HRP diluted in blocking buffer for 1 h at RT. Check with supplier for appropriate concentration.
10. Wash blot 4× for 5 min in washing buffer with shaking at RT.
11. Treat blot with ECL solution for 1 min.
12. Expose to X-ray film for 15 s to 30 min. A longer exposure time may be required to detect cleavage fragments.
13. Examine blots for the appearance of caspase cleavage products (*see* **Note 10**).

3.5. Caspase Protease Assay

1. Dilute caspase substrates to 1 mg/ml working solution in reaction buffer.
2. In a 96-well plate, add 100 μl of whole-cell lysis buffer per assay well.
3. Add 50 μl of cell lysates (∼ 1 mg/ml) prepared as described in **Subheadings 3.4.1, 3.4.2, and 3.4.3**. Add 50 μl of whole-cell lysis buffer to act as a blank. It is recommended that samples be added in triplicate wells.
4. Incubate at 37 °C for 15 min.
5. Add 4 μl of 1 mg/ml caspase substrate prepared in **step 1**.
6. Read fluorescence at time zero (*see* **Note 1**).
7. Incubate at 37 °C.
8. Read fluorescence at various time points (i.e., 0.25, 0.5, 1, 2, and 4 h).
9. Calculate caspase activity using the Michaelis–Menten equation. The Michaelis constant (K_m) is specific for each caspase–substrate combination and should be indicated by the supplier.

3.6. Measurement of Mitochondrial Membrane Potential

1. Seed cells on chambered slide.
2. Prepare working solution (1 mg/ml) of TMRM by diluting the stock solution in ' (*see* **Note 11**).
3. Add working TMRM solution to serum-free culture medium at a final concentration of 50–100 nM.
4. Incubate cells for 10–30 min at 37 °C and maintain cells in this media throughout the experiment.
5. Visualize TMRM accumulation in the mitochondria by fluorescence microscopy (excitation 550 nm and emission 575 nm; *see* **Notes 12–14**).

3.7. Detection of Cytochrome C Release from Mitochondria

1. Wash cells 2× with PBS.
2. Trypsinize cells with trypsin–EDTA.
3. Centrifuge cells at 1000 *g* for 5 min.
4. Discard supernatant.
5. Count cells using a hemocytometer and resuspend cells at 2×10^8/ml in homogenization buffer.
6. Homogenize cells with a Potter–Elvehjem homogenizer and Teflon pestle (*see* **Note 15**).
7. Centrifuge cells at 3500 *g* for 5 min to remove nuclei and intact cells.
8. Recentrifuge supernatant at 3500 *g* for 5 min.
9. Collect supernatant.
10. Centrifuge at 10,000 *g* for 15 min and collect the pellet (contains crude mitochondria).

11. Re-centrifuge the post-mitochondrial pellet at $100,000\,g$ for 30 min to obtain the cytosolic fraction.
12. Run samples on an SDS–PAGE and carry out Western blot analysis (*see* **Subheading 3.4.3.**).

3.8. Splenocyte Killing Assay on Adherent Cultured Cells

3.8.1. Harvesting and Activating Spleen Cells

1. Obtain spleens aseptically from C57BL/6 mice and chop up into small pieces.
2. Pass spleens through sterile wire mesh (Falcon 70 μm) with the blunt end of a 3-ml syringe to create a single-cell suspension into 5 ml of RPMI media +10% FCS.
3. Centrifuge the cell suspension for 5 min at $200\,g$.
4. Add 5 ml of RBC lysis buffer to pellet and incubate for 3 min.
5. Centrifuge for 5 min at $200\,g$.
6. Wash cells $2\times$ in 5 ml of PBS, centrifuging for 5 min at $1000\,g$.
7. Count splenocytes on hemocytometer.
8. Reconstitute 10^7 splenocytes in 10 ml of RPMI media +10% FCS containing 5 μg/ml Con A.
9. Incubate at 37 °C and 5% CO_2 for 72 h.
10. Count activated T blasts on hemocytometer.

3.8.2. Labeling Target Cells

1. Grow cells to approximately 70% confluence in 24-well cell culture plates.
2. Wash cells $2\times$ in PBS.
3. Treat cells with 10 μg/ml of calcein AM diluted in culture media.
4. Incubate for 30 min.
5. Wash cells $2\times$ in PBS.

3.8.3. Splenocyte Killing Assay

In this section, activated splenocytes prepared as described in **Subheading 3.8.1** are incubated with labeled target cells prepared as described in **Subheading 3.8.2** to quantify target cell apoptosis.

1. Add effector splenocytes to target cells at effector: target ratios of 100:1, 50:1, 25:1, 10:1, and 1:1 in culture medium for assay optimization. *See* **Note 16** for details on assay controls.
2. Incubate at 37 °C and 5% CO_2 for various time points (i.e., 4, 24, and 48 h) to determine the time course of probe release and cell death vulnerability.
3. Centrifuge for 5 min at 500–1000 g to remove cells and debris and read on a fluorescence microplate reader at excitation 485 nm and emission 535 nm.

4. Notes

1. Because caspases differ in their abilities to cleave specific sequences, caspase substrates should be selected for the caspase of interest. The excitation and emission wavelengths as well as the Michaelis constants will differ for each caspase substrate. This information should be obtained from the supplier prior to commencing the assay. It should also be noted that, for convenience, several caspase assay kits are commercially available.

2. Positive control slides or tissues must be included in every run. Normal spleen contains numerous apoptotic leukocytes and can be used for positive control. Negative control slides must also be included in every run. These can include spleen incubated with reaction mixture without the addition of TdT for 1 h at 37 °C. To ensure the specificity of staining, nuclear localization of staining and assessment of nuclear morphology are routinely employed. TUNEL positivity should be localized to nuclei. Cytoplasmic staining indicates either non-specific staining or detection of necrotic cells because of DNA fragmentation in the presence of nuclear disruption. TUNEL positivity in the cytoplasm of macrophages may also indicate the phagocytosis of apoptotic cells by these phagocytes.

3. Excessive background staining is typically related to the DAB development time. Although incubation with DAB for 5–10 min is recommended, this step should be optimized by the user. Certain tissues have high endogenous peroxidase activity. Therefore, increasing the incubation with 3% H_2O_2 may reduce background staining because of this process.

4. A lack of staining may result from inefficient digestion of DNA-associated proteins or low concentration/activity of TdT. The concentration of these components can be optimized by the operator.

5. If the density of total cells is reduced after the procedure, many of the apoptotic cells that become detached may have washed off. In this case, either can cells be washed more gently or the supernatant of the treated cells be cytospun with harvested (trypsinized) adherent cells and stained in the same manner as described.

6. If a known apoptotic inducer is utilized, and there remains a high frequency of Annexin V-positive and PI-positive cells, the procedure may need to be performed more rapidly. As this is a live staining method, maintenance of these cells outside a proper incubator may lead to cell damage. In addition, the concentration of Annexin V may need to be adjusted by the operator to ensure optimum staining of different cell types.

7. If limited staining is observed, the concentration of Annexin V can be increased—**Subheading 3.3, step 2**. In addition, the intensity of fluorescent dyes decreases with increased exposure time to excitation wavelengths (photobleaching). Therefore, the length of time the stained cells are exposed to light should be limited. Finally, anti-fade reagents that reduce the sensitivity of fluorescent dyes can be purchased. The cells can be maintained in solutions containing these reagents, **Subheading 3.3, step 4**, if need be.

8. The volume of whole-cell lysis buffer can be adjusted depending on the cell number. It is recommended that 100–200 µl be used per 10^6 pelleted cells. For adherent cells, the volume must be adjusted based on the surface area of the plate. Typically use 0.5–1 ml of buffer for 100-mm Petri dishes or 0.2–0.4 ml of buffer for 35-mm Petri dishes.

9. Many cells will detach upon apoptosis; therefore, it is important to collect these cells from the conditioned media. We recommend that pelleted cells from the media are lysed in the lysis buffer from their respective treatments. This ensures that all cells are lysed in an equal volume of buffer and are thus equally represented in the lysate.

10. Approximate sizes of caspase cleavage products. It is important to remember that some caspases also show intermediate cleavage products and that the sizes of these cleavage products can vary depending on which protease was responsible for their cleavage. Furthermore, antibodies typically only recognize the pro-form plus one cleavage product. It is important to check with the antibody supplier as to which cleavage fragments are recognized by each primary antibody (*see* **Table 1**).

11. Prepare stock solution of TMRM at 10 mg/ml diluted in DMSO and store at $-20\,°C$.

12. TMRM binds to polystyrene, so usage of this plastic should be avoided. TMRM fluorescence is temperature dependent (fluorescence decreases with increasing temperature), so a uniform temperature should be maintained for all samples.

13. A mitochondrial uncoupling reagent, carbonyl cyanide *p*-(trifluoromethoxy) phenyl-hydrazone (FCCP, Sigma), can be utilized as a positive control for mitochondrial membrane depolarization. Expose cells to this proton ionophore for 10 min at a concentration of 5 µmol/l prior to TMRM incubation.

14. The TMRM probe can be used for flow cytometry utilizing approximately 10^5 cells at 20 nM TMRM with an incubation of 10 min at 30 °C.

Table 1
The Molecular weights of the Caspases in the uncleared pro form and cleared active form

Caspases	Molecular weight of pro-form (kDa)	Molecular weight of active subunits (kDa)
Caspase-2	49	12, 13, 18
Caspase-3	32	12, 17
Caspase-6	33	11, 18
Caspase-7	34	11, 20
Caspase-8	55	10, 18
Caspase-9	46	10, 35
Caspase-10	59	12, 17/25

15. The number of strokes taken with the homogenizer is dependent on the operator and cell type and should be optimized to minimize disruption of the mitochondrial outer membrane. This procedure can be optimized by determining the least amount of strokes required for >90% of cells to uptake trypan blue.

16. Negative and positive controls are essential for this splenocyte assay. Controls described in (a) and (b) are required, whereas controls described in (c) and (d) are strongly recommended:

   a. Negative calcein release control: Treat target cells with media alone for the determination of "spontaneous calcein AM release."

   b. Positive calcein release control: Treat target cells with 0.1–2% Triton to determine "total calcein release."

   c. Cell-death control: Use calcein-labeled rat splenocytes as target cells and treat them with activated mouse effector splenocytes. This will serve as a positive control for cell death. Rat splenocyte controls are prepared as described in **Subheading 3.8.1, steps 1–7**, followed by the complete procedure in **Subheading 3.8.2**.

   d. Additional control: Treat target cells with inactivated splenocytes (and without Con A treatment).

Acknowledgments

This work was supported by grants from the Canadian Institutes for Health Research (CIHR), The Canada Foundation for Innovation (CFI), the Michael Smith foundation for Health Research (MSFHR), the St. Paul's Hospital Foundation, and the British Columbia Transplant Society. D.J.G. is a Canada Research Chair in Cardiovascular Biochemistry and a MSFHR scholar. W.A.B. is a recipient of a CIHR/MSFHR Transplantation Training Research Award. The authors thank Jonathon C. Choy and Arwen L. Hunter for their contributions toward several of these protocols.

References

1. Granville, D.J., C.M. Carthy, D.W. Hunt, and B.M. McManus (1998) Apoptosis: molecular aspects of cell death and disease. *Lab. Invest.* **78**, 893–913.

2. Granville, D.J., and R.A. Gottlieb (2002) Mitochondria: regulators of cell death and survival. *Scientific World J.* **2**, 1569–1578.

3. Tedgui, A., and Z. Mallat (2003) Apoptosis, a major determinant of atherothrombosis. *Arch. Mal. Coeur. Vaiss.* **96**, 671–675.

4. Gottlieb, R.A. (2003) Mitochondrial signaling in apoptosis: mitochondrial daggers to the breaking heart. *Basic Res. Cardiol.* **98**, 242–249.

5. Gonzalez, A., M.A. Fortuno, R. Querejeta, S. Ravassa, B. Lopez, N. Lopez, and J. Diez (2003) Cardiomyocyte apoptosis in hypertensive cardiomyopathy. *Cardiovasc. Res.* **59**, 549–562.

6. Miller, L.W., D.J. Granville, J. Narula, and B.M. McManus (2001) Apoptosis in cardiac transplant rejection. *Cardiol. Clin.* **19**, 141–154.

7. Choy, J.C., D.J. Granville, D.W. Hunt, and B.M. McManus (2001) Endothelial cell apoptosis: biochemical characteristics and potential implications for atherosclerosis. *J. Mol. Cell. Cardiol.* **33**, 1673–1690.

8. Thompson, R.W., S. Liao, and J.A. Curci (1997) Vascular smooth muscle cell apoptosis in abdominal aortic aneurysms. *Coron. Artery Dis.* **8**, 623–631.

9. Gold, R., M. Schmied, G. Giegerich, H. Breitschopf, H.P. Hartung, K.V. Toyka, and H. Lassmann (1994) Differentiation between cellular apoptosis and necrosis by the combined use of in situ tailing and nick translation techniques. *Lab. Invest.* **71**, 219–225.

10. Gold, R., M. Schmied, G. Rothe, H. Zischler, H. Breitschopf, H. Wekerle, and H. Lassmann (1993) Detection of DNA fragmentation in apoptosis: application of in situ nick translation to cell culture systems and tissue sections. *J. Histochem. Cytochem.* **41**, 1023–1030.

11. Martin, S.J., C.P. Reutelingsperger, A.J. McGahon, J.A. Rader, R.C. van Schie, D.M. LaFace, and D.R. Green (1995) Early redistribution of plasma membrane phosphatidylserine is a general feature of apoptosis regardless of the initiating stimulus: inhibition by overexpression of Bcl-2 and Abl. *J. Exp. Med.* **182**, 1545–1556.

12. Kohler, C., S. Orrenius, and B. Zhivotovsky (2002) Evaluation of caspase activity in apoptotic cells. *J. Immunol. Methods* **265**, 97–110.

13. Nicholson, D.W. (1999) Caspase structure, proteolytic substrates, and function during apoptotic cell death. *Cell Death Differ.* **6**, 1028–1042.

14. Choy, J.C., V.H. Hung, A.L. Hunter, P.K. Cheung, B. Motyka, I.S. Goping, T. Sawchuk, R.C. Bleackley, T.J. Podor, B.M. McManus, and D.J. Granville (2004) Granzyme B induces smooth muscle cell apoptosis in the absence of perforin: involvement of extracellular matrix degradation. *Arterioscler. Thromb. Vasc. Biol.* **24**, 2245–2250.

12

Endothelial Progenitor Cell and Mesenchymal Stem Cell Isolation, Characterization, Viral Transduction

Keith R. Brunt, Sean R. R. Hall, Christopher A. Ward, and Luis G. Melo

Summary

Endothelial progenitor cells (EPCs) and mesenchymal stem cells (MSCs) have emerged as potentially useful substrates for neovascularization and tissue repair and bioengineering. EPCs are a heterogeneous group of endothelial cell precursors originating in the hematopoietic compartment of the bone marrow. MSCs are a rare population of fibroblast-like cells derived from the bone marrow stroma, constituting approximately 0.001–0.01% of the nucleated cells in the marrow. Both cells types have been isolated from the bone marrow. In addition, EPC can be isolated from peripheral blood as well as the spleen, and MSC has also been isolated from peripheral adipose tissue. Several approaches have been used for the isolation of EPC and MSC, including density centrifugation and magnetic bead selection. Phenotypic characterization of both cell types is carried out using immunohistochemical detection and fluorescence-activated cell sorting analysis of cell-surface molecule expression. However, the lack of specific markers for each cell type renders their characterization difficult and ambiguous. In this chapter, we describe the methods that we use routinely for isolation, characterization, and genetic modification of EPC and MSC from human, rabbit, and mouse peripheral blood and bone marrow.

Key Words: Density centrifugation; Endothelium; Flow cytometry; Immunofluorescence; Neovascularization

1. Introduction

Since Asahara et al. *(1)* first reported the isolation of endothelial progenitor cell (EPC) from peripheral blood, several studies have suggested that circulating EPC may play a significant role in endogenous neovascularization of ischemic

From: *Methods in Molecular Medicine, Vol. 139: Vascular Biology Protocols*
Edited by: N. Sreejayan and J. Ren © Humana Press Inc., Totowa, NJ

tissues and in re-endothelialization of ischemic tissues *(2–4)*. Other studies have suggested the potential therapeutic usefulness of EPC as a cell substrate for neovascularization of ischemic tissues and bioengineering of prosthetic grafts *(3–5)*, and several small early phase clinical trials have demonstrated the safety of bone marrow cell transplantation in patients with myocardial or peripheral limb ischemia *(3,4)*. Mesenchymal stem cells (MSCs) have also emerged as substrate for tissue engineering and repair of infarcted tissues *(3,6)*. The cells are pluripotent and able to differentiate into a variety of cell types, including osteocytes, adipocytes, chondrocytes, and possibly myocytes *(6)*. Furthermore, the ease of handling and enormous expansion potential of MSC, together with their self-renewal, pluripotency, and immunotolerance, render them suitable for a wide range of therapeutic applications *(6)*.

EPCs are a heterogeneous group of cells originating from multiple precursors in the bone marrow and present at different stages of differentiation in the circulation *(4,7)*. The cells can be isolated from peripheral blood using density centrifugation and are differentiated and expanded in medium enriched with endothelial cell-growth factors *(1)*. As the cells differentiate, they acquire endothelial lineage markers such as VE-cadherin, PECAM-1 (CD31), and eNOS *(7)*. The function of EPC is reduced in a variety of cardiovascular diseases *(4,8)*, and an inverse relation was recently reported between the number circulating EPC and risk factors for cardiovascular disease *(9)*. MSCs are a rare population of fibroblast-like cells resident in the stroma of the bone marrow *(6)*. In addition, MSCs have been isolated from adipose tissue *(6)* and vessel wall *(10)*, and some recent evidence suggests that MSC can also be mobilized into the circulation *(11)*. In the bone marrow, MSCs represent approximately 0.001–0.01% of the nucleated cells and are readily separated from the hematopoietic stem cells (HSCs) in culture by their preferential attachment to plastic surfaces *(6)*.

The phenotypic characterization of EPC relies on immunohistochemical detection of cell-surface molecule markers *(12–14)* (*see* **Table 1**). However, many of these cell-surface markers are not unique to the EPC or MSC and are shared by HSCs and by adult endothelial-derived and mesenchyme-derived cells, rendering accurate characterization of EPC and MSC quite challenging. In humans, the expression of CD133 is believed to be an early marker of the angioblast, and the loss of this marker appears to coincide with EPC differentiation into cells with functional characteristics of adult endothelial cells *(3,4)*. Undifferentiated MSCs express CD29, CD106, and Sca-1 but do not express common hematopoietic and endothelial markers such as CD14, CD31, CD34, and CD45 (for review, *see* **ref.** *6*).

Table 1
Common Antigen Expression for Endothelial Progenitor Cell (EPC) and Mesenchymal Stem Cell (MSC)

Antigen	Alternate name(s)	EC	EPC	MSC	Other
CD10	CALLA, NEP, gp100	−	−	+	T, B, Fibro
CD13	APN, gp150	−	−	+	Mono, Gran
CD14	LPS-R	−	−	−	Mono, Mac, Gran
CD29	Integrin β1, gpIIa	−	−	+	Mono, Gran, Plt, T, B
CD31	PECAM1, Endocam	+	+	−	Mono, Plt, Gran, Lymph, B
CD34	Gp105–120, Mucosialin	+	+	−	HemSC
CD44	H-CAM, Pgp-1, EMCRIII, Ly-24	−	−	+	Leuko
CD45	LCA, T200, B220, Ly-5	−	−	−	T, B, HemP
CD54	ICAM1	+	+/−	†	Mono, Lympho
CD62e	E-Selectin, ELAM1, LECAM2	†	+/−	†	Leuko
CD90	Thy-1	+	+/−	+	Hem
CD105	Endoglin	+	+	+	Mono, Mac
CD106	VCAM1, INCAM	†	†/−	†	
CD117	c-kit, SCFR	−	+	+	HemSC
CD133	AC133[a], Prominin 1[a]	−	+/−	−	HemSC
CD141	Thrombomodulin	+	+/−	−	Mono, Plt, Neutro, SMC
CD144	VE-cadherin, cadherin 5	+	+	+	StemCell
CD146	P1H12, S-endo-1, Muc18	+	−	−	T
CD166	ALCAM	−	−	+	Mono, T, Fib
CD202b	Tie2, Tek, Angiopoietin-R	+	+	+	StemCell
CD309	VEGFR2, Flk1, KDR	+	+	+	StemCell
Sca1*		−	+	+	StemCell

−, antigen negative; +, antigen positive; +/−, subject/age/differentiation dependent; †, antigen activation dependent; B, B cells; EC, endothelial cells; EPC, endothelial progenitor cells; Fib, fibroblast; Gran, granulocyte; Hem, hematopoietic cells; HemP, hematopoeitic progenitors; HemSC, hematopoietic stem cells; Leuko, leukocytes; Lympho, lymphocytes; Mac, macrophage; Mono, monocyte; Neutro, neutrophil; Plt, platelet; T, T cells.

[a] Murine only.

2. Materials

2.1. Blood/Marrow Collection

1. Appropriate surgical/cell culture tools/facilities.
2. Povidone-iodine (PVP-I) U.S.P. prep pad.
3. Alcohol prep pads, 70%.
4. Insyte autoguard shielded I.V. catheter 20 G, 1.16″, 1.1 × 30 mm, 60 ml/min (BD Biosciences, Bedford, MA, USA).
5. Angiocath 20 G, 1.88″, 1.1 × 48 mm, 42 ml/min (BD Biosciences).
6. Double male luer-lock priming volume (PV), 0.1 ml (Medex, Dublin, OH, USA).
7. Minibore extension set (72″) with male/female luer-locks (Codan, Santa Ana, CA, USA).
8. Vacutainer sodium heparin, 6 ml (BD Biosciences).
9. Heparin sodium injection U.S.P. (1000 U/ml).
10. Evacuated container (250 ml) (Baxter, Deerfield, IL, USA).
11. 0.9% Sodium chloride injection U.S.P. (100 ml).
12. Ketamine hydrochloride injection U.S.P. (100 mg/ml) stored at 4–8 °C protected from light (Pfizer, Kirkland, Quebec, Canada).
13. Xylazine hydrochloride injection U.S.P. (20 mg/ml) stored at 4–8 °C protected from light (Bayer, Toronto, Ontario, Canada).

2.2. Isolation and Cell Culture

1. Ficoll-Paque™ Plus stored at 4–8 °C protected from light (Amersham Biosciences, Uppsala, Sweden).
2. Hanks' balanced salt solution (HBSS; Invitrogen, Burlington, Ontario, Canada).
3. BioCoat human fibronectin cellware (BD Biosciences) stored at 4–8 °C.
4. Rabbit fibronectin: aseptically prepare 1 mg/ml solution dissolved in HBSS stored stably at 4–8 °C for 1 week, otherwise at −20 °C for 3–6 months (Bioxys & Gentaur, Brussels, Belgium).
5. Mouse fibronectin: aseptically prepare 1 mg/ml solution dissolved in HBSS stored stably at 4–8 °C for 1 week, otherwise at −20 °C for 3–6 months (IMFBN Innovative Research, Southfield, MI, USA).
6. Endothelial growth media (EGM)-2-MV bullet kit (EGM-2 media; EGM Basal Media and Single Quots, Cambrex, Walkersville, MD, USA).
7. Endothelial growth supplement (EGS; Invitrogen).
8. L-Glutamine (Invitrogen).
9. Fetal bovine serum (FBS; Hyclone, Logan, UT, USA).
10. Penicillin–streptomycin (P/S; Sigma, Oakville, Ontario, Canada).
11. Alpha minimal essential medium (αMEM; Invitrogen) supplemented with 10% FBS and 1% P/S.
12. Medium 199 (growth medium) supplemented with 37.5 g EGS, 2 mM L-glutamine, 5% FBS, 1% P/S.

13. HyQTase® cell-detachment solution (Hyclone).
14. Dispase (BD Biosciences).
15. Hexadimethrine bromide (polybrene; Sigma-Aldrich, St. Louis, MO, USA).

2.3. Characterization

2.3.1. Human EPC

1. CD31-FITC (PECAM) antibody (BD Biosciences).
2. CD309-PE (VEGFR2) antibody (R&D Systems, Minneapolis, MN, USA).
3. CD144-PE (VE-cadherin) antibody (R&D Systems).
4. CD202b-APC (Tie2) antibody (R&D Systems).
5. CD34-FITC (Mucosialin) antibody (BD Biosciences).
6. CD117-PE (c-kit) antibody (R&D Systems).
7. CD133-APC (Prominin 1) antibody (Miltenyi Biotec, Auburn, CA, USA).
8. IgG1-FITC isotype (BD Biosciences).
9. IgG1-PE isotype (BD Biosciences).
10. IgG1-APC isotype (BD Biosciences).
11. IgG2-PE isotype (BD Biosciences).
12. IgG1-PE isotype (R&D Systems).
13. IgG1-APC isotype (R&D Systems).

2.3.2. Mouse EPC

1. CD31-FITC (PECAM) antibody (BD Biosciences).
2. CD309-PE (VEGFR2) antibody (BD Biosciences).
3. CD144-PE (VE-cadherin) antibody (BD Biosciences).
4. CD202b-APC (Tie2) antibody (R&D Systems).
5. CD34-FITC (Mucosialin) antibody (BD Biosciences).
6. CD117-PE (c-kit) antibody (R&D Systems).
7. CD133-APC (Prominin 1) antibody (Miltenyi Biotec).
8. IgG2a-FITC isotype (BD Biosciences).
9. IgG2a-PE isotype (BD Biosciences).
10. IgG2a-PE isotype (R&D Systems).
11. IgG1-APC isotype (R&D Systems).

2.3.3. Mouse MSCs

1. CD29-PE antibody (Biolegend, San Diego, CA, USA).
2. CD44-APC antibody (BD Biosciences).
3. CD105-FITC antibody (Biolegend).
4. CD34-PE antibody (Biolegend).
5. CD90-APC antibody (BD Biosciences).
6. CD14-FITC antibody (BD Biosciences).
7. CD45-FITC antibody (BD Biosciences).

8. IgG1-FITC isotype control (BD Biosciences).
9. IgG2b-FITC isotype control (BD Biosciences).
10. IgG2b-APC isotype control (BD Biosciences).
11. IgG2a-APC isotype control (BD Biosciences).
12. IgG2a-FITC isotype control (Biolegend).
13. IgG2a-PE isotype control (Biolegend).

2.3.4. Tube Formation

1. Growth factor-reduced Matrigel™ matrix (BD Biosciences).
2. BD Falcon four-well culture slides (BD Biosciences) or MatTek 35-mm glass bottom culture dishes (MatTek, Ashland, MA, USA).
3. PKH26 red fluorescent cell-linker kit (Sigma).

3. Methods

3.1. Human Blood Collection

1. Add 100 ml HBSS followed by 100 U/heparin to 250 ml evacuated container.
2. Assemble luer-lock extension set with a 18-G needle.
3. Wipe vigorously with PI U.S.P. prep pad to clean the articulation of the elbow. Clean respective area with alcohol pads allowing area to completely dry (*see* **Notes 1** and **2**).
4. Phlebotomize cephalic vein, place I.V. catheter, and quickly join assembled luer-lock extension set (18-G needle).
5. Tape i.v. to stabilize catheter and draw 5 ml to heparinized tube to discard (*see* **Note 3**).
6. Draw 100 ml blood to evacuated container containing heparin and 100 ml HBSS (*see* **Note 4**).
7. Remove to flow hood.

3.2. Rabbit Blood Collection

1. Fast a 4-kg to 5-kg rabbit 12–16 h prior to procedure(s).
2. Place rabbit in restriction bag or anesthesia cage (*see* **Note 5**).
3. Administer 35–50 mg/kg ketamine and 5–10 mg/kg xylazine IM (*see* **Note 6**).
4. Shave back of the right ear to identify the marginal vein and heparinize with 300 U of heparin.
5. Expose and stabilize the left groin, shave, wipe with PI U.S.P. prep pad, clean with alcohol pad, and drape area.
6. Surgical incision along the left leg and blunt dissection to femoral vessels.
7. Isolate femoral vein and dissect from surrounding connective tissue to the level of the inguinal ligament. Take care not to rupture side branches.
8. Clamp femoral vein proximally and distally with reflow hemoclamps.
9. Isolate the long femoral side branch and dissect it clean of connective tissues. Cannulate the side branch with angiocath; direct the catheter tip toward midline to proximal clamp site.

10. Place tie around angiocath and advance past clamp site while releasing clamp; quickly tie down; release distal clamp.
11. Slowly draw 1 ml blood and discard (*see* **Note 3**). Attach 60-cc syringe containing 0.5 cc heparin and slowly withdraw 50 cc or 10% blood volume (5-kg rabbit; ~ 500 ml blood).
12. Replenish blood volume with equal volume of sterile saline solution to prevent shock (*see* **Note 7**).
13. Clamp again proximally and distally. Remove angiocath and tie off side branch. Remove clamps and confirm good flow.
14. Clean wound and close skin with running 3-0 prolene suture.
15. Remove blood to flow hood and add an equal volume of HBSS.

3.3. Mouse Blood/Bone Marrow Collection for EPC and MSC

1. Place 10 ml αMEM at 37 °C in a 50-ml tube and have a small dish with sterile HBSS on ice for bone dissection.
2. Euthanize mice by CO_2 asphyxiation. Immerse in ethanol and dry animal with sterile pads.
3. Make a transverse abdominal incision to the midline. Firmly grip skin and pull rostrally and caudally to expose.
4. Remove section of abdominal muscle to expose descending vena cava and venopucture with 27-G intramuscular (IM) needle and 1-ml syringe containing 0.1 cc heparin.
5. Draw 500–800 μl blood and dispense to a 15-ml conical tube on ice. About 8–12 mice should provide 5 ml blood that is then diluted with equal volume of HBSS and kept on ice. Proceed to **Subheadings 3.4–3.5** after collection of bone marrow.
6. Remove tibia and femur bones to HBSS dish on ice.
7. Dissect muscle, tendon, and connective tissue.
8. Cut both ends of each bone shaft and flush with pre-warmed αMEM using a 27-G needle and 10-ml syringe. About 8–12 mice should provide adequate number of marrow cells for isolation.
9. Disaggregate the bone marrow with repeated passages through 27-G needle.
10. Add an additional 7 ml αMEM (37 °C) and 17 ml HBSS. Proceed to **Subheading 3.5**.

3.4. EPC Isolation and Culture

1. Coat 100-mm culture dishes with 100 μl species-specific fibronectin for 30 min in incubator.
2. Place media into dishes (1 dish/50 ml collected blood) and allow it to equilibrate for 30 min in incubator.
3. Split 35 ml blood/HBSS into 50-ml centrifuge tubes.
4. With a 25-ml serological pipette, take 15 ml Ficoll-Paque™, and with the pipette tip at the bottom of the 50-ml centrifuge, below the blood, underlay the Ficoll while slowly raising the pipette.

5. For mouse blood, underlay Ficoll with an extra long Pasteur pipette to the 5-ml gradation on the 15-ml tube.

6. Centrifuge with brake off at 400 g or 20 min at room temperature (*see* **Note 8**).

7. There will be four visible layers after centrifugation: (i) the bottom is predominately red blood cells, (ii) above is a clear Ficoll layer, (iii) above this is a slightly pink and hazy layer containing the majority of the mononuclear cells, and (iv) the uppermost is predominately phosphate-buffered saline (PBS).

8. Aspirate the uppermost layer taking care to avoid disturbing the mononuclear layer.

9. Remove the mononuclear layer and two-thirds of the Ficoll layer to new 50-ml tubes.

10. Add PBS to fill each tube and spin at 1500 g for 5 min to pellet cells. Wash pellet with PBS once.

11. Resuspend cells in EGM-2 media (5–10 ml) and count.

12. Plate 20–40 × 10⁶ cells, or the number of mononuclear cells from each 50 ml collected blood, onto one 100-mm fibronectin-coated dish.

13. Allow cells to equilibrate in the incubator for 1 h (*see* **Note 9**).

14. Transfer media to a new fibronectin-coated dish and incubate it for 48 h.

15. Change media on first dish and wash three times with PBS.

16. Transfer the media from the second dish to a 50-ml tube. Wash the second dish with EGM-2 media and collect in a 50-ml tube. Add new media to the second dish.

17. Pellet the collected cells from the second plate by centrifugation (500 g, 5 min). Resuspend pellet in 10 ml EGM-2 and plate on a third fibronectin-coated dish.

18. Change media every 48 h. Colonies should appear in either second or third dish by the 10th day.

19. When cells are approaching confluence, the cells can be passaged into normal (i.e., without fibronectin) cell culture dishes (*see* **Fig. 1**).

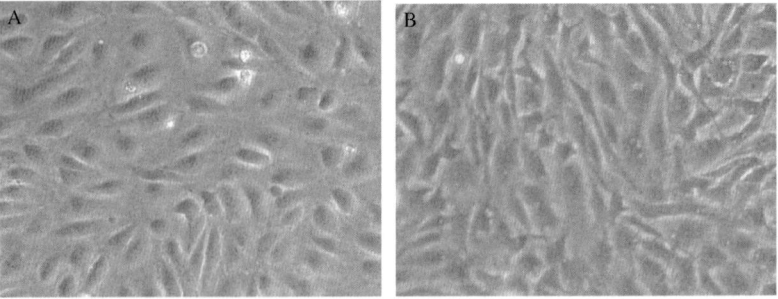

Fig. 1. Morphological features of confluent EPC and MSC. (A) EPC after 30 days in culture demonstrating distinct flat, spreadout, cobblestone morphology. (B) MSC after 20 days in culture demonstrating clustered, spindle shaped morphology.

3.5. MSC Isolation and Culture

1. With a 15-ml serological pipette, take 15 ml Ficoll-Paque™. With the pipette tip at the bottom of the 50-ml centrifuge, below the marrow, underlay the Ficoll while slowly raising the pipette.
2. Centrifuge at 2000 g (with brake off) for 25 min at room temperature (*see* **Note 8**).
3. There will be four visible layers after centrifugation: (i) the bottom is predominately red blood cells, (ii) above is a clear Ficoll layer, (iii) above this is a slightly pink hazy layer containing the majority of the mononuclear cells, and (iv) the uppermost is predominantly PBS.
4. Aspirate the uppermost layer taking care to avoid disturbing the mononuclear layer.
5. Remove the mononuclear layer and two-thirds of the Ficoll layer to new 50-ml tubes.
6. Centrifuge at 1000 g for 10 min.
7. Aspirate supernatant and wash pellet twice with HBSS. Centrifuge at 500 g for 5 min.
8. Resuspend cells in growth medium passing cells several times with a pipette to ensure homogeneous suspension.
9. Plate cells in 100-mm cell culture dishes with 10 ml growth medium.
10. Add 2 ml fresh pre-warmed media every day. Take care not to dislodge attached cells when adding fresh media.
11. Colonies of MSCs should appear 4–6 days after initial plating. When colonies appear, aspirate media and replace with growth media supplemented with additional serum to 10%. Repeat until the cells are confluent (*see* **Fig. 1**).

3.6. Characterization of Stem Cells

3.6.1. Colony Formation Assay

1. Perform mononuclear cell isolation as in **Subheading 3.4** or **3.5**.
2. With a 3% acetic acid/PBS solution, lyse red blood cells in a small aliquot of cell suspension.
3. Count mononuclear cells.
4. Plate 5×10^6 mononuclear cells in an uncoated six-well dish.
5. After 2 days, remove floating cells and replate at 1×10^6 cells/well. Use a fibronectin-coated dish with 24 wells at $2 \, cm^2$ per well.
6. Between days 3 and 15, count the number of colonies and outgrowth EPC colonies (*see* **Note 10**).

3.6.2. Tube Formation Assay

1. Detach adherent cells and count.
2. Pellet 1×10^6 cells by centrifugation at 400 g for 5 min.
3. Gently resuspend cell pellet in 1 ml Diluent C (kit).
4. In a separate tube, prepare PKH26 dye by adding 3 μl dye to 1 ml Diluent C. Mix well and protect from light.
5. Combine Diluent C–dye mix and cell suspension. Gently mix by pipetting. Do not vortex cells.

6. Incubate at room temperature for 5 min. Invert periodically to maintain cells in suspension.
7. Stop reaction by adding 2 ml FBS. Incubate for 1 min.
8. Add 4 ml of complete growth medium and pellet cells.
9. Aspirate the medium and wash cells three times with 10 ml of complete medium. Resuspend cells each time by gentle pipetting.
10. Resuspend cells in 1 ml medium. Cells may be examined by fluorescent microscopy or flow cytometry to determine level of labeling.
11. Plate cells and allow them to recover for 24 h in complete growth media.
12. Thaw Matrigel (*see* **Note 11**).
13. Add 50 μl/cm^2 Matrigel for thin layer or 150 μl/cm^2 for thick layer. Take care to avoid air bubbles.
14. Place plates in 37 °C incubator with 5% CO_2 for 30 min.
15. Remove culture dishes and cover gel with appropriate cell culture medium. Return culture dishes to incubator.
16. Wash once with HBSS to remove floating cells. Detach adherent cells and count.
17. Add 2×10^3 stained cells to the top of the gel in 200–300 μl media to cover the Matrigel. Allow cells to settle for 3–5 min in cell culture incubator (*see* **Note 12**).
18. With extreme care, bring the volume of media in each well to 1.5 ml by pipetting slowly down the side of well.
19. To recover cells for continuous culture or fluorescence-activated cell sorting (FACS) analysis, aspirate media and wash with HBSS.
20. Add 500 μl dispase to each dish/well. Place in cell culture incubator for 5–15 min or until cells are freely floating.
21. Pellet cells and wash once with complete medium.
22. Pellet cells and resuspend in appropriate complete medium for plating or 5% FBS–FACS medium.
23. Document the formation of vascular structures using differential interference contrast (DIC) and epifluorescent microscopy (*see* **Fig. 2**).

3.6.3. Immunophenotyping

1. Detach cells, count, and resuspend $1–5 \times 10^6$ cells per ml in PBS with 5% FBS (*see* **Note 13**).
2. Wash cells with serum PBS once.
3. Pellet cells and gently resuspend in 90 μl serum PBS/5×10^5 cells.
4. Add 10 μl antibody/isotype control/5×10^5 cells.
5. Shake on ice protected from light for 1 h.
6. Wash cells two to three times with serum PBS.
7. Resuspend in 500 μl serum PBS/5×10^5 cells.
8. Use forward-scatter and side-scatter analysis with flow cytometer for live gate without debris.

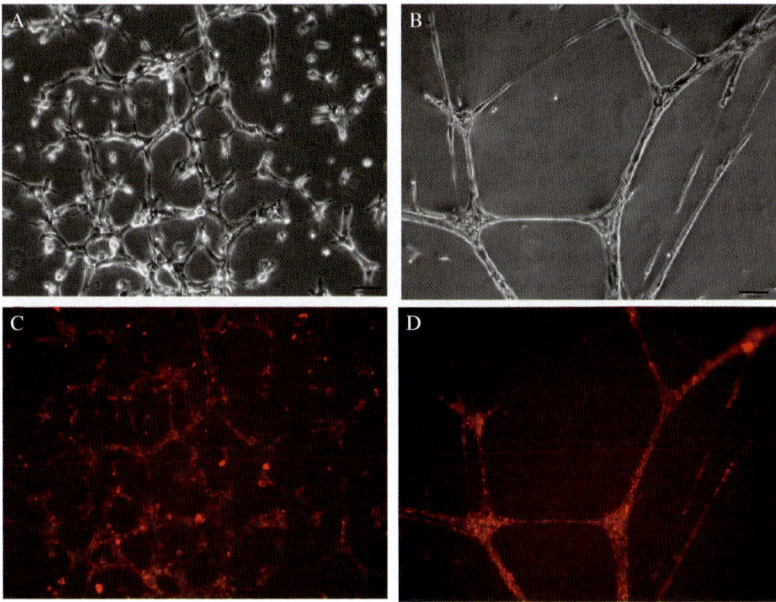

Fig. 2. DIC contract images of EPC plated on matrigel. (A) EPC show initial capillary tube formation after 4 hours in matrigel. (C) Same field as in (A) demonstrating PKH26 labelling of EPC and early colocalization of multiple EPC (bright spots). (B) EPC show profound capillary formation matrigel after 24 hours in matrigel. (D) Same field as in (B) demonstraint PKH26 labelling of EPC and nodal clustering of EPC with distinct internodal tube formation.

9. Use isotype controls to calibrate photomultiplier tubes, compensations, and analysis gates (*see* **Note 13**).
10. Early populations of EPC (less than 2 weeks from isolation) will be CD34/ CD117/CD133 positive. Longer cultures will be CD31/CD144/CD202b/CD309. All cultures should be negative for CD19/CD45 (*see* **Table 1**).
11. MSC populations will be positive for CD29/CD44/CD105/CD90 and negative for CD14/CD34/CD45 (*see* **Table 1**).

3.7. Viral Transduction

3.7.1. Viral Transduction of EPC

1. Seed EPC in a dish of desired size at 30–40% confluence (usually $20,000$ cells/cm^2 in EGM-2 medium).
2. Incubate with $4\,\mu g/ml$ 3,4-hexadimethrine bromine (polybrene).
3. Add 1–20 multiplicities of infection (MOI) or viral particles/cell.

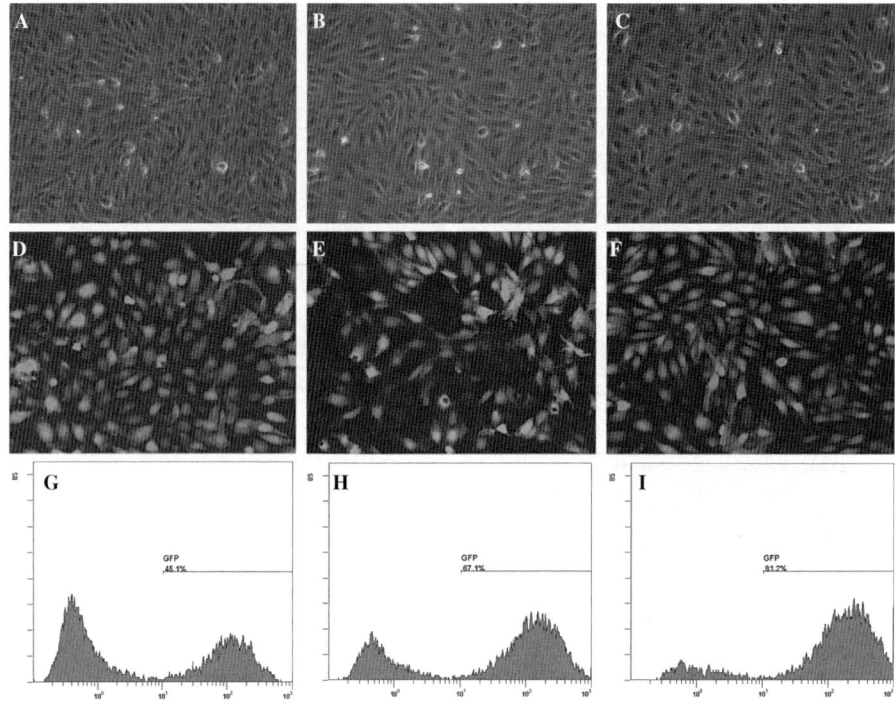

Fig. 3. Confluent EPC tranduced with MSCV-IGFP. (A) Phase contract of EPC trans-
duced with 5 MOI MSCV-IGFP (B), 20 MOI MSCV-IGFP (C). (D,E,F) same field under
green fluorescene respectively to (A,B,C). (G,H,I) Flow cytometry histograms of green.

4. Incubate cells for 24 h in full media. Examine cells by fluorescent microscopy and
   flow cytometry (*see* **Fig. 3**).

3.7.2. Viral Transduction of MSC

1. Seed MSCs in a dish of desired size at 10–20% confluence (usually 5000 cells/cm^2
   in growth medium).
2. Incubate with 8 μg/ml 3,4-hexadimethrine bromine (polybrene).
3. Add 1–20 MOI.
4. Incubate cells for 24 h in growth media.
5. Ninety-six hours after transduction, examine by fluorescent microscopy and flow
   cytometry to determine efficiency (*see* **Fig. 3**).

4. Notes

1. All procedures should be conducted in conformance with institutional requirements
   for use of human research volunteers and in accordance with animal care policies.

2. Unless otherwise stated, all procedures should be conducted with the utmost adherence to aseptic technique.
3. Utilizing a soft stabilized catheter and bleeding a small amount of blood reduces the incidence of contaminating mature endothelial cells. Take care during bleeding not to move the catheter.
4. Avoid hemolysis, as this can have adverse effects on isolation.
5. Take care to ensure rabbit is secured snuggly. Failure to do so may result in fracture of the rabbit's spine.
6. Observe lack of withdrawal to ear pinch and slowed respiratory rate. If procedures exceed 45 min, use a supplemental 1/3 dose. Xylazine can induce profound bradycardia and hypothermia. To avoid complications, we recommend a warm heating pad. Atropine can be used to prevent bradycardia and fluid collection in lungs; however, rabbits often demonstrate reduced effectiveness of atropine because of circulating atropinase. Yohimbine can reverse effects of xylazine on alpha2 adrenergic receptors and will also assist in recuperation.
7. Xylazine carries analgesic effects. Buprenorphine (0.01–0.05 mg/kg s.c.) should be administered upon first signs of recuperation to avoid analgesic overdose.
8. It is very important to allow the centrifuge to accelerate/decelerate slowly to avoid disrupting proper separation.
9. The first hour of plating will collect monocytes/macrophages and any endothelial cells that were sloughed off from the vessel wall during blood collection. These cells should be monitored until colonies form. Monocytes/macrophages will wash away after 2–3 days.
10. Colonies appear as a central core of round cells surrounded by more elongated cells at the periphery. Outgrowth colonies have similar characteristics but are smaller and appear near or adjacent to larger early forming colonies. The number of colonies is an index of EPC functional capacity.
11. Matrigel solidifies at room temperature. To ensure uniform consistency, liquefy Matrigel on ice at 4 °C. Pre-incubate all cell culture dishes and tips at −20 °C for 1 h. Keep all culture dishes and Matrigel on ice during coating procedure.
12. Take extreme care not to disturb or agitate the culture plates when plating on Matrigel, as this can adversely affect the growth characteristics.
13. We recommend a minimum of 5×10^5 cells per antibody or isotype control and a minimum of 50,000 events for characterization. It is best to first run cells with isotype control, then a second set of cells with only one antigen alone. When doing multiple antigen labeling, these controls can be used to verify there is no spectral overlap in multiparameter measurements.

Acknowledgments

This work is supported by the Canadian Institute of Health Research (CIHR) and Heart and Stroke Foundation of Ontario Grants awarded to Luis G. Melo and Christopher A. Ward. Sean R. R. Hall is supported by a Merck Frosst

Postdoctoral Fellowship. Keith R. Brunt is supported by a CIHR doctoral award from the Gasotransmitter Research and Training Program (G.R.E.A.T.).

References

1. Asahara T., Murohara T., Sullivan A., Silver M., van der Zee R., Li T., Witzenbichler B., Schatteman G., and Isner J.M. (1997) Isolation of putative progenitor endothelial cells for angiogenesis. *Science* **275**, 964–967.
2. Rafii S., and Lyden D. (2003) Therapeutic stem cell and progenitor cell transplantation for organ vascularization and regeneration. *Nat. Med.* **9**, 702–712.
3. Dimmeler S., Zeiher A.M., and Schneider M.D. (2005) Unchain my heart: the scientific foundations of cardiac repair. *J. Clin. Invest.* **115**, 572–883.
4. Dzau V.J., Gnecchi M., Pachori A.S., Morello F., and Melo L.G. (2005) Therapeutic potential of endothelial progenitor cells in cardiovascular diseases *Hypertension* **46**, 7–18.
5. Griese D.P., Ehsan A., Melo L.G., Kong D., Zhang L., Mann M.J., Pratt R.E., Mulligan R.C., and Dzau V.J. (2003) Isolation and transplantation of autologous circulating endothelial cells into denuded vessels and prosthetic grafts: implications for cell-based vascular therapy. *Circulation* **108**, 2710–2715.
6. Pittenger M.F., and Martin B.J. (2004). Mesenchymal stem cells and their potential as cardiac therapeutics. *Circ. Res.* **95**, 9–20.
7. Urbich C., and Dimmeler S. (2004) Endothelial progenitor cells. Characterization and role in vascular biology. *Circ. Res.* **95**, 343–353.
8. Khakoo A.Y., and Finkel T. (2005) Endothelial progenitor cells. *Annu. Rev. Med.* **56**, 79–101.
9. Vasa M., Fichtlscherer S., Aicher A., Adler K., Urbich C., Martin H., Zeiher A.M., and Dimmeler S. (2001) Number and migratory activity of circulating endothelial progenitor cells inversely correlates with risk factors and coronary artery disease. *Circ. Res.* **89**, e1–e7.
10. Abedin M., Tintut Y., and Demer L.L. (2004) Mesenchymal stem cells and the artery wall. *Circ. Res.* **95**, 671–676.
11. Roufosse C.A., Direkze N.C., Otto W.R., and Wright N.A. (2004). Circulating mesenchymal stem cells. Intern. *J. Biochem. Cell. Biol.* **36**, 585–594.
12. Bellik L., Ledda F., and Parenti A. (2005) Morphological and phenotypical characterization of human endothelial progenitor cells in an early stage of differentiation. *FEBS Lett* **579**, 2731–2736.
13. Khan S.S., Solomon M.A., and McCoy J.P. (2004) Detection of circulating endothelial cells and endothelial progenitor cells by flow cytometry. *Cytometry* **64**, 1–8.
14. Kong D., Melo L.G., Gnecchi M., Zhang L., Mostoslavsky G., Liew C.C., Pratt R.E., and Dzau V.J. (2004) Cytokine-induced mobilization of circulating endothelial progenitor cells enhances repair of injured arteries. *Circulation* **110**, 2039–2046.

13

Assessment of Endothelial Damage/Dysfunction: A Focus on Circulating Endothelial Cells

Christopher J. Boos, Andrew D. Blann, and Gregory Y. H. Lip

Summary

Endothelial injury represents a major initiating step in the pathogenesis of vascular disease and atherosclerosis. The identification and quantification of circulating endothelial cells (CECs) has evolved as a novel marker of endothelial function. As a technique, it correlates with other markers of endothelial function such as flow-mediated dilation, the measurement of von Willebrand factor, and tissue plasminogen activator. Quantification of CECs is difficult due to low numbers, variable morphology, and a lack of standardization in current techniques used. CECs appear to be a different population of cells to endothelial progenitor cells. Increased CECs have been noted in a number of disease states and is evolving as a novel method of assessment of both disease severity and response to treatment.

Key Words: Endothelial dysfunction; Circulating endothelial cells.

1. Introduction

The endothelium is the largest organ in the body, occupying a surface exceeding $1000 \, m^2$. Although originally thought of simply as an inert lining to blood vessels, it is now clear that it is a dynamic and metabolically active organ, which has a crucial role in several physiological and pathological processes *(1)*. The functions of the endothelium include inflammation control, the regulation of the coagulation and local blood flow, vascular homeostasis controlling the passage of fluid into the media, as well as the regulation of cell migration through the vascular intima *(2)*. Endothelial injury is a major initiating step in the pathogenesis of vascular disease and atherosclerosis *(3,4)*.

From: *Methods in Molecular Medicine, Vol. 139: Vascular Biology Protocols*
Edited by: N. Sreejayan and J. Ren © Humana Press Inc., Totowa, NJ

Current methods used to assess endothelial function include physiological assessment by flow-mediated dilatation (FMD), the use of laser Doppler flowmetry, and the measurement of specific plasma indices of endothelial damage/dysfunction, such as von Willebrand factor (vWF), soluble thrombomodulin, soluble endothelial cell protein C receptor, soluble E selectin, endothelin-1, soluble thrombomodulin, and tissue plasminogen activator (tPA) *(5–8)*. However, a recent development has been the measurement of immunologically defined circulating endothelial cells (CECs) in the peripheral blood, a method rapidly gaining ground as an important and novel technique for assessment of endothelial injury. These CECs are felt to represent those which have detached (or have been driven from) from the mural endothelium in response to particular injurious process(es). In this chapter, we present a brief overview of CECs, their origins, methods of identification, and potential use in clinical practice.

1.1. Historical Perspective

The founders of the concept of CECs were Bouvier and colleagues, who, in 1970, were the first to report the discovery of non-hemopoietic cells of possible endothelial origin in the blood of rabbits after endotoxin injection *(9)*. Since then, several authors described similar circulating cells, associated with denuded vascular areas on biopsies in different experimental situations *(10–12)*. However, the methods of CEC identification by vital light microscopy, using morphology, Giemsa staining, and separation by Ficoll density centrifugation, failed to unequivocally identify CECs from blood samples, highlighting the need for a reliable endothelial-specific cell surface molecule that could be used as an appropriate tool.

In 1991, two groups independently developed two endothelial cell monoclonal antibodies (mAbs) (HEC 19 and S-Endo 1) recognizing an antigen present on all types of endothelial cells and used these antibodies to quantify blood-borne CECs *(13–14)*. Dignat-George and colleagues subsequently characterized their antibody as recognizing the CD146 molecule and used it to demonstrate CECs in a variety of conditions *(15–17)*.

1.2. The Preparation and Quantification of CECs

The precise quantification of CEC is difficult. This is explained in part by the low numbers present in the circulation and because of their differing morphological appearances. However, their detection has been improved by cell-enrichment techniques with labeling of the cells with specific markers *(18,19)*. Current techniques of cell enrichment include the elimination of red blood cells and concentration by centrifugation, the use of an Fc-blocking agent to prevent non-specific leukocyte binding, the preparation of a mononuclear cell suspension by density centrifugation (Ficoll, Percoll, Lymphoprep, Histopaque, etc) and capture by immunomagnetic beads bearing CD146 *(20)*. A variety of mAbs to cell-surface glycoproteins on endothelial cells (such as indirect immunofluoresence, immunocytochemistry, and flow cytometry) are then used to identify the CECs *(20)*.

CD146 has clearly evolved as the most popular marker for the identification of CECs. However, this molecule has also been described on trophoblast, mesenchymal stem cells, periodontal tissue, and malignant (prostatic cancer and melanoma) tissues so that some caution is demanded, especially in cancer *(21–23)*.

There is a marked variability in the detected CEC counts reported in disease states, with the number of CECs ranging from an average of 1 to 39,000 cells/ml of whole blood *(20)*. This variance is partly related to the diversity of disease processes and to differences in the methods of quantification and preparation. However, the wide variation of counted CECs among supposedly healthy controls does raise serious questions regarding their identification and quantification as number have ranged from 0 to 7900 cells/ml. The immunobead and density centrifugation methods appear to be the most specific method of CEC identification (with values in the order of 10 cells/ml), whereas flow cytometry seems more sensitive, often reporting greater numbers, up to 1000-fold higher than the immunobead method *(20)*.

Sections 2&3 show our method for measuring CECs.

1.3. Differentiation of CECs From Endothelial Progenitor Cells

Accumulating evidence has indicated that the peripheral blood of adults contains bone marrow-derived large non-leukocyte progenitor cells with properties similar to embryonal angioblasts *(24)*. These precursor cells have the capacity to differentiate into mature endothelial cells and line the endothelial monolayer, and therefore, they have been termed endothelial progenitor cells (EPCs). These cells appear to be different to CECs, which are mature cells that appear to detach from the vessel wall in response to a damaging stimulus. Differentiation of these two cell populations has been difficult due to a lack of clear consensus regarding their characterization *(25)*.

EPCs represent a subset of cells at varying stages of development commencing with the early progenitor cell arriving in the peripheral blood stream bearing immaturity surface markers, such as CD34 and CD133 not present on CECs (*see* **Table 1**) *(24,25)*. Later, after migrating into the

Table 1
Differences Between CECs and EPCs

	CECs	EPCs
Origin	Blood vessel wall	Bone marrow
Morphology	Mature cells of diameter 20–50 μM	Immature cells of diameter less than 20 μM
Phenotype	CD133−ve/CD146+ve	CD133+ve/CD146−ve
Capacity to form colonies with high proliferative potential	No	Yes
Pathophysiology	Reflective of damage	Neovascularization

CECs, circulating endothelial cells; EPCs, endothelial progenitor cells.
Reproduced with permission and adapted from *(20,25)*.

circulation, EPCs gradually lose their progenitor properties and start to express endothelial markers such as vWF *(26)*.

The general thrust of the literature seems to support the concept that bone marrow-derived EPCs are viable cells that have the ability to promote endothelialization and form colonies in vitro *(20,27,28)*. In contrast, it is uncertain as to whether CECs are truly viable, with marked conjecture in the literature *(20)*. CECs do not appear to form colonies often appearing in the circulation as irregular carcasses, as clumped cells, or as filamentous cell remnants.

Current evidence would suggest that CD-146-dependent isolation does not capture EPCs to any relevant extent. Hence, most groups working on CECs first capture cells with CD146 immunobeads, then perform additional phenotyping with various stains such as CD31, vWF immunocytochemistry, and Ulex europaeus lectin-1. Others prepare mononuclear cells by density centrifugation, and then stain for CD146.

1.4. The Origins of CECs

It would seem logical to hypothesize that CECs detected in blood are the result of disease process leading to endothelial damage. This theory is supported by the finding of increased CECs in a variety of disease processes that are known to cause endothelial disruption. The mechanisms responsible for the structural integrity of the endothelium are not precisely known but include many factors such as interendothelial receptors, cytoskeletal components, pro-angiogenic and anti-angiogenic growth factors, and endothelial adhesive molecules (such as vitronectin and fibronectin) *(15,29,30)*.

Although not fully understood, it would appear that CEC detachment from the endothelium involves many factors such as mechanical injury, alteration of endothelial cellular adhesion molecules (such as Integrin $\alpha V \beta 3$), defective binding to anchoring matrix proteins (such as fibronectin, laminin, or type IV collagen), and cellular apoptosis with decreased survival of cytoskeletal proteins *(29,31)*. The net effect of these processes is the reduced interaction between the endothelial cell with basement membrane proteins, weakening their tethering to the substratum, and subsequent cellular detachment from the basement membrane.

1.5. The Relationship Between CECs and Other Markers of Endothelial Function

Impaired FMD and increased plasma markers of endothelial injury/activation (e.g., vWf, tPA, and to a lesser extent soluble E selectin) have acted as indirect surrogate markers of endothelial integrity and have been shown to predict adverse cardiovascular outcomes *(32–36)*. There has been evidence to support the correlation of CECs with other markers of endothelial dysfunction. For example, an inverse correlation between CECs and FMD has been demonstrated *(37)*. There appears to be a strong correlation between CECs and vWf and tPA *(31,37–39)*. Studies looking into a possible correlation between CECs and E-selectin have been conflicting, with one study supporting the correlation *(31)*, whereas another refuting it *(38)*.

Recently, it has been shown in vitro that endothelial cells release endothelial microparticles (EMPs) on activation or apoptosis and that an assay of EMPs can provide useful information on EC status in patients with thrombotic disorders *(40,41)*. It would thus seem logical that there might be a relationship between microparticles and CECs; however, as yet, this has not been studied.

1.6. The Relationship Between CECs and Disease States

CECs are rarely found in normal healthy individuals, that is, in order of < 3 cells/ml. Elevated numbers of CEC have been identified in a wide array of disease processes (*see* **Table 2**) such as chronic venous insufficiency *(42)*, aortitis *(43)*, pulmonary hypertension *(44)*, post-coronary angioplasty *(45)*, Behcets disease *(46)*, diabetes mellitus *(19)*, septic shock *(47)*, thalassaemia *(48)*, acute sickle cell crises *(49)*, and thrombotic thrombocytopenia *(50)*.

Table 2
Reports of CECs in Case/Control Studies of Human Disease and Particular Methodology

Condition (reference)	Method	Number of CECs/ml[a]	
		Cases	Controls
Acute coronary syndromes *(57,58)*	IB	7.5	
Aortoarteritis *(43)*	SC	58	16 *(4)*
Behcet's disease *(46)*	IB	0–25	<3
Bone marrow transplantation *(59,60)*	IB	16–44	8
Breast cancer and lymphoma *(63)*	FC	6800–39,100	1200–7900
Various cancers *(64)*	IB	399	121
Coronary angioplasty *(45)*	IB	6–10	<3
Chronic venous insufficiency *(42)*	SC	1001	514
Diabetes *(19)*	IB	69	10
Inflammatory vasculitis *(18,52)*	IB	136	5
Kawasaki disease *(53)*	IB	15	6
Peripheral vascular disease *(39)*	IB	1.1–3.5	0.9
Pulmonary hypertension *(44)*	DC+IB	30	3.5
Renal transplantation *(59)*	IB	24–72	6
Rickettsial infection *(51)*	IB	5–1600	<3
Septic shock *(47)*	DC	16.1	1.9
Sickle cell anemia *(52)*	IB	13.2–22.8	2.6
Systemic lupus erythematosus *(57)*	DC/FC	32–89	5–10
Systemic sclerosis *(49)*	FC	243–375	77
Thalassaemia *(48)*	DC	45	4
Thrombotic thrombocytopenia *(50)*	IB	6–220	<3

IB, immunobeads; DC, density centrifugation; FC, flow cytometry; SC, standard centrifugation.

[a] As the distribution of CECs is often non-normal, median number is appropriate although some report the mean.

Reproduced with permission and adapted from *(20)*.

Elevated CECs have also been demonstrated to correlate with disease severity in Mediterranean spotted fever *(51)*, inflammatory vasculitis *(18,52)*, Kawasaki's disease *(53)*, systemic lupus erythmatosis *(54)*, systemic sclerosis *(55)*, peripheral vascular disease *(39)*, and acute coronary artery disease *(56,57)*. Our group were the first to demonstrate that CECs are an independent predictor of both death and major adverse cardiovascular events at 30 days and 1 year among patients presenting with acute coronary syndromes *(56)*.

In transplantation medicine, CECs have been identified as a useful marker of endothelial damage and potential vascular rejection *(58)*. Popa et al. *(59)* were able to demonstrate that the presence of donor CECs in recipient blood related to post-transplantation injury. And, more recently, Woywodt et al. *(60)* demonstrated that in allogenic hematopoietic stem cell transplantation, those patients who received reduced-intensity conditioning had significantly lower CEC numbers.

The endothelium plays an important role in the spread and propagation of cancer cells. It is actively involved in angiogenesis (new vessel formation) and metastasis *(61)*. Increased CECs have been identified in both breast cancer and lymphoma, and their numbers have been shown to significantly correlate with plasma vascular cell adhesion molecule-1 (VCAM)-1 and vascular endothelial growth factor (VEGF) levels *(62,63)*.

In various clinical conditions, the longitudinal quantification of CEC has shown that their levels vary according to the clinical position. Levels among patients who are acutely ill are higher than those in patients who are in clinical remission or in a recovery phase of the disease. From a pharmacological point of view, several studies have also shown that CECs may be useful in monitoring therapeutic efficacy. For example, Woywodt et al. *(52)* were able to demonstrate a fall in CECs during 6 months of successful treatment of anti-nuclear cytoplasmic antibody (ANCA)-associated small-vessel vasculitis, whereas George et al. *(64)* demonstrated a fall in CECs with treatment of *Rickettsial conorri* infection in familial Mediterranean fever. In another study involving the treatment of disseminated human cytomegalovirus (CMV) infection, there was a dramatic fall in the CEC count from blood within just a few days of treatment with ganciclovir and foscarnet *(65)*. In cancer, numbers of CECs are reported to be higher in those patients with progressive disease compared to those with stable disease and healthy controls *(63)*. Interestingly, these CECs appear not to express tissue factor *(66)*.

2. Estimating CECs by Immunobead Capture

2.1. Materials

1. Stock solution of immunomagnetic beads (diameter $4.5 \mu M$) coated with anti-CD146 antibody: Mix $100 \mu l$ anti-CD146 mAb (Biocytex, Marseille, France) with $125 \mu l$ Dynabead at 4×10^8 beads/ml (themselves coated with an anti-pan mouse IgG) (Dynal Biotech ASA, Oslo, Norway) at room temperature for 30 min. Unbound CD146 is washed off with three washes with phosphate-buffered saline (PBS), and the stock-coated beads are stored at $4°C$.

2. PBS.
3. Acridine orange (5 μg/ml PBS; Sigma, Poole, UK) or Ulex europaeus lectin-1.
4. Fluorescence microscope and hemocytometer.

2.2. Method

1. The subject is venesected from an ante-cubital vein: the first 4 millilitre (e.g., a vacutainer) is discarded as it may contain mural ECs removed by trauma.
2. Fifteen microliters from a preparation of 7×10^7 immunobeads/μl is mixed with venous blood diluted 1 ml with 3 ml in PBS in a head-over-head mixer for 30 min at 4 °C.
3. Excess bound and unbound cells are washed out with PBS four times inside a magnet at 4 °C.
4. Rosetted cells are recovered and divided into fractions allowing staining with acridine orange (5 μg/ml PBS; Sigma) or Ulex europaeus lectin-1 for counting with fluorescence microscopy and hemocytometer and for subsequent cell characterization.
5. CECs are defined (i) morphologically as CD146 rosetted cells, bearing more than 10 beads (diameter 20–50 μm) or bearing less than 10 beads but with a clear nucleus in a well-delineated cytoplasm and (ii) immunologically by the expression of endothelial markers and the lack of expression of leukocyte antigens. Sheets or clumps of cells can often be found. For these aggregates, the number of cells is deduced from the number of nuclei or from the number of spherical rosetted features discriminated in the aggregate.

3. Conclusions

It is clear that increased numbers of CECs are present in the blood of patients with a wide variety of clinical diseases that lead to vascular injury. CEC counts have correlated well with many traditional methods of endothelial injury detection. Current methods of CEC quantification are variable and difficult, and blood samples have not been cross-referenced by other laboratories utilizing either similar or different techniques of quantification. CECs appear to be a valuable tool for assessment of both disease severity and treatment efficacy in a number of clinical conditions.

References

1. Cines, D.B., Pollak, E.S., Buck, C.A., Loscalzo, J., Zimmerman, G.A., McEver, R.P., Pober, J.S., Wick, T.M., Konkle, B.A., Schwartz, B.S., Barnathan, E.S., McCrae, K.R., Hug, B.A., Schmidt, A.M., and Stern, D.M. (1998) Endothelial cells in physiology and in the pathophysiology of vascular disorders. *Blood* **91**, 3527–3561.

2. Galley, H.F., and Webster, N.R. (2004) Physiology of the endothelium. *Br. J. Anaesth.* **93**, 105–113.

3. Gonzalez, M.A., and Selwyn, A.P. (2003) Endothelial function, inflammation, and prognosis in cardiovascular disease. *Am. J. Med.* **115**, 99S–106S

4. Landmesser, U., Hornig, B., and Drexler, H. (2004) Endothelial function: a critical determinant in atherosclerosis? *Circulation* **109**, II27–II33

5. Hinderliter, A.L., and Caughey, M. (2003) Assessing endothelial function as a risk factor for cardiovascular disease. *Curr. Atheroscler. Rep.* **5**, 506–513.

6. Faulx, M.D., Wright, A.T., and Hoit, B.D. (2003) Detection of endothelial dysfunction with brachial artery ultrasound scanning. *Am. Heart. J.* **145**, 943–951.

7. Morris, S.J., and Shore, A.C. (1996) Skin blood flow responses to the iontophoresis of acetylcholine and sodium nitroprusside in man: possible mechanisms. *J. Physiol.* **496**, 531–542.

8. Poredos, P. (2003) Endothelial dysfunction in the pathogenesis of atherosclerosis. *Int. Angiol.* **21**,109–116.

9. Bouvier, C.A., Gaynorm E., Cintron, J.R., Bernhardt, B., and Spaet, T.H. (1970) Circulating endothelium as an indicator of vascular injury. *Thromb. Diath. Haemorrh.* **40**, 163–168.

10. Hladovec, J., and Rossman, P. (1973) Circulating endothelial cells isolated together with platelets and the experimental modification of their counts in rats. *Thromb. Res.* **3**, 665–674.

11. Wright, H.P., and Giacometti, N.J. (1972) Circulating endothelial cells and arterial endothelial mitosis in anaphylactic shock. *Br. J. Exp. Pathol.* **53**, 1–4.

12. Gaynor, E., Bouvier, C.A., and Spaet, T.H. (1968) Circulating endothelial cells in endotoxin treated rabbits. *Clin. Res.* **16**, 535 (abstract).

13. Sbarbati, R., de Boer, M., Marzilli, M., Scarlattini, M., Rossi, G., and van Mourik, J. (1991) Immunologic detection of endothelial cells in human whole blood. *Blood* **77**, 764–769.

14. George, F., Brisson, C., Poncelet, P., Laurent, J.C., Massot, O., Arnoux, D., Ambrosi, P., Klein-Soyer, C., Cazenave, J.P., and Sampol, J. (1992) Rapid isolation of human endothelial cells from whole blood using S-Endo1 monoclonal antibody coupled to immuno-magnetic beads: demonstration of endothelial injury after angioplasty. *Thromb. Haemost.* **67**, 147–151.

15. Dignat-George, F., and Sampol, J. (2000) Circulating endothelial cells in vascular disorders: new insights into an old concept. *Eur. J. Haematol.* **65**, 215–220.

16. Bardin, N., George, F., Mutin, M., Brisson, C., Horschowski, N., Frances, V., Lesaule, G., and Sampol, J. (1996) S-Endo 1, a pan-endothelial monoclonal antibody recognizing a novel human endothelial antigen. *Tissue Antigens* **48**, 531–539.

17. Bardin, N., Frances, V., Lesaule, G., Horschowski, N., George, F., and Sampol, J. (1996) Identification of the S-Endo 1 endothelial-associated antigen. *Biochem. Biophys. Res. Comm.* **218**, 210–216.

18. Woywodt, A., Goldberg, C., Scheer, J., Regelsberger, H., Haller, H., and Haubitz, M. (2004) An improved assay for enumeration of circulating endothelial cells. *Ann. Hematol.* **83**, 491–494.

19. McClung, J.A., Naseer, N., Saleem, M., Rossi, G.P., Weiss, M.B., Abraham, N.G., and Kappas, A. Circulating endothelial cells are elevated in patients with type 2 diabetes mellitus independently of HbA(1)c. *Diabetologia* (2005) **48**, 345–350.

20. Blann, A.D., Woywodt, A., Bertolini, F., Bull, T.M., Buyon, J.P., Clancy, R.M., Haubitz, M., Hebbel, R.P., Lip, G.Y., Mancuso, P., Sampol, J., Solovey, A., and Dignat-George, F. (2005) Circulating endothelial cells. Biomarker of vascular disease. *Thromb. Haemost.* **93**, 228–235.

21. Liu, Q., Yan, X., Li, Y., Zhang, Y., Zhao, X., and Shen, Y. (2004) Pre-eclampsia is associated with the failure of melanoma cell adhesion molecule (MCAM/CD146) expression by intermediate trophoblast. *Lab. Invest.* **84**, 221–228.

22. Wu, G.J., Peng, Q., Fu, P., Wang, S.W., Chiang, C.F., Dillehay, D.L., and Wu, M.W. (2004) Ectopical expression of human MUC18 increases metastasis of human prostate cancer cells. *Gene* **327**, 201–213.

23. Seo, B.M., Miura, M., Gronthos, S., Bartold, P.M., Batouli, S., Brahim, J., Young, M., Robey, P.G., Wang, C.Y., and Shi, S.(2004) Investigation of multipotent postnatal stem cells from human periodontal ligament. *Lancet* **364**, 149–155.

24. Hristov, M., and Weber, C. (2004) Endothelial progenitor cells: characterization, pathophysiology, and possible clinical relevance. *J. Cell. Mol. Med.* **8**, 498–508.

25. Woywodt A, Blann AD, Kirsch T, Erdbruegger U, Banzet N, Haubitz M, Dignat-George F. Isolation and enumeration of circulating endothelial cells by immuno-magnetic isolation: proposal of a definition and a consensus protocol. *J Thromb Haemost.* 2006;4:671–7.

26. Bompais, H., Chagraoui, J., Canron, X., Crisan, M., Liu, X.H., Anjo, A., Tolla-Le Port, C., Leboeuf, M., Charbord, P., Bikfalvi, A., and Uzan, G. (2004) Human endothelial cells derived from circulating progenitors display specific functional properties compared with mature vessel wall endothelial cells. *Blood* **103**, 2577–2584.

27. Gehling, U.M., Ergun, S., Schumacher, U., Wagener, C., Pantel, K., Otte, M., Schuch, G., Schafhausen, P., Mende, T., Kilic, N., Kluge, K., Schafer, B., Hossfeld, D.K., and Fiedler, W. (2000) In vitro differentiation of endothelial cells from AC133-positive progenitor cells. *Blood* **95**, 3106–3112.

28. Ingram, D.A., Mead, L.E., Tanaka, H., Meade, V., Fenoglio, A., Mortell, K., Pollok, K., Ferkowicz, M.J., Gilley, D., and Yoder, M.C. (2004) Identification of a novel hierarchy of endothelial progenitor cells using human peripheral and umbilical cord blood. *Blood* **104**,2752–2756

29. Chavakis, E., and Dimmeler, S. (2002) Regulation of endothelial cell survival and apoptosis during angiogenesis. *Arterioscler. Thromb. Vasc. Biol.* **22**, 887–893.

30. Ruegg, C., Yilmaz, A., Bieler, G., Bamat, J., Chaubert, P., and Lejeune, F.J. (1998) Evidence for the involvement of endothelial cell integrin alphaVbeta3 in

the disruption of the tumor vasculature induced by TNF and IFN-gamma. *Nat. Med.* **4**, 408–414.

31. Rajagopalan, S., Somers, E.C., Brook, R.D., Kehrer, C., Pfenninger, D., Lewis, E., Chakrabarti, A., Richardson, B.C., Shelden, E., McCune, W.J., and Kaplan, M.J. (2004) Endothelial cell apoptosis in systemic lupus erythematosus: a common pathway for abnormal vascular function and thrombosis propensity. *Blood* **103**, 3677–3683.

32. Thompson, S.G., Kienast, J., Pyke, S.D., Haverkate, F., and van de Loo, J.C. (1995) Hemostatic factors and the risk of myocardial infarction or sudden death in patients with angina pectoris. European Concerted Action on Thrombosis and Disabilities Angina Pectoris Study Group. *N. Engl. J. Med.* **332**, 635–641.

33. Jansson, J.H., Nilsson, T.K., and Johnson, O. (1998) von Willebrand factor, tissue plasminogen activator, and dehydroepiandrosterone sulphate predict cardiovascular death in a 10-year follow up of survivors of acute myocardial infarction. *Heart* **80**, 334–337.

34. Montalescot, G., Philippe, F., Ankri, A., Vicaut, E., Bearez, E., Poulard, J.E., Carrie, D., Flammang, D., Dutoit, A., Carayon, A., Jardel, C., Chevrot, M., Bastard, J.P., Bigonzi, F., and Thomas, D. (1998) Early increase of von Willebrand factor predicts adverse outcome in unstable coronary artery disease: beneficial effects of enoxaparin. French Investigators of the ESSENCE Trial. *Circulation* **98**, 294–299.

35. Blann AD. Plasma von Willebrand factor, thrombosis, and the endothelium: the first 30 years. Thromb Haemost. 2006;95:49–55.

36. Roldan V, Martin F, Lip GYH, Blann AD. Soluble I selectin in cardiovascular disease and its risk factors. Thromb Haemostas 2003;90;1007–1020.

37. Chong, A.Y., Blann, A.D., Patel, J., Freestone, B., Hughes, E., and Lip, G.Y. (2004) Endothelial dysfunction and damage in congestive heart failure: relation of flow-mediated dilation to circulating endothelial cells, plasma indexes of endothelial damage, and brain natriuretic peptide. *Circulation* **110**, 1794–1798.

38. Kas-Deelen, A.M., Harmsen, M.C., De Maar, E.F., Oost-Kort, W.W., Tervaert, J.W., Van Der Meer, J., Van Son, W.J., and The, T.H. (2000) Acute rejection before cytomegalovirus infection enhances von Willebrand factor and soluble VCAM-1 in blood. *Kidney Int.* **58**, 2533–2542.

39. Makin, A., Chung, N.A.Y., Silverman, S.H., Blann, A., and Lip, G.Y. (2004) Assessment of endothelial damage in atherosclerotic vascular disease by quantification of circulating endothelial cells. *Eur. Heart. J.* **25**, 371–376.

40. Bernal-Mizrachi, L., Jy, W., Jimenez, J.J., Pastor, J., Mauro, L.M., Horstman, L.L., de Marchena, E., Ahn, Y.S. (2003) High levels of circulating endothelial microparticles in patients with acute coronary syndromes. *Am. Heart. J.* **145**, 962–970.

41. Martinez, M.C., Tesse, A., Zobairi, F., and Andriantsitohaina, R. (2005) Shed membrane microparticles from circulating and vascular cells in regulating vascular function. *Am. J. Physiol. Heart Circ. Physiol.* **288**, H1004–H1009.

42. Janssens, D., Michiels, C., Guillaume, G., Cuisinier, B., Louagie, Y., and Remacle, J. (1999) Increase in circulating endothelial cells in patients with primary chronic venous insufficiency: protective effect of Ginkor Fort in a randomized double-blind, placebo-controlled clinical trial. *J. Cardiovasc. Pharmacol.* **33**, 7–11.

43. Dang, A., Wang, B., Li, W., Zhang, P., Liu, G., Zheng, D., Ruan, Y., and Liu, L. (2000) Plasma endothelin-1 levels and circulating endothelial cells in patients with aortoarteritis. *Hypertens. Res.* **23**, 541–544.

44. Bull, T.M., Golpon, H., Hebbel, R.P., Solovey, A., Cool, C.D., Tuder, R.M., Geraci, M.W., and Voelkel, N.F. (2003) Circulating endothelial cells in pulmonary hypertension. *Thromb. Haemost.* **90**, 698–703.

45. George, F., Brisson, C., Poncelet, P., Laurent, J.C., Massot, O., Arnoux, D., Ambrosi, P., Klein-Soyer, C., Cazenave, J.P., and Sampol, J. (1992) Rapid isolation of human endothelial cells from whole blood using S-Endo1 monoclonal antibody coupled to immuno-magnetic beads: demonstration of endothelial injury after angioplasty. *Thromb. Haemost.* **67**, 147–153.

46. Camoin-Jau, L., Kone-Paut, I., Chabrol, B., Sampol, J., and Dignat-George, F. (2000) Circulating endothelial cells in Behcet's disease with cerebral thrombophlebitis. *Thromb. Haemost.* **83**, 631–632.

47. Mutunga, M., Fulton, B., Bullock, R., Batchelor, A., Gascoigne, A., Gillespie, J.I., and Baudouin, S.V. (2001) Circulating endothelial cells in patients with septic shock. *Am. J. Respir. Crit. Care Med.* **163**, 195–200.

48. Butthep, P., Rummavas, S., Wisedpanichkij, R., Jindadamrongwech, S., Fucharoen, S., and Bunyaratvej, A. (2002) Increased circulating activated endothelial cells, vascular endothelial growth factor, and tumour necrosis factor in thalassemia. *Am. J. Hematol.* **70**, 100–106.

49. Solovey, A., Lin, Y., Browne, P., Choong, S., Wayner, E., and Hebbel, R.P. (1997) Circulating activated endothelial cells in sickle cell anemia. *N. Engl. J. Med.* **337**, 1584–1590.

50. Lefevre, P., George, F., Durand, J.M. and Sampol, J. (1993) Detection of circulating endothelial cells in thrombotic thrombocytopenic purpura. *Thromb. Haemost.* **69**, 522.

51. Drancourt, M., George, F., Brouqui, P., Sampol, J., and Raoult, D. (1992) Diagnosis of Mediterranean spotted fever by indirect immunofluorescence of Rickettsia conorii in circulating endothelial cells isolated with monoclonal antibody-coated immunomagnetic beads. *J. Infect. Dis.* **166**, 660–663.

52. Woywodt, A., Streiber, F., de Groot, K., Regelsberger, H., Haller, H., and Haubitz, M. (2003) Circulating endothelial cells as markers for ANCA-associated small-vessel vasculitis. *Lancet* **361**, 206–210.

53. Nakatani, K., Takeshita, S., Tsujimoto, H., Kawamura, Y., Tokutomi, T., and Sekine, I. (2003) Circulating endothelial cells in Kawasaki disease. *Clin. Exp. Immunol.* **131**, 536–540.

54. Clancy, R., Marder, G., Martin, V., Belmont, H.M., Abramson, S.B., and Buyon, J. (2001) Circulating activated endothelial cells in systemic lupus erythematosus: further evidence for diffuse vasculopathy. *Arthritis Rheum.* **44**, 1203–1208.

55. Del Papa, N., Colombo, G., Fracchiolla, N., Moronetti, L.M., Ingegnoli, F., Maglione, W., Comina, D.P., Vitali, C., Fantini, F., and Cortelezzi, A. (2004) Circulating endothelial cells as a marker of ongoing vascular disease in systemic sclerosis. *Arthritis Rheum.* **50**, 1296–1304.

56. Lee, K.W., Lip, G.Y., Tayebjee, M., Foster, W., and Blann, A.D. (2005) Circulating endothelial cells, von Willebrand factor, interleukin-6, and prognosis in patients with acute coronary syndromes. *Blood* **105**, 526–532.

57. Quilici, J., Banzet, N., Paule, P., Meynard, J.B., Mutin, M., Bonnet, J.L., Ambrosi, P., Sampol, J., and Dignat-George, F. (2004) Circulating endothelial cell count as a diagnostic marker for non-ST-elevation acute coronary syndromes. *Circulation* **110**, 1586–1591.

58. Woywodt, A., Schroeder, M., Gwinner, W., Mengel, M., Jaeger, M., Schwarz, A., Haller, H., and Haubitz, M. (2003) Elevated numbers of circulating endothelial cells in renal transplant recipients. *Transplantation* **76**, 1–4.

59. Popa, E.R., Kas-Deelen, A.M., Hepkema, B.G., Van Son, W.J., The, T.H., and Harmsen, M.C. (2002) Donor-derived circulating endothelial cells after kidney transplantation. *Transplantation* **74**, 1320–1327.

60. Woywodt, A., Scheer, J., Hambach, L., Buchholz, S., Ganser, A., Haller, H., Hertenstein, B., and Haubitz, M. (2004) Circulating endothelial cells as a marker of endothelial damage in allogeneic hematopoietic stem cell transplantation. *Blood* **103**, 3603–3605.

61. Mancuso, P., Calleri, A., Cassi, C., Gobbi, A., Capillo, M., Pruneri, G., Martinelli, G., and Bertolini, F. (2003) Circulating endothelial cells as a novel marker of angiogenesis. *Adv. Exp. Med. Biol.* **522**, 83–97.

62. Mancuso, P., Burlini, A., Pruneri, G., Goldhirsch, A., Martinelli, G., and Bertolini, F. (2001) Resting and activated endothelial cells are increased in the peripheral blood of cancer patients. *Blood* **97**, 3658–3661.

63. Beerepoot, L.V., Mehra, N., Vermaat, J.S., Zonnenberg, B.A., Gebbink, M.F., and Voest, E.E. (2004) Increased levels of viable circulating endothelial cells are an indicator of progressive disease in cancer patients. *Ann. Oncol.* **15**, 139–145.

64. George, F., Brouqui, P., Boffa, M.C., Mutin, M., Drancourt, M., Brisson, C., Raoult, D., and Sampol, J. (1993) Demonstration of Rickettsia conorii-induced endothelial injury in vivo by measuring circulating endothelial cells, thrombo-modulin, and von Willebrand factor in patients with Mediterranean spotted fever. *Blood* **82**, 2109–2116.

65. Percivalle, E., Revello, M.G., Vago, L., Morini, F., and Gerna, G. (1993) Circulating endothelial giant cells permissive for human cytomegalovirus (HCMV) are detected in disseminated HCMV infections with organ involvement. Circulating endothelial giant cells permissive for human cytomegalovirus (HCMV) are

detected in disseminated HCMV infections with organ involvement. *J. Clin. Invest.* **92**, 663–670.

66. Beerepoot, L.V., Mehra, N., Linschoten, F., Jorna, A.S., Lisman, T., Verheul, H.M., and Voest, E.E. (2004) Circulating endothelial cells in cancer patients do not express tissue factor. *Cancer Lett.* **213**, 241–248.

14

Evaluation of Endothelial Function by In Vivo Microscopy

Rosario Scalia

Summary

This chapter describes a method that permits simultaneous measurement of leukocyte–endothelium interactions and endothelial nitric oxide (NO) levels in the microcirculation in vivo. The method is also useful to study the effect of NO replenishing therapy on adhesion of leukocytes to the vascular endothelium in acute and chronic inflammatory states of the cardiovascular system. This research approach requires the combination of two well-established physiology techniques, that is, intravital microscopy and real-time measurement of NO with microelectrodes. Intravital microscopy is considered the method of choice to monitor leukocyte–endothelial cell interactions in intact vascular beds of live animals. In vivo microscopy is currently used to study the endothelial cell phenotype of mice carrying mutations or deletion of targeted genes. Intravital microscopy is also used to study endothelial cell function in acute (e.g., ischemia-reperfusion injury) and chronic (e.g., hypercholesterolemia, hyperglycemia, and diabetes) inflammatory states of the cardiovascular system. NO sensors allow for continuous, amperometric quantification of NO levels in cells and organ tissues. Coupling of NO electrode technology with intravital microscopy has recently permitted to measure NO bioavailability in the normal and inflamed microcirculation. The method described here can be used to study in vivo how acute and chronic inflammatory states of the cardiovascular system alter endothelial function resulting in endothelial cell activation and damage.

Key Words: Intravital microscopy; Leukocyte rolling; Leukocyte adhesion; Leukocyte extravasation; NO electrode; Endothelial dysfunction; Post-capillary venules; Shear rates; Inflammation.

From: *Methods in Molecular Medicine, Vol. 139: Vascular Biology Protocols*
Edited by: N. Sreejayan and J. Ren © Humana Press Inc., Totowa, NJ

1. Introduction

The discovery of the role exerted by the vascular endothelium in the homeostasis of the cardiovascular system has generated the growing need of studying endothelial cell function in vivo. Studies in the live, intact animal offer the obvious advantage of monitoring the function of the vascular endothelium in a physiologic context, without disrupting the anatomic relationship between the blood and the vascular wall. The information provided is of tremendous value as it clarifies and validates the physiological meaning of results obtained with reductionistic studies in vitro. Thus, intravital microscopy has become an invaluable research tool, instrumental for translating information obtained with in vitro systems into potential clinical applications.

Because of its strategic location within the structure of parenchymatous organs, the microvascular endothelium is a critical checkpoint for inflammation and immunity. Microvascular endothelial cells play at least three roles in the interaction with leukocytes. First, they are gatekeepers in leukocyte recruitment to inflammatory foci and lymphocyte homing to secondary lymphoid organs. Second, they modulate activation of circulating leukocytes. Finally, they are targets of leukocyte-derived molecules, resulting either in endothelial cell activation or in endothelial cell death. The normal endothelium maintains a non-thrombogenic surface and prevents adhesion of circulating leukocytes to the vessel wall *(1)*. Nitric oxide produced by endothelial cells (eNO) is an important mediator of the anti-inflammatory properties of the normal vascular endothelium. Several studies have demonstrated that loss of eNO causes a rapid activation of endothelial cells leading to up-regulation of cell adhesion molecules *(2–4)*. Infiltration of circulating leukocytes damages the vascular wall and organ tissue *(5,6)*. Indeed, infiltrating leukocytes become activated and release free radicals able to rapidly quench NO further reducing the levels of eNO at the vascular wall *(7,8)*. This widely accepted paradigm has been confirmed in acute and chronic disease states of the cardiovascular systems including ischemia-reperfusion injury, shock, hypercholesterolemia, atherosclerosis, hyperglycemia, and diabetes. Hence, methods that permit to study leukocyte–endothelium interactions and simultaneously monitor eNO bioavailability in vivo become a powerful research tool in vascular biology. It should be noted that more recently, intravital microscopy (IVM) techniques have also been employed to study mechanisms of cancer metastasis based on the observation that metastatic diffusion of cancer cells largely occurs through tumor microvessels and that it involves adhesion of cancer cells to the vascular endothelium of invaded organs.

Modern video-based IVM allows for the visualization of leukocytes as they slow down, roll along the vascular endothelium, adhere, and infiltrate the surrounding tissue. At acquisition speed of 30 frames/s, this technology captures all the stages of leukocyte–endothelium interactions in live animals. IVM observations are usually made in post-capillary venules. Post-capillary venules are the beginning of the venous system, and they basically resemble capillaries but with a wider lumen. Structurally, post-capillary venules are tubes of endothelial cells supported by a basal membrane and an adventitia of collagen fibers and fibroblasts *(9)*. The lack of a significant smooth muscle cell layer and the high expression of cell adhesion molecules make post-capillary venules the preferred site of migration of blood cells by diapedesis *(10,11)*.

To quantify levels of endothelial NO during IVM studies, an amperometric NO sensor of 100 nm diameter has been recently used in our laboratory *(12,13)*. Our team has previously demonstrated that NO sensors are clearly sensitive, highly selective, avoid problems with contaminations of oxidized derivatives of NO or oxygen, and allow for real-time measurement of NO release from biological systems *(14)*. Continuos monitoring of NO release in intact microvessels in vivo using NO sensors provides a useful means for studying the role of NO in inflammation on a rapid analysis bases. Also, coupling of IVM with this method is useful in testing the effectiveness of NO donors to release NO and to modulate inflammatory signals in the microcirculation.

2. Materials

2.1. Instrumentation

2.1.1. Intravital Microscopy

1. For basic IVM experiments, the investigator should consider the acquisition of a commercially available microscope, high-resolution intensified video camera, high-resolution video monitor, and digital DVD recorder. Several computer-based video acquisition systems are also available in the market. Depending on the configuration, computerized systems can be indeed very expensive. Albeit most of the instrumentation required is commonly available in the market, the investigator must also be prepared to acquire a few specialized devices. In addition, a certain degree of modification to the stage of commercially available microscopes must be made. All of these special needs are listed in the points given below.
2. Different laboratories have successfully employed both upright and inverted microscopes for IVM studies. However, the use of an upright microscope is preferred as it permits the placement of NO sensors over the microvessel under study (*see* **Fig. 1**). The investigator must be able to visualize the reading segment of the

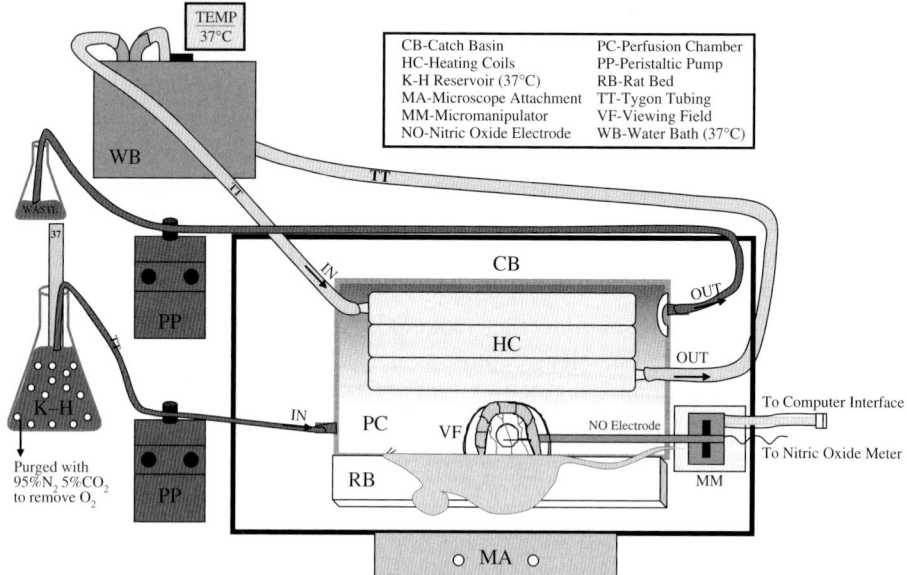

Fig. 1. Schematic of intravital microscopy experimental setup. CB, catch base; HC, heating coils; K–H Reservoir (37 °C); MA, microscope attachment; MM, micromanipulator; NO, nitric oxide electrode; PC, perfusion chamber; PP, peristaltic pump; RB, rat bed; TT, Tygon tubing; VF, viewing field; WB, water bath (37 °C).

NO sensor during microscopy so that its precise location and distance from the vessel wall can be accurately determined. NO sensors are usually driven over the microvessels with the aid of a micromanipulator attached to the microscope stage (*see* **Fig. 1**).

3. The microscope stage must be able to hold a perfusion chamber while being fully operative. In the case of the Nikon microscopes used in our laboratory, perfusion chambers of different sizes can be effectively screwed on the microscope stage in place of the standard specimen holder. The perfusion chamber is necessary to preserve the physiological integrity of the IVM preparation for the entire duration of the experiment. It is recommended that the investigator devise the perfusion chamber keeping in mind the experimental design. Thus, perfusion chambers can vary in size and shape to accommodate laboratory rodents of different weight, such as rats and mice. As an example, a schematic of the perfusion chamber used in our laboratory for studying the rat mesentery is illustrated in **Fig. 1**. Perfusion chambers are usually made of Plexiglas, and they must be individually tailored to fit different microscopes. With a volume capacity of approximately 250 ml, the perfusion chamber used in our laboratory permits to completely immerse the

exposed mesenteric tissue in physiological buffer. It should be noted that even a brief drying of surgically exposed tissue causes massive microvascular dysfunction, which must be prevented with superfusion with physiological buffer. The buffer in the perfusion chamber is circulated from a reservoir with the aid of two peristaltic pumps and through Tygon tubing. A water-based heating coil maintains the temperature of the perfusion buffer at 37 °C. Accurate control of the temperature in the perfusion chamber is crucial not only for the viability of the tissue but also because the sensitivity of NO sensors is greatly affected by changes in temperature *(15)*. The microcirculation is visualized with ×20 or ×40 water-immersion objectives during bright-field or fluorescence microscopy. Detailed discussion on the opportunity of using bright-field or fluorescence microscopy can be found elsewhere *(16)*. Inflammatory agents or eNOS inhibitors can be added to the perfusion buffer to study relevant mechanisms of endothelial dysfunction in vivo.

4. An optical Doppler velocimeter should be attached to one available optical port of the microscope. This instrument is required to measure online the velocity of blood flow in the post-capillary venules *(17)*. The optical Doppler velocimeter instrument used in our laboratory can be obtained from the Cardiovascular Research Institute (Texas A&M College of Medicine, College Station, TX, USA).

5. Video caliper to measure vessel diameter in a real-time video image. Several models are available on the market.

6. NO microelectrodes and NO meters are now commercially available. Currently, World Precision Instruments (Saratosa, FL, USA) manufactures NO sensors of different sizes and specific NO meters to which the sensor can be coupled (e.g., Apollo 4000).

2.2. Reagents and Buffers

1. Pentobarbital or other appropriate anesthetic of choice (*see* **Note 1**).
2. NaCl (saline) solution, 0.9%.
3. Modified Krebs–Henseleit solution for IVM: 118 mM NaCl, 4.74 mM KCl, 2.45 mM $CaCl_2$, 1.19 mM KH_2PO_4, 1.19 mM $MgSO_4$, and 12.5 mM $NaHCO_3$. This buffer is warmed to 37 °C and bubbled with 95% N_2 and 5% CO_2 (*see* **Note 2**).
4. Saturated solution of cuprous chloride. This solution is used to calibrate daily the NO sensor. It is prepared by adding 150 mg cuprous chloride to 500 ml distilled deoxygenated water. The distilled water can be deoxygenated by purging with pure nitrogen or argon gas for 15 min. The saturated cuprous chloride solution will have a concentration in the range 2.4 mM, and it should be stored in the dark.
5. *S*-nitroso-*N*-acetyl-D,L-penicillamine (SNAP): To prepare a standard SNAP solution, dissolve 5 mg ethylenediaminetetraacetic acid (EDTA) in 250 ml high-performance liquid chromatography (HPLC) pure water (HPLC grade; Sigma, St. Louis, Mo). Deoxygenate the solution using the method described in **step 4** above. Add 5.6 mg SNAP to the solution. SNAP solution must be refrigerated and stored in dark until use to prevent degradation.

3. Methods

3.1. Surgical Preparation of Experimental Animals

Small laboratory rodents such as mice and rats are used for intravital microscopy studies, albeit larger species can also be considered (e.g., rabbit and cat). A surgical plane of anesthesia is induced with administration of 80–100 mg/kg pentobarbital injected intraperitoneally. Intraperitoneal injections should be made very carefully to avoid traumatic injury to the peritoneum and/or intestine. If bleeding or hematomas are found in the peritoneal organs, the animal should be excluded from the study. Animals are also tracheotomized and cannulated at the left carotid artery and the right jugular vein to maintain patent airways and for continuous blood pressure monitoring and fluids/drugs administration, respectively.

3.2. Preparation of the Mesentery for Intravital Microscopy

Membranous tissues are ideal for intravital microscopy studies and simultaneous measurement of NO levels in microvessels in vivo because of their translucency and thinness. The most common tissues issued in the mouse and the rat are the mesentery, the cremaster muscle, and the meningeal membranes. We will focus on the preparation of the rat mesentery (*see* **Note 3**). A midline incision laparotomy is performed to exteriorize four distal loops of ileal tissue. If bleeding of the abdominal wall at the site of laparotomy occurs, it should be stopped using a battery-operated small vessel cauterizer. The intestinal and mesenteric tissue should be handled gently to avoid stretching or trauma. The tissue can be sandwiched between sterile dressing gauzes moistened with saline. At the end of the surgical procedures, the anesthetized rat is carefully positioned on the microscope stage as shown in **Fig. 1**. The exteriorized intestinal loops with the mesenteric tissue is carefully freed of the gauzes and positioned on the viewing field of the perfusion chambers. During the experiments, the tissue is superfused with the modified K–H buffer listed in step 3 of section 2.2, which also preserves its suitability for ex vivo biochemical analyses (*see* **Note 4**).

3.3. Evaluation of Leukocyte–Endothelial Cell Interactions

Following a 30-min stabilization of the microcirculation, three to four relatively straight, unbranched segments of post-capillary venules with lengths of $> 100 \mu m$ and diameters between 25 and $40 \mu m$ are randomly selected from each rat to monitor leukocyte–endothelial cell interactions. The numbers of rolling leukocyte, firmly adhering leukocyte, and extravasated leukocyte

extravasation can be quantified over a period of time of 120 min. Longer observation periods are not recommended because of the risk of tissue deterioration.
During this time, recordings are made at 15-min intervals. Video recordings are
stored on DVD. The number of rolling and adherent leukocytes is determined
off-line during video playback analysis. Baseline leukocyte rolling (number of
cells moving past a designated point per minute), adherence (cells that remain
stationary for >30 s in 100 μm of vessel length), and extravasation (number of
leukocytes found in a defined interstitial area; e.g. an area of 20×100 μm at a
distance of 10 μm from the vessel wall) are quantified.

3.4. Evaluation of Shear Rate Values

Evaluation of shear rate values is important for the overall evaluation of the
results. The risk exists that as result of anesthesia, surgery, and tissue manipulation, macrohemodynamic and microhemodynamic conditions become altered,
which would affect non-specifically leukocyte–endothelium interactions. The
range of acceptable shear rates values in intravital microscopy studies has been
previously published *(18)*. Centerline red blood cell velocity (V_{RBC}) is measured
on-line using an optical Doppler velocimeter reported in **Subheading 2.1.1**
(step 4). V_{RBC} and Venular diameter (D_{ven}) were used to calculate the venular
wall interfacial shear rate (γ) by using the formula: $= 4.9 \times 8 \times (V_{mean}/D_{ven})$
where mean red blood cell velocity (V_{mean}) is equal to $V_{RBC}/1.6$ (19). Venular
diameter (D_{ven}) is measured on-line using a video caliper calibrated using a
stage micrometer.

3.5. NO Electrode Calibration Procedure

The NO electrodes must be connected to the NO meter at least 12 h before
use. This allows for polarization of the probe. Calibration of the sensor can
be made in a 20-ml vial containing 10 ml of the CuCl solution. The vial
should be capped and purged of residual oxygen with nitrogen or argon (*see*
Note 5). Drop a small stirring bar into the solution and place the vial on top
of a magnetic stirring plate. The NO probe is then immersed in the purged
cuprous chloride solution. The background current detected after 10–15 min of
stabilization is recorded. As soon as the background current becomes stable,
increasing concentrations of SNAP are added to the cuprous chloride solution to
generate a calibration curve (*see* **Note 6**). It is recommended that the calibration
range be kept close to the anticipated experimental concentrations of NO (*see*
Note 7). These concentrations are in the nanomolar range in vivo. Following
addition of a given concentration of SNAP to the solution, the current output
of the electrode increases rapidly. Once the response reaches a plateau, the

second aliquot of SNAP can be added. For every addition, the responses should reach a plateau within seconds. The calibration curve is constructed by plotting the signal output versus the concentration of SNAP. Each addition of SNAP corresponds to equivalent NO concentration. The dose–response should be linear from 10 to 1000 nM.

3.6. Positioning of the NO Electrode Under Microscopy

With the aid of the computerized micromanipulator, the calibrated micro-electrode is driven directly over a mesenteric post-capillary venule at a distance of approximately 3 μm from the venular wall (*see* **Fig. 2**). The distance is calculated using the video-caliper. To establish a 0 nM reference for NO during in vivo conditions, the microelectrode tip is moved 0.5–1 mm above the tissue surface, as previously described *(20)*. Subsequently, the electrode is placed in a non-vascular region of the mesentery (interstitial space). Typically, the NO concentration decreases sharply in the interstitial space, and a stable background current is obtained at a distance of 200 μm from the venular wall. The background current is then subtracted from the current measured at the venular

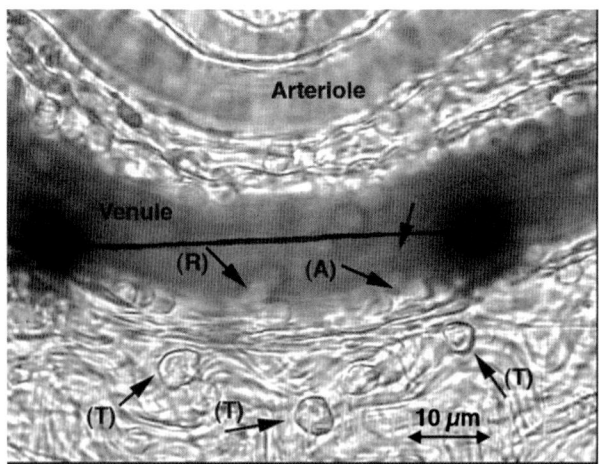

Fig. 2. Photomicrograph of a typical IVM video recording taken from an inflamed rat mesenteric microcirculation. Nitric oxide (NO) released from rat mesenteric post-capillary venules is measured using the ISO-NO-NM microelectrode (100 nm diameter, WPI, Sarasota, FL, USA). Large black dots are the optical projections of the optical Doppler velocimeter used to measure the blood flow velocity during intravital microscopy. Arrows indicate Rolling (R), adherent (A), and transmigrated (T) leukocytes can be simultaneously measured during the intravital microscopy.

wall to calculate the amount of NO released by the vascular endothelium. NO values are normalized according to the size of blood vessel studied and reported as nanomolar $NO/1000\,\mu m^2$ vessel area. It is recommended that a minimum of five randomly selected venules be studied in each rat.

4. Notes

1. Anesthetics are known to alter respiratory and cardiovascular functions. This could result in non-specific changes in the homeostasis of the microcirculation. Thus, the use of anesthetic should be tightly monitored and the possible occurrence of non-specific adverse reactions caused by anesthesia should always be considered in the interpretation of the results.

2. The buffer's composition, pH, dissolved gas content, and potential bacterial endotoxin contaminants must be carefully controlled.

3. A viable IVM preparation must produce an interpretable optical signal from the focal plane of the microscope objective. This can only be achieved through dissection and surgical manipulation of the tissue of interest. These procedures almost inevitably cause a certain degree of trauma that can result in local or even systemic inflammation. Investigator's experience and carefully standardized laboratory procedures are important in establishing acceptable baseline values for control IVM preparations. Based on published data, a good rat mesenteric preparation should have a very low number of rolling and adhering leukocytes in the range of 20 cell/min and 2 cell/100 μm, respectively.

4. At the end of intravital microscopy, the mesenteric tissue can be isolated and quickly frozen or fixed for biochemistry studies or immunohistochemical studies, respectively. Western blot analyses can be for instance used to evaluate expression/activity of relevant protein such as eNOS. Similarly, immunohisto-chemistry or immunofluorescence can be used to study expression of endothelial cell adhesion molecules on the vascular endothelium. The use of biochemical techniques greatly enhances the significance of results obtained with IVM, in that it may help provide important mechanistic information to explain the results in vivo. It is advised that the mesentery and intestine be freed of the circulating blood before isolation. Blood-free intestinal tissue can be prepared by K–H perfusion of the intestinal circulation through the superior mesenteric artery and drainage of the blood perfusate through the portal vein.

5. The described NO electrode calibration procedure requires the use of cuprous(I) chloride that acts as catalyst for the conversion of SNAP to NO. Cu(I) could be readily oxidized to Cu(II) if the compound is accidentally exposed to air or improperly stored. Cu(II) catalyzes SNAP to NO less efficiently, which could affect the result of the calibration procedure.

6. SNAP is relatively stable at low temperature, in dark, and in absence of trace metal ions. The presence of the chelating reagent, EDTA, guarantees the absence of free metal ions in the solution. Because of this, the standard SNAP solution can

be used for many calibrations of NO probes throughout the day. Nonetheless, it is advised that a fresh standard stock solution of SNAP is made fresh every day to secure the accuracy of the calibration procedure. In addition, only high-grade SNAP with minimal purity of 98% should be used to obtain a correct calculation of the NO levels generated.

7. Keep in mind that most NO probes are sensitive to changes in temperature. Accordingly, the calibration procedure should be performed at the same temperature used for the in vivo experimental setting.

Acknowledgments

This work was supported by NIH grant R01 DK064344. The author is thankful to Mr. Brett Berzins, BS, for his help with the graphic work.

References

1. Lefer, A.M., and Scalia, R. (2001) Nitric oxide in inflammation. In *Physiology of Inflammation*, Ley, K., Ed. New York, Oxford University Press, 447–472.
2. Davenpeck, K.L., Gauthier, T.W., and Lefer, A.M. (1994) Inhibition of endothelial-derived nitric oxide promotes P-selectin expression and actions in the rat microcirculation. *Gastroenterology* **107**, 1050–1058.
3. Granger, D.N., and Kubes, P. (1996) Nitric oxide as antiinflammatory agent. *Methods Enzymol.* **269**, 434–442.
4. Gaboury, J., Woodman, R.C., Granger, D.N., Reinhardt, P., and Kubes, P. (1993) Nitric oxide prevents leuocyte adherence: role of superoxide. *Am. J. Physiol.* **265**, H862–H867.
5. Hayward, R., Campbell, B., Shin, Y.K., Scalia, R., and Lefer, A.M. (1999) Recombinant soluble P-selectin glycoprotein ligand-1 protects against myocardial ischemic reperfusion injury in cats. *Cardiovasc. Res.* **41**, 65–76.
6. Scalia, R., Armstead, V.E., Minchenko, A.G., and Lefer, A.M. (1999) Essential role of P-selectin in the initiation of the inflammatory response induced by hemorrhage and reinfusion. *J. Exp. Med.* **189**, 931–938.
7. Stokes, K.Y., Cooper, D., Tailor, A., and Granger, D.N. (2002) Hypercholesterolemia promotes inflammation and microvascular dysfunction: role of nitric oxide and superoxide. *Free Radic. Biol. Med.* **33**, 1026–1036.
8. Rodenas, J., Mitjavila, M.T., and Carbonell, T. (1998) Nitric oxide inhibits superoxide production by inflammatory polymorphonuclear leukocytes. *Am. J. Physiol.* **274**, C827–C830.
9. Kierszenbaum, A.L. (2002) *Histology and Cell Biology.* St. Louis, Mosby.
10. Eriksson, E.E., Karlof, E., Lundmark, K., Rotzius, P., Hedin, U., and Xie X. (2005) Powerful inflammatory properties of large vein endothelium in vivo. *Arterioscler. Thromb. Vasc. Biol.* **25**, 723–728.
11. Granger, D.N., and Kubes, P. (1994) The microcirculation and inflammation: modulation of leukocyte-endothelial cell adhesion. *J. Leukoc. Biol.* **55**, 662–675.

12. Stalker, T.J., Gong, Y., and Scalia, R. (2005) The calcium-dependent protease calpain causes endothelial dysfunction in type 2 diabetes. *Diabetes* **54**, 1132–1140.

13. Stalker, T.J., Skvarka, C.B., and Scalia, R. (2003) A novel role for calpains in the endothelial dysfunction of hyperglycemia. *FASEB J.* **17**, 1511–1513.

14. Guo, J.P., Panday, M.M., Consigny, P.M., and Lefer, A.M. (1995) Mechanisms of vascular preservation by a novel NO donor following rat carotid artery intimal injury. *Am. J. Physiol.* **269**, H1122–H1131.

15. Zhang, X., Cardosa, L., Broderick, M., Fein, H., and Davies, I.R. (2000) Novel calibration method for nitric oxide microsensors by stoichiometrical generation of nitric oxide from SNAP. *Electroanalysis* **12**, 425–428.

16. Mempel, T.R., Scimone, M.L., Mora, J.R., and von Andrian, U.H. (2004) In vivo imaging of leukocyte trafficking in blood vessels and tissues. *Curr. Opin. Immunol.* **16**, 406–417.

17. Borders, J.L., and Granger, H.J. (1984) An optical doppler intravital velocimeter. *Microvasc. Res.* **27**, 117–127.

18. Perry, M.A., and Granger, D.N. (1991) Role of CD11/CD18 in shear rate-dependent leukocyte-endothelial cell interactions in cat mesenteric venules. *J. Clin. Invest.* **87**, 1798–1804.

19. Smith, M.L., Long, D.S., Damiano, E.R., and Ley, K. (2003) Near-wall micro-PIV reveals a hydrodynamically relevant endothelial surface layer in venules in vivo. *Biophys J* **85**, 637–645.

20. Bohlen, H.G. (1998) Mechanism of increased vessel wall nitric oxide concentrations during intestinal absorption. *Am. J. Physiol.* **275**, H542–H550.

15

Haptotaxis of Endothelial Cell Migration Under Flow

Steve Hsu, Rahul Thakar, and Song Li

Summary

Endothelial cell (EC) migration plays an important role in embryonic vasculogenesis, angiogenesis, and wound healing. EC migration can be regulated by extracellular matrix (ECM) and hemodynamic forces through haptotaxis (induced by an ECM gradient) and mechanotaxis (induced by mechanical forces). Previously, the effects of haptotaxis or mechanotaxis alone on EC migration have been studied; however, the dual effect of haptotaxis and mechanotaxis on EC migration is not known. We developed a micropatterning technique to generate step changes of collagen surface density to monitor haptotactic EC migration. To investigate the crosstalk between haptotaxis and mechanotaxis on EC migration, we used an in vitro flow system to apply a well-defined fluid shear stress on ECs cultured on the micropatterned collagen. The study on the effects of haptotaxis and mechanotaxis on EC migration will provide a rational basis for promoting vascular wound healing, angiogenesis, and vascularization in engineered tissues.

Key Words: Endothelial cell; Migration; Micropatterning; Haptotaxis; Mechanotaxis; Flow; Time-lapse microscopy; Molecular dynamics; Focal adhesion; Cytoskeleton.

1. Introduction

Endothelial cell (EC) migration plays an important role in embryonic vasculogenesis, angiogenesis, and wound healing. Under pathological conditions such as atherosclerosis and denudation injury following angioplasty and bypass grafting, the loss of endothelial integrity leads to EC dysfunction and thrombosis *(1–3)*. The recovery of endothelial integrity and vascular wall homeostasis requires endothelial healing, which involves EC migration at the wound edges. EC migration is also necessary for vascular assembly in engineered and native

From: *Methods in Molecular Medicine, Vol. 139: Vascular Biology Protocols*
Edited by: N. Sreejayan and J. Ren © Humana Press Inc., Totowa, NJ

tissues, which promotes the healing of damaged tissue and the survival of tissue-engineered implants *(4,5)*. Understanding the environmental factors that modulate EC migration will advance our understanding of vascular repair and remodeling in vivo and lead to the development of improved strategies to promote vascularization in tissue-engineered constructs.

EC migration can be modulated by environmental factors through different mechanisms, such as haptotaxis and mechanotaxis. Haptotaxis refers to the directed migration of ECs from a less adherent to a more adherent surface and plays a role in dictating EC migration during angiogenesis *(6,7)*. With respect to vascular tissue engineering, haptotaxis may be used to enhance EC migration and angiogenesis by controlling the spatial distribution of extracellular matrix (ECM) and its derived peptides in vascular grafts and scaffolds.

Mechanotaxis refers to the directed migration of ECs induced by a mechanical force *(2,8)*. The focal adhesions (FAs), cytoskeleton, and signaling molecules inside cells need to respond to diverse extracellular signals and translate them into coordinated intracellular responses to mediate cell migration. ECs are constantly subjected to a fluid shear stress, the tangential component of hemodynamics forces, due to blood flow. By using an in vitro flow system, we and others have shown that physiological shear stress levels ($\sim 12\,\mathrm{dyn/cm^2}$) increase EC wound healing *(9,10)* and promote lamellipodial protrusion and EC migration in the flow direction *(8,10–12)*. However, the crosstalk between haptotaxis and mechanotaxis during EC migration is not known.

We developed a system to investigate the dual effects of fluid flow and ECM gradients on EC migration. We used soft lithography to create micropatterned, parallel strips of collagen and generate step changes in collagen surface density in a quantifiable and controlled manner. ECs cultured on these micropatterned surfaces were then subjected to a well-defined physiological fluid shear stress using an in vitro parallel plate flow system. Phase-contrast time-lapse video microscopy generated quantitative information with respect to the percentage of ECs exhibiting haptotactic or mechanotactic behavior. By transfecting green fluorescent protein (GFP)-tagged paxillin in ECs, cell FAs were visualized, and the dynamics of FAs during EC migration on micropatterned surfaces was observed using fluorescent confocal time-lapse microscopy. Lastly, the dual effects of collagen step gradients and mechanical forces on FA formation in EC migration were visualized by immunofluorescent staining of vinculin and fluorescent microscopy.

2. Materials

2.1. EC Culture

1. Bovine aortic ECs (BAECs).
2. Dulbecco's modified Eagle's medium (DMEM), high glucose, with L-glutamine, and sodium pyruvate (store at 4 °C and protect from light; Gibco/BRL, Bethesda, MD, USA).
3. Fetal bovine serum (FBS; store in 50 ml aliquots at −20 °C; Gibco/BRL).
4. Penicillin–streptomycin solution (store in 5 ml 100× aliquots at −20 °C; Gibco/BRL).
5. Trypsin–ethylenediaminetetraacetic acid (EDTA), 10×: 0.5% trypsin and 5.3 mM EDTA (store in 5 ml 10× aliquots at −20 °C; Gibco/BRL).
6. Phosphate-buffered saline (PBS), 1× (autoclaved and stored at room temperature; Gibco/BRL).
7. Collagenase (stored at 4 °C; Worthington, Lakewood, NJ, USA).
8. Vascular endothelial (VE)-cadherin antibody (stored in 20 μl aliquots at −20 °C; Santa Cruz Biotechnologies, Santa Cruz, CA, USA).
9. Acetylated low-density lipoprotein conjugated with DiI (DiI ac-LDL; Molecular Probes, Eugene, OR, USA).

2.2. Microfabrication and Micropatterning

1. Silicon wafers.
2. Photomasks (either glass plate with emulsion pattern or plastic transparencies using a very high-resolution printer).
3. Photoresist (OIR 897-101, Arch Chemicals, Norwalk, CT, USA).
4. Poly(dimethylsiloxane) (PDMS; Sylgard 184, Dow Corning, Midland, MI, USA).
5. PDMS stamps.
6. Rat tail collagen type I (BD Biosciences, Bedford, MA, USA).
7. Fluorescein isothiocyanate (FITC)-conjugated collagen type I (Sigma-Aldrich, St. Louis, MO, USA).
8. Acetic acid, 0.1% (v/v) (EM Sciences, Darmstadt, Germany).
9. Filter, 0.20 μm and 20-ml syringe (Fisher Scientific, Santa Clara, CA, USA).
10. Bovine serum albumin (BSA), 1% (w/v) (Sigma-Aldrich).
11. Glass slides (25 mm × 75 mm × 1 mm; Fisher Scientific).
12. Aluminum foil.
13. Scalpel or razor blade.
14. Ethanol, 70%.

2.3. In Vitro Parallel Plate Flow System

1. Polycarbonate flow chamber base.
2. Silicone gasket.
3. Metal plates and screws for the assembly of flow chamber.

4. Glass slides and cover glass.
5. Peristaltic pump and pump head for flow system (Cole-Parmer Instrument Co., Vernon Hills, IL, USA).
6. Medium reservoirs for flow system.
7. Clamps and clamp holders for medium reservoirs.
8. Tubing for flow system (Cole-Parmer Instrument Co.).
9. Three-way stopcocks (Baxter Healthcare Corp., Deerfield, IL, USA).
10. Tube connectors and tube adaptors for flow system (Cole-Parmer Instrument Co.).
11. Syringes (20 ml).
12. Temperature chamber.
13. Heater, fan, and temperature controller.
14. Compressed air with 5% CO_2.
15. Flowmeter for air flow control (Cole-Parmer Instrument Co.).
16. Ethanol, 70%.
17. Sonicator (VWR Scientific Products, West Chester, PA, USA).

2.4. Phase-Contrast Time-Lapse Video Microscopy

1. Inverted microscope (Nikon, Melville, NY, USA) with image acquisition system for time series (Hamamatsu Orca100 digital charge-coupled device camera; Hamamatsu Photonics, Hamamatsu City, Japan).
2. Motorized X–Y microscope stage with Z motor for auto-focusing.
3. Image analysis software with particle tracing algorithm (C-Imaging System Software; Compix Inc., Cranberry Township, PA, USA).

2.5. Immunofluorescent Staining Analysis of FAs

1. Paraformaldehyde (Fisher Scientific): 50 ml aliquots of 4% (w/v) paraformaldehyde in PBS stored at 4 °C.
2. Triton X-100: 50 ml aliquots of 0.5% Triton X-100 in PBS stored at 4 °C.
3. BSA (Sigma-Aldrich): 50 ml aliquots of 1% BSA in PBS stored at 4 °C.
4. Primary vinculin mouse antibody (Sigma-Aldrich): 20 µl aliquots of 1:100 dilution from stock in PBS stored at −20 °C.
5. Rhodamine donkey anti-mouse secondary antibody (Jackson ImmunoResearch, West Grove, PA, USA): 20 µl of 15 µg/ml aliquots stored at −20 °C.
6. VectorShield antifade solution (Vector Laboratories, Burlingame, CA, USA): stored at 4 °C.
7. Confocal microscope with argon laser, appropriate filters, and image acquisition system (Leica Microsystems, Wetzlar, Germany).

2.6. Dynamics of FAs in BAEC Migration on Micropatterned ECM Step Gradients

1. LipofectAMINE 2000 transfection reagent (stored at 4 °C; Gibco/BRL).
2. DNA plasmids encoding GFP-tagged paxillin (stored at −20 °C).

3. Chamber coverslip (Fisher Scientific).
4. Rhodamine collagen I (stored at −20 °C).
5. Opti-MEM1 for transfection (stored at 4 °C and protect from light; Gibco/BRL).
6. Confocal microscope with argon laser, appropriate filters, and image acquisition system (Leica Microsystems).

3. Methods

3.1. EC Culture

1. Isolate BAECs from bovine aorta obtained from a local slaughterhouse (*see* **Note 1**).
2. Wash the aorta in PBS, cut into pieces ($\sim$ 3 cm ×3 cm), and place in 100-mm culture dishes.
3. Incubate the inner surface of the vessel in DMEM with 0.5% collagenase. After 30 min, collect the detached cells, spin down, and re-suspend in DMEM supplemented with 10% FBS and penicillin–streptomycin [(penicillin: 10,000 units/ml) streptomycin: 10,000 µg/ml, complete medium)].
4. Seed cells in six-well plates and culture in complete medium until confluency is reached. Change medium every 2–3 days.
5. Select wells that display a nice cell monolayer with cobblestone-like cell morphology for further characterization of EC-like properties. Determine EC characterization by positive staining for VE-cadherin using a VE-cadherin antibody and by cell incorporation of DiI ac-LDL.
6. Following characterization, isolate and culture BAECs in 100-mm cell culture dishes in complete medium. Upon confluency, passage BAECs by incubating with a 0.5× trypsin solution for 2–3 min at 37 °C. Maintain expanded BAEC cultures in a humidified 5% CO_2−95% air incubator at 37 °C prior to use in experiments. Use only low-passage BAEC cultures (usually the first 10 passages of BAECs) for the experiments.

3.2. Microfabrication and Micropatterning (see Fig. 1)

1. Using a software program such as AutoCAD, design the pattern (e.g., an array of strips) required for the experiments.
2. Convert the design into a photomask. (In microfabrication facility, silicon wafers, photoresist, and the photomask were used to create silicon wafer molds of the desired pattern using standard microfabrication procedures and equipment; *see* **Note 2**.)
3. Mix PDMS pre-polymer and initiator in a 10:1 (pre-polymer: initiator) ratio. Mix the solution well to a homogenous consistency so as to avoid non-uniform properties of the PDMS.
4. Place the silicon wafer on a piece of aluminum foil and fold the extra foil on the edges of the wafer to make a reservoir. This reservoir ensures the PDMS stamp is thick enough to handle and provides a low cost and easy and clean way of polymerizing the PDMS on the silicon wafer molds.

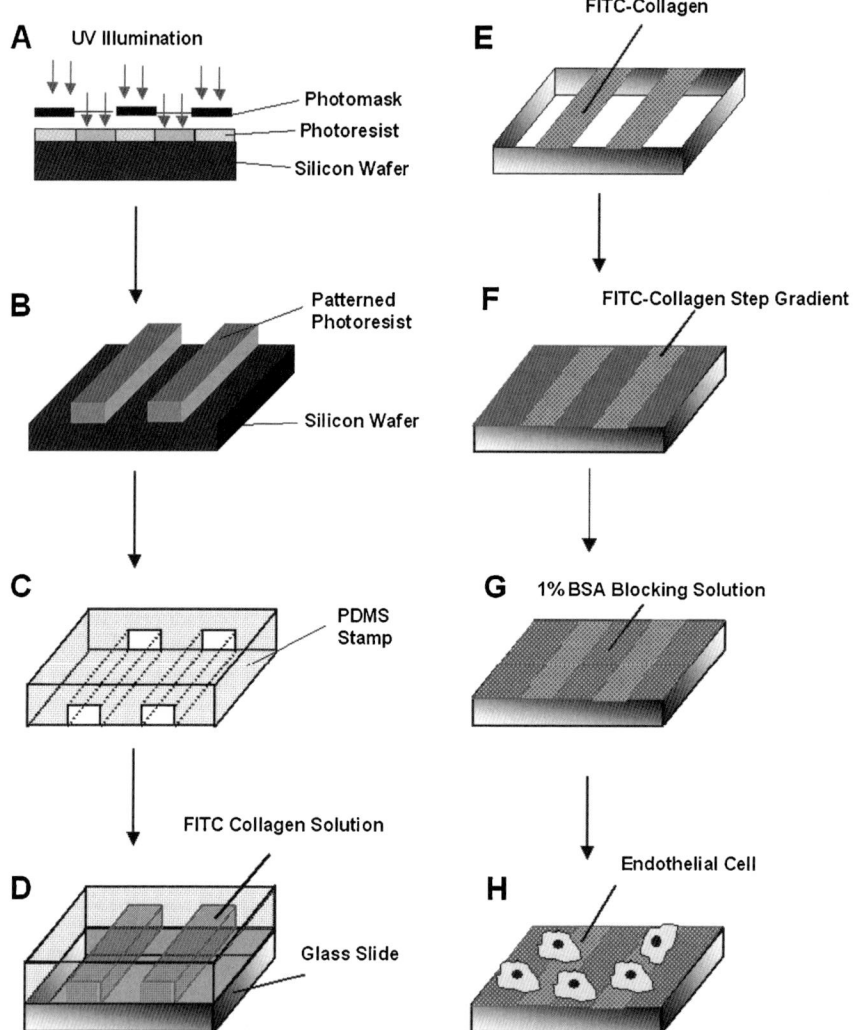

Fig. 1. Micropatterning procedures. (**A**) The negative photoresist was spin-coated on silicon wafer and exposed to UV light through a photomask. (**B**) Photoresist without UV polymerization was developed away, leaving a patterned surface. Then, the silicon wafer was etched using the STS Deep Reactive Ion Etcher, and the photoresist was washed away. (**C**) Poly(dimethylsiloxane) (PDMS) was cast onto the silicon wafer and cured, and the PDMS mold was removed and cleaned. (**D**) PDMS mold was put onto the surface to be patterned to form microfluidic channels, and FITC collagen solution was introduced into the channels and incubated overnight.

5. Pour the PDMS onto the silicon wafer in the aluminum foil reservoir so that the PDMS is approximately 2–2.5 cm thick. Leave the poured PDMS aside at room temperature for 10–15 min to allow bubbles to rise to the surface. To remove bubbles below the surface of the PDMS, degas under vacuum for another 10–15 min. Once bubbles are eliminated, place the degassed PDMS in an oven at 70 °C for 4 h (*see* **Note 3**).

6. Cut the solidified PDMS into stamps using a scalpel or a razor.

7. Place the PDMS stamps in a beaker containing 70% ethanol and clean it in a sonicator for 15 min, followed by a rinsing in dIH$_2$O.

8. Dry the PDMS stamps using an aspirator and then place in an oven at 37 °C for 15–20 min to ensure complete drying.

9. Use a mixture of rat-tail collagen I and FITC collagen I (with known concentration) as the patterning solution. Dilute collagen solution to desired concentration (e.g., 0.8 mg/ml) using 0.1% (v/v) acetic acid.

10. Place the PDMS stamp in conformal contact with the glass slide giving a complete seal (*see* **Note 4**).

11. Drip the collagen solution on one end of the microchannels at the end of PDMS stamp. On the other side of the stamp, introduce a vacuum or negative pressure to draw the solution through the channels (*see* **Note 5**). Incubate the patterning solution in the stamps overnight at room temperature.

12. Remove the PDMS stamp and wash the micropatterned surface with 1× PBS. Coat the entire surface with a lower concentration (e.g., 0.008 mg/ml) of collagen solution for 1 h at room temperature to create step density changes on the collagen surface.

13. Coat the entire surface with 1% BSA in PBS for 30 min to block non-specific adhesion sites.

14. Wash the surface gently with 1× PBS.

3.3. Quantification of ECM Surface Density

1. To generate standard curve for collagen surface density, dilute the mixture of FITC collagen I and collagen I serially in 0.1% (v/v) acetic acid, for example, with concentrations 0.8, 0.08, and 0.008 mg/ml.

2. Spot 1 μl collagen solutions after a serial dilution, for example, 0.8, 0.08, and 0.008 mg/ml, onto a glass slide and let dry. Calculate the collagen surface density at each spot.

Fig. 1. (**E**) The PDMS mold was removed, leaving a surface with patterned collagen strips. (**F**) Low concentration of FITC collagen was applied to the surface to create a step gradient. (**G**) The surface was blocked with 1% bovine serum albumin (BSA). (**H**) The cells were seeded onto the patterned surface, and the non-adherent cells were washed away.

3. Visualize the spot with a fluorescence microscope. Quantify pixel intensity with SCION Image software, where pixel values are 0–255 (0 is black and 255 is white).
4. Plot the collagen surface density and pixel intensity and curve fit to generate a standard curve.
5. Micropattern the collagen solution onto a glass slide using the procedure described in **Subheading 3.2**.
6. With the same microscope settings, collect the images of micropatterned collagen and quantify the intensity of the pixel.
7. Using the standard curve, calculate the collagen surface density in the high-density and low-density areas on micropatterned surfaces.

3.4. Cell Seeding

1. Aspirate the medium from BAEC culture and wash BAECs with 1× PBS.
2. Add 0.5× trypsin (2 ml for a 100-mm dish) and incubate BAECs with trypsin in the incubator at 37 °C for 1 min.
3. Check detachment of BAEC under a light microscope and add 10% FBS DMEM (8 ml for a 100-mm dish) to de-activate trypsin.
4. Centrifuge the cell suspension for 3 min at 200 g and 25 °C to pellet the BAECs.
5. Aspirate the fluid above the BAEC pellet and re-suspend BAECs in 2 ml 0.5% FBS DMEM.
6. Dilute BAECs and seed at 10% confluency on the micropatterned surface.
7. Allow BAECs to attach and spread for 3 h at 37 °C on the micropatterned surface.

3.5. Setting Up the In Vitro Parallel Plate Flow System for Phase-Contrast Time-Lapse Microscopy

We set up our flow system inside a custom-made temperature chamber built around the microscope to maintain the temperature at 37 °C (*see* **Note 6**) as shown in **Fig. 2**. In this way, we were able to observe cell migration over ECM gradients under fluid flow by using phase-contrast time-lapse video microscopy. The temperature chamber also allowed us to ventilate the medium in circulation within the flow system with air/5% CO_2 to keep the medium at a physiological pH level (7.4–7.5) (*see* **Note 7**). A beaker of autoclaved deionized water was placed within the chamber to maintain humidity.

1. Sonicate the polycarbonate flow chamber base and silicone gaskets for 20 min, followed by dipping in a 70% ethanol bath for 10 min, and allow to sit overnight for air drying in a tissue culture hood. Autoclave all other flow system components (e.g., tubing, tube connectors/adaptors, and medium reservoirs). Wipe clean the temperature chamber with 70% ethanol.

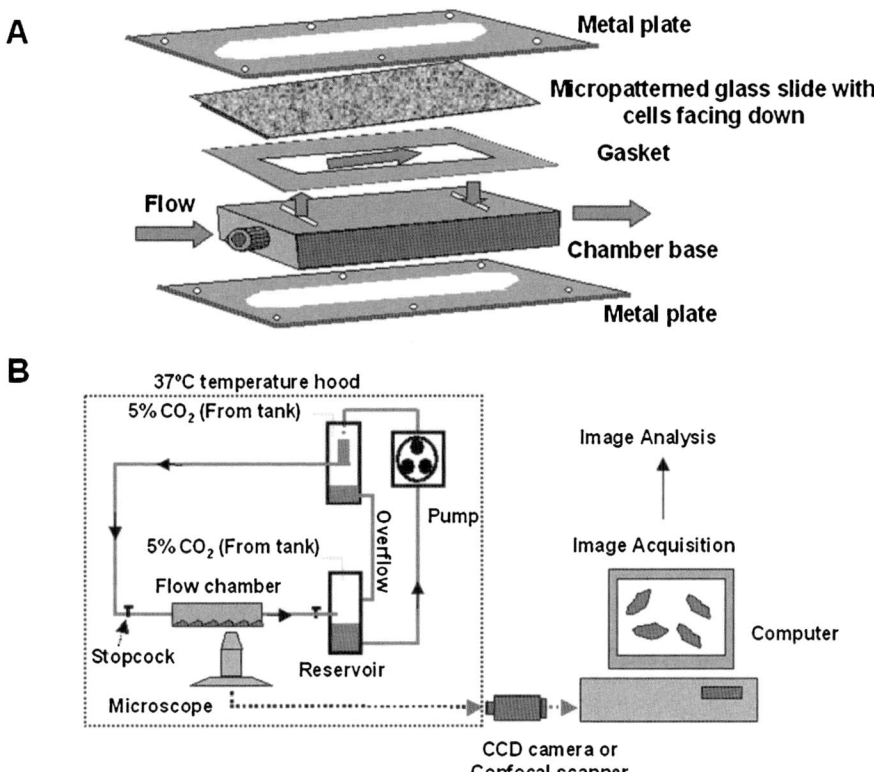

Fig. 2. In vitro flow system. (**A**) In vitro parallel plate flow chamber system set-up. A silicone gasket was placed on top of the polycarbonate flow chamber base, leaving the central area of the base and the two slits at the ends exposed. The base was then wetted with media, and the micropatterned glass slide with cells was placed face down onto the gasket. The flow chamber was then sandwiched by two metal plates and tightened with six screws. (**B**) Diagram of flow system setup for phase-contrast time-lapse video microscopy. After flow chamber assembly, tubing was used to connect the flow chamber to the flow system. Cells were exposed to a well-defined steady laminar flow generated by the hydrostatic pressure between two medium reservoirs, with the hydrostatic pressure maintained by a peristaltic pump. The system was set up in a custom-made 37 °C temperature hood built around a phase-contrast microscope. Circulating media within the flow system was ventilated by air with 5% CO_2. For time-lapse video microscopy, the stage was motorized in X, Y, and Z directions. In this way, multiple images at different locations across the slide could be visualized by a Hamamatsu Orca100 digital charge-coupled device (CCD) camera. Images were stored on a computer and analyzed.

2. Place a beaker of deionized water in the temperature chamber. Turn on the heater, fan, and temperature controller to warm the temperature chamber to 37 °C. Turn on the compressed air with 5% CO_2 tank and flowmeter so as to maintain the pH level.

3. Assemble the flow system in the temperature chamber as shown in **Fig. 2**. Fix the two custom-made glass medium reservoirs at different heights using clamps and clamp holders.

4. Assemble the tubing, tube connectors, three-way stopcocks, peristaltic pump and pump head, and polycarbonate flow chamber base to complete the circulation system (*see* **Note 8**).

5. Place an autoclaved glass slide without cells on the top of the silicone gasket and the polycarbonate chamber base to form a chamber. Sandwich the chamber between two metal plates and fix it with screws (*see* **Fig. 2A**).

6. Attach a 20-ml syringe to the circulation system at the side port of the three-way stopcock upstream of the flow chamber. Infuse slowly low serum medium (DMEM with 0.5% FBS) into the upstream and downstream tubing and medium reservoirs to avoid air bubbles.

7. Turn on the peristaltic pump and adjust the speed to maintain the hydrostatic pressure between the two medium reservoirs and the flow through the system.

8. Calculate the desired flow rate for a specific shear stress using the following equation: $Q = \tau Wh^2/6\mu$, where Q is the volumetric flow rate across the flow chamber, τ is the desired physiological shear stress level (e.g., $12\,dyn/cm^2$), W is the inner width of the silicone gasket (1.5 cm), h is the height of the flow chamber (i.e., the gasket thickness, 250 μm), and μ is the medium viscosity ($0.008\,dyn \times s/cm^2$ at 37 °C).

9. Measure the flow rate across the flow chamber. Stop the flow by turning the stopcock (at the same height as the entry of the lower reservoir downstream of the flow chamber) and allow the medium to flow from the side port into a 15-ml centrifuge tube. After collecting medium for 30 s, determine the weight of the medium using a weigh balance. Calculate the volume of the collected medium by dividing the weight by the medium density ($\sim 1\,g/ml$). Calculate the flow rate in milliliter per second units.

10. Adjust the relative heights of the two medium reservoirs according to the flow rate measurement. Repeat this step until the desired flow rate is obtained.

11. After the desired flow rate is obtained, stop the flow across the flow chamber by turning off the peristaltic pump and closing the three-way stopcocks. Disassemble the flow chamber, remove the glass slide without cells, and wet the surface of the chamber base with medium. Assemble the micropatterned slide with cells in the wetted chamber base surface (avoid air bubbles during assembly) and sandwich the flow chamber between the two metal plates. After assembly, turn the flow chamber upside down so that the cells are facing up and the slide is adjacent to the objective.

3.6. Phase-Contrast Time-Lapse Video Microscopy and Analysis of BAEC Migration on ECM Gradients Under Fluid Flow

1. Secure tightly the flow chamber on the microscope stage with adhesive tape and initiate fluid flow.
2. Adjust the exposure time for image acquisition under a 10× phase objective (*see* **Note 9**).
3. Using the image acquisition software, select the BAECs located near micropatterned ECM strips (with FITC fluorescent signal) and generate a list of points by scanning the slide manually. Store X, Y, and Z coordinates of each point in the computer (*see* **Note 10**).
4. After creating a list of points, set the time interval between cycles of image acquisition (e.g., 10 min) and define the number of cycles (i.e., 91 cycles for 15 h). After testing a cycle of images, start the time series of image acquisition.
5. After completing all image cycles, use a fluorescent image of each point to generate an image of the ECM gradient at each point. Disassemble the flow chamber and use the micropatterned slide with BAECs for immunofluorescent staining analysis.
6. Construct the time-series images using a avi movie file or extract as individual image files for further analysis.
7. For each point, open the computer file containing the fluorescent image of the ECM gradient. Tape a piece of transparency film onto the computer screen and trace the micropatterned ECM strip onto the transparency so that the exact location of the ECM gradient would be known when visualizing the phase-contrast time-lapse videos of BAEC migration.
8. Construct a time-lapse avi movie for each point and play it to view BAEC migration. BAECs near the edge of an ECM strip will appear pronounced, and their migration can be followed during video re-playing under slow motion.
9. Quantify the number of BAECs that displayed haptotactic (i.e., BAECs migrating from low to high ECM density against the fluid flow direction) or mechanotactic (i.e., BAECs migrating in the flow direction against the ECM gradient). Use statistical analysis (i.e., chi-square analysis) to determine whether there was a significant difference in terms of the percentage of BAECs exhibiting haptotactic behavior compared with static control samples (*see* **Note 11**).

3.7. Immunofluorescent Staining Analysis of FAs

1. After flow experiments, fix BAECs in 4% paraformaldehyde in PBS for 15–20 min at room temperature.
2. Remove paraformaldehyde and wash the slide with PBS three times.
3. Permeabilize BAECs with 0.5% Triton X-100 in PBS for 15–30 min at room temperature.
4. Wash with PBS three times. Incubate BAECs with 1% BSA in PBS for 1 h at room temperature to minimize background signal during staining.

5. Wash 3× with PBS. Incubate BAECs with a primary vinculin mouse antibody for 2 h at room temperature.

6. Wash BAECs with PBS three times and incubate with a rhodamine donkey anti-mouse secondary antibody (15 μg/ml in PBS) for 1–1.5 h at room temperature.

7. After washing with PBS, mount the BAEC samples in VectorShield antifade solution to help maintain fluorescent intensity. Analyze the effect of micropatterned ECM gradients and fluid shear stress on BAEC FA formation using confocal microscopy images from the stained samples. Visualize the micropatterned collagen (with FITC fluorescent signal) and FA staining (rhodamine) at different wavelengths.

3.8. Dynamics of FAs in BAEC Migration on Micropatterned ECM Step Gradients

1. Seed BAECs into wells of a six-well cell culture plate at 50–60% confluency the day before transfection in complete medium.

2. Replace complete medium with plain DMEM right before transfection.

3. According to manufacturer's instructions, mix LipofectAMINE 2000 transfection reagent in Opti-MEM1 and incubate for 5 min at room temperature with gentle swirling.

4. Mix gently DNA plasmids encoding GFP-tagged paxillin with Opti-MEM1 to achieve a concentration of 100 nM and incubate with LipofectAMINE 2000 transfection reagent for 20 min at room temperature.

5. Incubate DNA plasmids/LipofectAMINE 2000 solution with BAEC cultures for 5–6 h in a humidified incubator at 37 °C.

6. Replace DNA plasmids/LipofectAMINE PLUS solution with complete medium, and BAECs were allowed to recover overnight in humidified incubator.

7. Plate BAECs transfected with DNA plasmids encoding GFP-tagged paxillin onto a coverslip chamber cover glass (for ×40 oil immersion objectives) coated with micropatterned rhodamine collagen (0.8 mg/ml) strips in complete medium.

8. Load chamber cover glass onto a confocal microscope scanning stage inside a temperature hood. Track the dynamics of FAs in BAEC migration on micropatterned ECM step gradients through time-lapse fluorescence microscopy under a ×40 oil immersion objective using a confocal microscope, with the laser intensity, gain, focus plane, and digital zoom adjusted to optimize the image of GFP-tagged paxillin (*see* **Note 12**). Set laser intensity and scanning time as low as possible to minimize the adverse effect on the cell.

9. After the experiment, view the images (gray-scale .tif files), and the time-lapse movie of FA dynamics using an image analysis software to observe how ECM step gradients affect the dynamics of FAs during BAEC migration from low to high ECM density.

4. Notes

1. Aortic bovine, porcine, or human ECs may also be obtained commercially from Cambrex Inc. (East Rutherford, NJ, USA) and Cascade Biologics Inc. (Portland, OR, USA).

2. Unless we have routine access to a microfabrication facility, we may save time and effort if we place an order and have the silicon wafer molds made at a microfabrication service company.

3. PDMS will be polymerized when its surface is slightly sticky to the touch. The stickiness arises from weak interactions. PDMS is inherently sticky, but if it is wet, then it should be left in the oven for a longer period of time.

4. The conformal seal can be observed under a microscope if one is unsure. The seal should be resistant to a slight push by one's finger against the PDMS stamp. Without a strong seal, the patterning solution in step 11 of **Subheading 3.2** will leak in between channels yielding a poor pattern.

5. The solution can be seen traveling through the channels. Channels that are filled with solution are darker than empty ones.

6. To minimize the chances of contamination, the flow system can be assembled in a tissue culture hood and then transferred to the temperature chamber-enclosed microscope by wrapping the flow system components in autoclaved aluminum foil.

7. If compressed air with 5% CO_2 tanks is unavailable, CO_2-independent DMEM (Gibco/BRL) may be used to perfuse the flow system. The CO_2-independent medium is composed of monobasic and dibasic sodium phosphate and a small amount of sodium bicarbonate and can maintain the pH of the flow system for at least 24 h without a supply of CO_2.

8. Transparent Tygon LFL tubing (Cole-Parmer Instrument Co.) is used throughout the circulation system except for the tubing loaded in the pump head. Pharmed tubing (Cole-Parmer Instrument Co.) is loaded in the pump head, as this tubing is tougher and more resistant to wearing and tearing by the pump head.

9. After adjusting the exposure time for images, do not alter the light condition in the room significantly, as the changes in light will affect image quality.

10. When selecting the first point, choose a "blank" spot at a point outside of the flow chamber. In this way, between each cycle of images, the light will not remain focused on the cells on the slide, as prolonged exposure of cells to light may have adverse effects. Alternatively, an automatic shutter can be used to block light between image cycles.

11. For static controls, the same procedure outlined in **Subheading 3.6** was carried out, except that BAECs were grown on micropatterned chamber slides and no flow was applied to the cells. Analysis of movies was then carried out, and BAECs that touched the border of the ECM strips during migration were used for analysis to determine the percentage of BAECs exhibiting haptotactic behavior (i.e., migrate from low to high ECM density). Analysis of time-lapse videos of static control samples using Dynamic Image Analysis System (DIAS) software

with particular tracing function (Solltech, Oakdale, IA, USA) was used to show that BAEC migration speed was slower in regions of high ECM density compared with regions of low ECM density. Cells contacting each other during migration were excluded. The DIAS program was used to determine the centroid position of each cell from the cell outline at each time point, and the cell migration speed was quantified based on the generated cell migration paths.

12. When using the ×40 oil immersion objective, the samples might go out of focus due to stage or temperature fan vibrations. To minimize these vibrations, the microscope should be placed on an anti-vibration air table. The heater/fan should be placed in the temperature hood but not come in contact with microscope stage. If the cell is out of focus, manual correction with confocal scanning is needed.

References

1. Ross, R. (1993) The pathogenesis of atherosclerosis: a perspective for the 1990s. *Nature* **362**, 801–809.
2. Davies, P.F. (1995) Flow-mediated endothelial mechanotransduction. *Physiol. Rev.* **75**, 519–560.
3. Toborek, M., and Kaiser, S. (1999) Endothelial cell functions. Relationship to atherogenesis. *Basic Res. Cardiol.* **94**, 295–314.
4. Griffith, L.G., and Naughton, G. (2002) Tissue engineering–current challenges and expanding opportunities. *Science* **295**, 1009–1014.
5. Nugent, H.M., and Edelman, E.R. (2003) Tissue engineering therapy for cardio-vascular disease. *Circ. Res.* **92**, 1068–1078.
6. Carter, S.B. (1967) Haptotactic islands: a method of confining single cells to study individual cell reactions and clone formation. *Exp. Cell Res.* **48**, 189–193.
7. Herbst, T.J., McCarthy, J.B., Tsilibary, E.C., and Furcht, L.T. (1988) Differential effects of laminin, intact type IV collagen, and specific domains of type IV collagen on endothelial cell adhesion and migration. *J. Cell Biol.* **106**, 1365–1373.
8. Li, S., Butler, P., Wang, Y., Hu, Y., Han, D.C., Usami, S., Guan, J.L., and Chien, S. (2002) The role of the dynamics of focal adhesion kinase in the mechanotaxis of endothelial cells. *Proc. Natl. Acad. Sci. U. S. A.* **99**, 3546–3551.
9. Albuquerque, M.L., Waters, C.M., Savla, U., Schnaper, H.W., and Flozak, A.S. (2000) Shear stress enhances human endothelial cell wound closure in vitro. *Am. J. Physiol. Heart Circ. Physiol.* **279**, H293–H302.
10. Hsu, P.P., Li, S., Li, Y.S., Usami, S., Ratcliffe, A., Wang, X., and Chien, S. (2001) Effects of flow patterns on endothelial cell migration into a zone of mechanical denudation. *Biochem. Biophys. Res. Commun.* **285**, 751–759.
11. Hu, Y.L., Li, S., Miao, H., Tsou, T.C., del Pozo, M.A., and Chien, S. (2002) Roles of microtubule dynamics and small GTPase Rac in endothelial cell migration and lamellipodium formation under flow. *J. Vasc. Res.* **39**, 465–476.
12. Wojciak-Stothard, B., and Ridley, A.J. (2003) Shear stress-induced endothelial cell polarization is mediated by Rho and Rac but not Cdc42 or PI 3-kinases. *J. Cell Biol.* **161**, 429–439.

16

Isolation, Culture, and Functional Analysis of Adult Mouse Cardiomyocytes

Ronglih Liao and Mohit Jain

Summary

The pathogenesis of heart disease and the development of myocardial failure are highly dependent upon the cardiomyocyte—the basic contractile cell within the heart. Understanding and elucidating the complex networks that regulate cardiomyocyte function are central to the development of specific target-based therapeutic interventions. The relative recent advances in molecular genetics and generation and routine usage of transgenic and knockout mouse models have further necessitated that previously established cardiomyocyte methods be adapted for the isolation, culture, and study of primary adult murine cardiomyocytes, both freshly isolated and in culture. Such adaptation is based not only upon scalability of established techniques, as the mouse heart is an order of magnitude smaller than the rat heart, but also upon properties unique to the mouse. This chapter therefore describes current methods for the isolation, culture, and functional analysis of adult murine cardiomyocytes.

Key Words: Adult mouse cardiomyocytes; Cardiomyocyte culture; Cardiomyocyte contractility; Intracellular calcium transients.

1. Introduction

Cardiovascular disease remains the single greatest cause of morbidity and mortality in the developed world *(1)*. The pathogenesis of heart disease and the development of myocardial failure are highly dependent upon the cardiomyocyte—the basic contractile cell within the heart. In response to hemodynamic stress, ischemic or drug injury, infection, neurohormonal stimulants, or primary genetic mutations, cardiomyocytes undergo a complex remodeling process with resultant molecular, biochemical, structural, and functional

From: *Methods in Molecular Medicine, Vol. 139: Vascular Biology Protocols*
Edited by: N. Sreejayan and J. Ren © Humana Press Inc., Totowa, NJ

changes *(2)*. Understanding and elucidating the complex networks that regulate cardiomyocyte function are central to the development of specific target-based therapeutic interventions. In contrast to many other human diseases, the study of cardiovascular disease, in particular cardiomyocyte biology, is complicated by the lack of a representative and reliable cell line for intensive in vitro examination; as such, primary adult ventricular cardiomyocytes are of particular importance. The relative recent advances in molecular genetics and generation and routine usage of transgenic and knockout mouse models have further necessitated that previously established cardiomyocyte methods be adapted for the isolation, culture, and study of primary adult murine cardiomyocytes, both freshly isolated and in culture. Such adaptation is based not only upon scalability of established techniques, as the mouse heart is an order of magnitude smaller than the rat heart, but also upon properties unique to the mouse. This chapter therefore describes current methods, modified from earlier reports *(3,4)*, for the isolation, culture, and functional analysis of adult murine cardiomyocytes.

2. Materials

2.1. Equipment

2.1.1. Cardiomyocyte Isolation and Culture

1. CO_2 (2%) tissue culture incubator.
2. Isolated heart perfusion system (*see* **Fig. 1**).
3. Fine surgical tools (small scissors, curve-tip forceps, and fine-tip forceps; *see* **Fig. 2**).
4. Filter, $0.22 \mu m$ (cellulose acetate membrane).
5. Polystyrene tissue culture dish ($60 mm \times 15 mm$ or $35 mm \times 10 mm$).
6. Micro-vessel clip (Miltex, York, PA).
7. Waxed, non-sterile silk suture (5-0).
8. Sterile plastic transfer pipettes.
9. Polypropylene conical tube (50 and 15 ml).
10. Filter top, $250 \mu m$ (*see* **Fig. 3**).

2.1.2. Isolated Heart Perfusion System (see **Fig. 1**)

1. Glassware (water-jacketed spiral condense).
2. Temperature probe.
3. Water circulators.
4. Cannulas (blunt-end 20-G needle).
5. Pump.
6. Tubing.
7. Three-way stopcock.

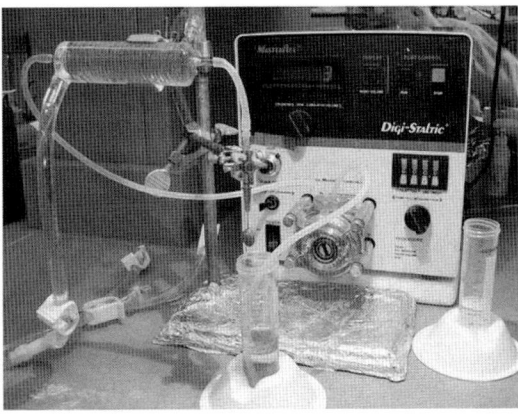 

Fig. 1. Isolated heart perfusion apparatus. Digestion buffer is pumped at a rate of 3 ml/min through a water-jacketed spiral condenser (attached to a heated circulating water bath) through a three-way stopcock and cannula and into an isolated murine heart. Higher magnification figure illustrates an isolated heart secured to a cannula tip with suture.

2.1.3. Cardiomyocyte Contractility and Calcium Measurements

1. Cardiomyocyte culture/stimulation chamber (Cell MicroControls, Norfolk, VA, Culture Camber System).
2. Inverted fluorescence microscope (Nikon, TS100, *see* **Note 1**).

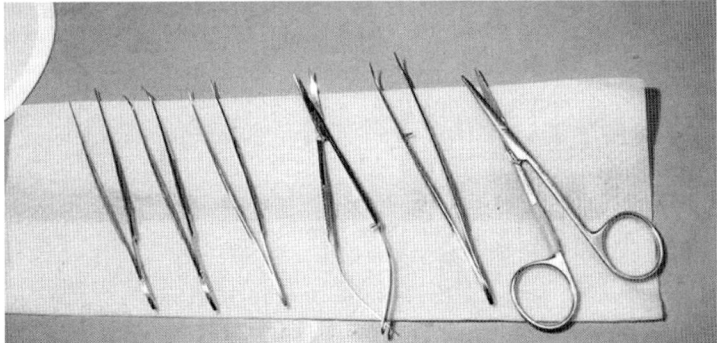

Fig. 2. Fine surgical tools. Displayed are required fine surgical tools, including (from left to right) three fine-tipped forceps, a fine-tipped surgical scissor, a curved-tip forceps, and a small scissor.

Fig. 3. Filter apparatus. Displayed is a 50-ml polypropylene conical tube with a 250-μm filter top for straining the cell suspension following enzymatic digestion.

3. Hardware and software for the assessment of cell contractility and intracellular calcium measurements (Ionoptix, Milton, MA, Myocyte System, *see* **Note 2**).

2.2. Buffers for Cardiomyocyte Isolation

1. Perfusion buffer: 135 mM NaCl, 4 mM KCl, 1 mM $MgCl_2$, 10 mM HEPES, 0.33 mM NaH_2PO_4, 10 mM glucose, 10 mM 2,3-butanedione monoxime (Sigma St. Louis, MO), 5 mM taurine (Sigma), pH 7.2 at 37 °C.
2. Digestion buffer: dissolve 0.3 mg/g body weight collagenase D (Roche, Indianapolis, IN), 0.4 mg/g body weight collagenase B (Roche), 0.05 mg/g body weight protease type XIV (Sigma) in 25 ml perfusion buffer.
3. Transfer buffers: Transfer buffer A—135 mM NaCl, 4 mM KCl, 1 mM $MgCl_2$, 10 mM HEPES, 0.33 mM NaH_2PO_4, 5.5 mM glucose, 10 mM 2,3-butanedione monoxime (Sigma), 5 mg/ml bovine serum albumin (Sigma), pH 7.4 at 37 °C. Transfer buffer B—137 mM NaCl, 5.4 mM KCl, 1.8 mM $CaCl_2$, 0.5 mM $MgCl_2$, 10 mM HEPES, 5.5 mM glucose, pH 7.4 at 37 °C.
4. Mix transfer buffers A and B to achieve a series of transfer solutions with increasing calcium concentrations of 0.06, 0.24, 0.6, and 1.2 mM as follows: 0.06 mM—29 parts transfer buffer A, 1 part transfer buffer B; 0.24 mM—13 parts transfer buffer

A, 2 parts transfer buffer B; 0.6 mM—2 parts transfer buffer A, 1 part transfer buffer B; 1.2 mM—1 part transfer buffer A, 2 parts transfer buffer B.

2.3. Media for Cardiomyocyte Culture

1. Laminin (Invitrogen, Carlsbad, CA).
2. Plating medium: minimal essential medium (MEM; GIBCO, Carlsbad, CA), 100 U/ml penicillin–streptomycin (Invitrogen), 2 mM L-glutamine, 10 mM 2,3-butanedione monoxime (Sigma), 5% fetal calf serum (Invitrogen).
3. Culture medium A (for less than 24-h culture): MEM, 100 U/ml penicillin–streptomycin, 2 mM L-glutamine, 0.1 mg/ml bovine serum albumin.
4. Culture medium B (for greater than 24-h culture): MEM, 100 U/ml penicillin–streptomycin, 2 mM L-glutamine, 0.1 mg/ml bovine serum albumin, 10 mM 2,3-butanedione monoxime, 10 μg/ml insulin, 5.5 μg/ml transferrin, 5 ng/ml selenium, insulin–transferrin–sodium selenite media supplement.

2.4. Buffers for Cardiomyocyte Contractility and Intracellular Calcium Measurements

1. Fura 2-AM stock: dissolve Fura 2-AM (Molecular Probes, Carlsbad, CA) in dimethyl sulfoxide (Sigma) to a final concentration of 1 mM Fura 2-AM. Aliquot and store in opaque tubes at −20 °C. Note: Fura 2-AM should not be exposed to direct light and should be handled only under dim light conditions.
2. Cardiomyocyte perfusion buffer: 137 mM NaCl, 5.4 mM KCl, 1.8 mM $CaCl_2$, 0.5 mM $MgCl_2$, 10 mM HEPES, 5.5 mM glucose, 0.5 mM probenecid (Sigma), pH 7.4 at 37 °C.
3. Fura 2-AM loading buffer: identical composite as "cardiomyocyte perfusion buffer", with the exception of pH 7.4 at room temperature.

3. Methods

3.1. Adult Mouse Cardiomyocytes Isolation and Culture

3.1.1. Setup Procedures

1. Prior to the isolation, clean and sterilize all dissection tools as well as prepare perfusion buffer, digestion buffer, and transfer buffers. Filter all buffers and media using 0.22-μm cellulose acetate membrane filters. Perfusion and digestion buffers should be equilibrated to 37 °C prior to use. Plating and culture media should be equilibrated in a 2% CO_2 culture incubator at 37 °C for at least 2 h prior to use.
2. Clean and sterilize the perfusion system with 50 ml 70% ethanol and then thoroughly rinse with another 100 ml sterile water. Finally, prime the perfusion system with 20 ml perfusion buffer for another 5 min. Remove any residual air bubbles from the perfusion system.

3. Adjust the temperature of the water circulator to achieve a perfusate temperature of 37 °C (measured at cannula tip) at a constant perfusion flow rate of 3 ml/min.
4. Place 10 ml perfusion buffer in a 60-mm × 15-mm tissue culture dish and chill to 4 °C.

3.1.2. Preparation of Laminin-Coated Culture Dishes

1. Add 10 ml cold phosphate-buffered saline to the laminin concentrate stock (Invitrogen) to achieve a final laminin concentration of 10 μg/ml.
2. Add 1 ml laminin to a 35-mm × 10-mm culture dish (2 ml for 60-mm × 15-mm culture dish), gently shake to ensure coating of the entire culture dish surface, and incubate at 37 °C for a minimum of 2 h prior to use. Laminin-coated dishes must be freshly prepared on the day of use.

3.1.3. Isolation and Cannulation of the Heart

1. Inject the mouse with 200 IU heparin i.p.
2. Ten minutes after heparin injection, anesthetize the mouse with pentabarbital (65 mg/kg body weight i.p.) (*see* **Note 3**).
3. Once the mouse is fully anesthetized, clip excess animal chest hair and wipe the chest region gently with 70% ethanol.
4. Expose the rib cage using a fine scalpel. Cut the rib cage along the mid-sternal axis using fine surgical scissors and separate the ribs to expose the thoracic cavity. Using fine forceps and surgical scissors, lift the heart and cut the aorta 2–3 mm from the aortic valve, freeing the heart from the thoracic cage. Immediately place the heart in a 60-mm × 15-mm culture dish containing the 10 ml pre-chilled perfusion buffer. Ensure that the entire heart and aortic root is entirely submerged to prevent air from entering the aorta. Trim excess soft tissue from the aorta.
5. Start the perfusion pump and adjust the perfusion buffer flow rate to 0.5 ml/min. Using fine-tip forceps, attach the heart to the perfusion cannula through the aorta (may be performed with the aid of a magnifying lens or dissection microscope). Visually ensure the tip of the cannula remains 0.5–1 mm above aortic valve, to allow for proper coronary perfusion. Use the micro-vessel clip to secure the aorta onto the cannula and then firmly tie the aorta to the cannula using suture (*see* **Fig. 1**). Once the heart is secured, increase the perfusion buffer flow rate to 3 ml/min. Note: Following excision, cannulation and perfusion of the heart should be completed in less than 60 s to prevent significant myocardial ischemia.

3.1.4. Cardiomyocyte Dissociation and Culture

1. Perfuse the heart with perfusion buffer for 3 min and switch to perfusion with the digestion buffer. Note: Digestion buffer may be re-circulated to conserve collagenase/protease.
2. Following 7–10 min of perfusion with the digestion buffer, the heart will become progressively edematous and appear pale and flaccid. Once flaccid, the digestion

procedure is complete. Note: The greatest degree of variability within the isolation procedure is the time to proper digestion.

3. Remove the heart from cannula using sterile scissors and place in a 60-mm × 15-mm culture dish containing 5 ml room temperature transfer buffer A. All subsequent steps should be performed in a strictly sterile manner under a laminar flow culture hood.

4. Trim the atria and great vessels from the ventricles and dispose of the extraneous tissue. Cut the ventricles into 4–5 sections and gently tease the tissue apart with sterile-opposed forceps (*see* **Fig. 4**), dissociating the cells. Pipette the suspension several times with a sterile plastic transfer pipette to further allow cardiomyocyte dissociation (*see* **Fig. 4**).

5. Transfer the cell suspension with a sterile plastic transfer pipette to a 50-ml polypropylene conical tube, straining through a 250-μm filter top (*see* **Fig. 3**). Rinse the dish with another 3 ml transfer buffer A and transfer the remaining solution to the same 50-ml conical tube. Place the conical tube upright and allow the cells to settle by gravity for 15 min.

6. Pipette the cells pellet with a sterile plastic transfer pipette and transfer to a 15-ml polypropylene conical tube containing 5 ml 0.06 mM calcium transfer buffer.

7. Place the conical tube upright and allow the cells to settle by gravity for 15 min.

8. Again pipette the cell pellet with a sterile plastic transfer pipette and transfer to a 15-ml polypropylene conical tube containing 5 ml 0.24 mM calcium transfer buffer.

9. Repeat **steps 7** and **8** using 0.6 and 1.2 mM calcium transfer buffer.

10. Several minutes after the transfer of cells to 1.2 mM calcium transfer buffer, place a droplet of cell suspension on a glass slide and examine under ×40 light microscopy for estimation of the percentage of rod-shaped cardiomyocytes (procedure typically yields $\sim$ 80% calcium-tolerant rod-shaped cardiomyocytes).

3.1.5. Cardiomyocyte Plating and Culture (see *Fig. 5*)

1. Using a sterile plastic transfer pipette, transfer the cardiomyocyte pellet from the 1.2 mM calcium transfer buffer to a new 15-ml conical tube containing 12 ml plating media (pre-equilibrated in 2% CO_2 incubator at 37 °C for 2 h). To ensure cells are properly re-suspended in the plating media, gently pipette the suspension several times.

2. Plate the cardiomyocyte suspension on the laminin-coated culture dishes, using 2 ml cell suspension per 35-mm × 10-mm culture dish or 3 ml suspension per 60-mm × 15mm culture dish. Immediately after plating, place the culture dishes in a 2% CO_2 incubator at 37 °C for 1 h to allow cardiomyocytes to adhere to the laminin-plated culture dish.

3. After 1 h, gently (as to not displace adhered cardiomyocytes) aspirate and discard the plating medium with a sterile Pasteur pipette. Wash each culture dish with

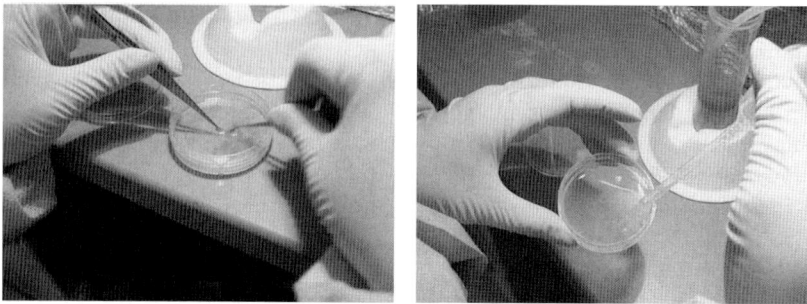

Fig. 4. Cardiomyocyte dissociation. Following enzymatic digestion, cardiomyocytes are dissociated by teasing apart the tissue with sterile-opposed forceps (left) followed by gently pipetting the suspension with a plastic transfer pipette (right).

1.5 ml culture media (culture media A or B, depending on desired length of culture), pre-equilibrated in 2% CO_2 incubator at 37 °C for 2 h. Again aspirate and discard the cell media, to ensure unattached cardiomyocytes and debris are removed. Add 1.5 ml culture media (culture media A or B, depending on desired length

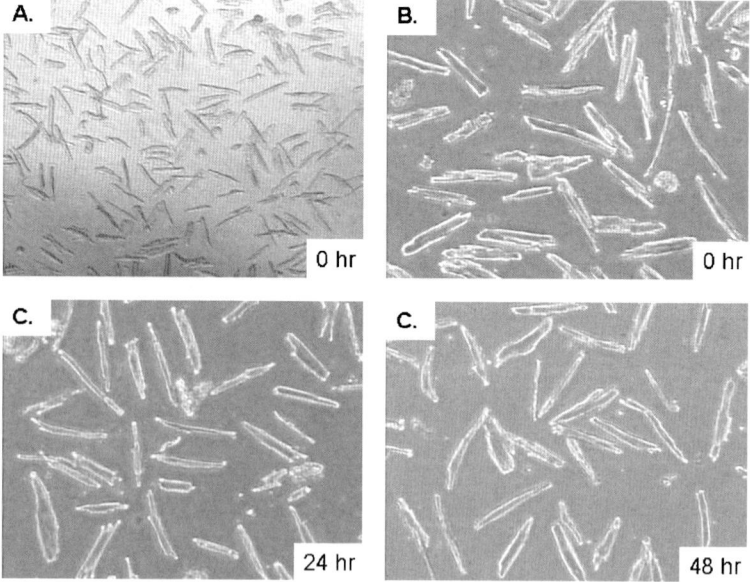

Fig. 5. Murine cardiomyocytes in culture. Adult murine cardiomyocytes are shown after (**A**) 0 h (×10), (**B**) 0 h (×20), (**C**) 24 h (×20), and (**D**) 48 h (×20) in culture.

of culture) to the culture dish and place back in a 2% CO_2 incubator at 37 °C. Note: During wash or media change, always gently add the media to the side of a dish, rather than pipetting media directly on the cells, to prevent displacing cardiomyocytes.

3.2. Cardiomyocyte Contractility and Intracellular Calcium Measurements

Analysis of in vitro cellular contractility and intracellular calcium transients provides a unique opportunity to investigate cardiomyocyte function, in the absence of neuro-hormonal stimulation or the influence of non-myocyte cardiac cells. Such analyses of murine cardiomyocytes have been instrumental in determining cardiomyocyte function—both basal performance and in response to pharmacologic and pacing stress *(5,6)*—in defining the cardiomyocyte phenotype in genetically modified animals *(7,8)*, and in elucidating specific cardiomyocyte signaling pathways *(9)*.

To perform measurements of cellular contractility or intracellular calcium transients, isolated cardiomyocytes must be plated or cultured on a suitable chamber. Our laboratory currently utilizes a commercially available cardiomyocyte stimulation chamber system (Cell MicroControls; http://www.cellmc.com) that consists of a cell-perfusion chamber, field electrodes, in-line heater, and temperature probe to allow for continuous regulation of temperature as well as field stimulation. More economical custom-made perfusion systems may also be employed, provided that they allow for close temperature regulation, continuous cell perfusion, field stimulation, and can be securely mounted on an inverted microscope. For clear assessment of cardiomyocyte shortening and intracellular calcium transients, it is recommended that cells are cultured at a density that allows for a minimum of 200 μm between adjacent cells.

3.3. Measurement of Cell Shortening

1. The instrument should be turned on according to the specific manufacturer's instructions prior to cell experiments.
2. Mount and secure the cell-stimulation chamber on the inverted fluorescence microscope stage. Electrically stimulate cells at a given frequency (1–10 Hz) and superfuse cells with cardiomyocyte perfusion buffer (pH 7.4 and 37 °C).
3. Isolated and cultured cells exhibit a degree of heterogeneity; therefore, the identical inclusion criteria should be applied to all groups for cell physiology experiments. Currently, employed inclusion criteria include (i) rod shaped with a clear striation pattern and defined cell edges (no blebs or cauliflower-shaped cell ends), (ii) quiescent when unstimulated, and (iii) stable contractility at desired stimulation

rate for at least 15 min. Collecting data from cells obeying the above criterion, despite the amplitude of contractility, will yield a representative and reproducible contractile phenotype (*see* **Note 4**).

3.4. Fura 2-AM Loading

1. Remove the myocyte stimulation chamber from culture incubator, aspirate culture media, and replace with 0.5 ml Fura 2-AM loading buffer. Add 0.5 µl Fura 2-AM stock (1 mM) to the chamber to achieve a final concentration of 1 µM and allow for incubation for 15 min at room temperature (*see* **Note 5**).
2. After 15 min, wash the cells twice with Fura 2-AM loading buffer to remove excess free Fura 2-AM. Allow another 40 min for the desertification of the Fura 2-AM ester in the cells and change buffer every 10 min during this time.

3.5. Measurement of $[Ca^{2+}]_i$

1. In addition to Fura 2-AM, cytosolic calcium can be determined using alternative calcium-sensitive fluorescence probes. A panel of fluorescence probes is available from Molecular Probes (http://probes.invitrogen.com/).
2. Intracellular calcium transients are usually collected simultaneously with cell shortening. To remove background from the fluorescence measurements, background measurements are assessed in four cell-free fields adjacent to the cardiomyocyte. The average fluorescence signal from the background is then subtracted from the cardiomyocyte's fluorescence signal.

3.6. Special Considerations

1. Isolation of cardiomyocytes from mice with structural cardiac disease—including ventricular hypertrophy, chamber dilation, and/or pathologic fibrosis—may require modification of isolation procedures *(7,8)*. Several approaches have been utilized to optimize isolation of cardiomyocytes from diseased hearts. In our experience, we have found that increasing the concentration of collagenase B and protease type XIV and/or extending the digestion time period yields the best results for calcium-tolerant cardiomyocytes.
2. Isolation of cardiomyocytes from aging mouse hearts may be performed using the above-described techniques without modification *(6)*.
3. Genetic transfection and the biochemical study of isolated adult murine cardiomyocytes have been previously well described *(3,4)*.

4. Notes

1. Any commercially available inverted fluorescence microscope with a ×40 fluorescence objective is suitable for the experiments.
2. We currently utilize commercially available equipment (Ionoptix, http://www.ionoptix.com/) to assess cardiomyocyte contractility and intracellular calcium

transients. Although several commercially available hardware and software packages are available, in our experience, the Ionoptix equipment provides particular easy of use, with both simultaneous measurement of cell shortening and intracellular calcium and off-line data analysis. Furthermore, the Ionoptix system allows for the assessment of cardiomyocyte contractility through cellular edge detection or sarcomere length.

3. Alternative methods of anesthesia can be used in accordance with approved individual Institutional Animal Care and Use Committee's guidelines.
4. Details of data collection are dependent upon the equipment used. Reader should refer to manufacturer instructions accordingly.
5. It is very important that the loading procedure be conducted at room temperature in the dark.

Acknowledgments

The authors thank all members of Cardiac Muscle Research Laboratory at Brigham and Women's Hospital, Harvard Medical School, and, in particular, Drs Lei Cui and Bo Wang for their technical expertise. This work was supported by National Institutes of Health grants HL-73756, HL-67297, and HL-71775 (RL).

References

1. American Heart Association. (2005) *Heart disease and stroke statistics – 2005 update*. American Heart Association. Dallas, Texas.
2. Braunwald, E., and Bristow, M. R. (2000) Congestive heart failure: fifty years of progress. *Circulation* **102**, 14–23.
3. Zhou, Y. Y., Wang, S. Q., Zhu, W. Z., Chruscinski, A., Kobilka, B. K., Ziman, B., Wang, S., Lakatta, E. G., Cheng, H., and Xiao, R. P. (2000) Culture and adenoviral infection of adult mouse cardiac myocytes: methods for cellular genetic physiology. *Am. J. Physiol. Heart Circ. Physiol.* **279**, H429–H436.
4. O'Connell, T. D., Ni, Y. G., Lin, K.-M., Han, H., and Yan, Z. (2003) Isolation and culture of adult mouse cardiac myocytes for signaling studies. *AFCS Research Reports [online]* **1**(5), http://www.signaling-gateway.org/reports/v1/CM0005/CM0005.htm.
5. Jain, M., Lim, C. C., Nagata, K., Davis, V. M., Milstone, D. S., Liao, R., and Mortensen, R. M. (2001) Targeted inactivation of Galpha(i) does not alter cardiac function or beta-adrenergic sensitivity. *Am. J. Physiol. Heart Circ. Physiol.* **280**, H569–H575.
6. Lim, C. C., Apstein, C. S., Colucci, W. S., and Liao, R. (2000) Impaired cell shortening and relengthening with increased pacing frequency are intrinsic to the senescent mouse cardiomyocyte. *J. Mol. Cell. Cardiol.* **32**, 2075–2082.
7. Michael, A., Haq, S., Chen, X., Hsich, E., Cui, L., Walters, B., Shao, Z., Bhattacharya, K., Kilter, H., Huggins, G., Andreucci, M., Periasamy, M.,

Solomon, R. N., Liao, R., Patten, R., Molkentin, J. D., and Force, T. (2004) Glycogen synthase kinase-3beta regulates growth, calcium homeostasis, and diastolic function in the heart. *J. Biol. Chem.* **279**, 21383–21393.

8. Matsui, T., Li, L., Wu, J. C., Cook, S. A., Nagoshi, T., Picard, M. H., Liao, R., and Rosenzweig, A. (2002) Phenotypic spectrum caused by transgenic overexpression of activated Akt in the heart. *J. Biol. Chem.* **277**, 22896–22901.

9. Nagata, K., Ye, C., Jain, M., Milstone, D. S., Liao, R., and Mortensen, R. M. (2000) Galpha(i2) but not Galpha(i3) is required for muscarinic inhibition of contractility and calcium currents in adult cardiomyocytes. *Circ. Res.* **87**, 903–909.

17

Mechanical Measurement of Contractile Function of Isolated Ventricular Myocytes

Loren E. Wold and Jun Ren

Summary

Isolation of ventricular myocytes from all species of animals has revolutionized the field of cardiovascular research, allowing the assessment of true cardiac effects of drugs, treatments, and so on. With recent advances in physiology at the cellular level, direct assessment of isolated ventricular myocyte mechanics has become an increasingly powerful and important area of research. This technique provides important information on excitation–contraction coupling of the heart, either when investigating the effect of a drug on the heart or mechanical function under certain pathophysiological states. The goal of this chapter is to provide a detailed list of methods for isolation of ventricular myocytes from both rats and mice, as well as a list of certain "tricks" that have proven useful in enhancing the yield of viable cells from the whole heart. Mastering the isolation of myocytes will provide a much better framework for real-time beat-to-beat recording of myocyte contraction as well as intracellular Ca^{2+} transients.

Key Words: Myocyte; Contractility; Intracellular Ca^{2+} transients.

1. Introduction

The ventricular myocyte is the essential contractile unit of the heart and is the main cell comprising the myocardium. Measurement of contractile function of the individual myocyte provides the most direct information on whether cardiac contractile function is altered by a certain drug or under a certain pathological state. Several new techniques have been developed for use by the physiologist for assessment of contractile function, including video-based edge-detection systems, photodiode arrays, and spatial video-based edge imaging *(1,2)*. The

From: *Methods in Molecular Medicine, Vol. 139: Vascular Biology Protocols*
Edited by: N. Sreejayan and J. Ren © Humana Press Inc., Totowa, NJ

most commonly utilized technique, however, is edge-detection. This technique measures changes in isotonic contraction of the myocyte in real time.

The ability to delineate the exact function of isolated ventricular myocytes is extremely important, particularly because the results obtained from multicellular preparations such as papillary muscles are difficult to interpret due to the presence of a heterogenous mixture of cell types. Mechanical function of the myocardium may be affected by non-myocyte factors such as fibroblasts and nerve terminals that can complicate interpretation of heart function under certain disease states. For example, under ethanol exposure, heart function is altered. However, it is unclear whether the alteration is due to enhanced interstitial fibrosis versus changes in myocyte function *(3)*. Therefore, it is important that a tool exists for the physiologist to study isolated ventricular myocyte function, without the presence of other tissue types. This is available through video-based edge-detection of isolated ventricular myocytes.

2. Materials

2.1. Equipment and Reagents

1. IonOptix MyoCam System. (We use an Olympus IX-70 inverted microscope.)
2. Rat or mouse heart.
3. Forceps ($\times 2$), plastic weigh dishes, and suture (for securing the heart to the cannula).
4. Retrograde perfusion setup with water-jacketed tubing.
5. Krebs–Henseleit bicarbonate (KHB) buffer containing 118 mM NaCl, 4.7 mM KCl, 1.25 mM $CaCl_2$, 1.2 mM $MgSO_4$, 1.2 mM KH_2PO_4, 25 mM $NaHCO_3$, 10 mM N-[2-hydro-ethyl]-piperazine-N'-[2-ethanesulfonic acid] (HEPES), and 11.1 mM glucose, equilibrated with 5% CO_2 and 95% O_2.
6. Calcium-free KHB buffer: same as in **step 7** below, without calcium.
7. Ca^{2+}-free KHB containing 223 U/ml collagenase (Worthington Biochemical Corp., Freehold, NJ, USA) and 0.1 mg/ml hyaluronidase (Sigma).
8. Trypsin solution: 0.02 mg/ml (Sigma Chemical, Inc., St. Louis, MO, USA).
9. Ca^{2+}-free Tyrode's buffer containing 131 mM NaCl, 4 mM KCl, 1 mM $MgCl_2$, 10 mM HEPES, and 10 mM glucose, supplemented with 2% bovine serum albumin, with a pH of 7.4 at 37 °C.
10. Modified Tyrode solution (pH 7.4) containing 135 mM NaCl, 4 mM KCl, 1 mM $MgCl_2$, 10 mM HEPES, 0.33 mM NaH_2PO_4, 10 mM glucose, and 10 mM butane-dione, equilibrated with 5% CO_2 and 95% O_2.
11. Type II collagenase (for rat), collagenase D, 0.9 mg/ml (for mouse), and hyaluronidase.
12. Buffer for cell shortening/relengthening assay: 131 mM NaCl, 4 mM KCl, 1 mM $CaCl_2$, 1 mM $MgCl_2$, 10 mM glucose, and 10 mM HEPES, at pH 7.4.
13. High-speed centrifuge.
14. Sterile incubator for cell culture.

3. Methods

3.1. Animal Preparation

1. Inject the animals with heparin (1000 U/kg, i.p.) approximately 20 min prior to commencing the experimental protocol (*see* **Note 1**).
2. Anesthetize the animal by injecting ketamine (75 mg/kg, i.p.) and xylazine (5 mg/kg, i.p.).
3. Shave the dorsal side of the animal.
4. Make sure that the animal is free of any pedal reflexes before continuing.
5. Make a surgical incision posterior to the xiphoid process. Cut superiorly until reaching the xiphoid process and make another incision running down the chest cavity along the rib cage. Continue cutting through the sternum until you reach the diaphragm.
6. Cut down along the sides of the diaphragm close to the heart. Be careful not to puncture the heart.
7. To remove the heart, pick it up between the index finger and thumb and cut the aorta below.
8. Place the heart in a weigh bath with a small amount of Ca^{2+}-containing buffer.
9. Carefully remove all remnant non-heart tissue and discard. Make sure to expose the aorta properly, leaving enough of the vessel for cannulation (preferably to the carotid bifurcation).
10. While holding the aorta with two forceps, lift the heart from the dish and place on the perfusion cannula that should have buffer running through at a slow rate.
11. Clamp the aorta to the surgical needle and secure it with a piece of suture.
12. Increase the flow of buffer until the drops fall at a rate of 5–10 times per minute. If the cannulation is performed successfully, buffer will fill the carotid arteries, making them appear translucent.
13. The heart should begin to beat rhythmically and the drops will become clear (*see* **Notes 2** and **3**).

3.2. Rat Ventricular Myocyte Isolation Procedure

1. Once the heart is successfully cannulated, perfuse it with KHB buffer at 37°C.
2. Continue perfusing the hearts with a nominally Ca^{2+}-free KHB buffer for 2–3 min until spontaneous contractions cease.
3. Depending on the physiological condition of the animal, continue perfusion for another 15–20 min with Ca^{2+}-free KHB containing collagenase and hyaluronidase.
4. After perfusion, remove the ventricles and mince, under sterile conditions, and incubate with the Ca^{2+}-free KHB with collagenase solution for 3–5 min.
5. Filter the digested tissue through a nylon mesh (300 μm).
6. Centrifuge at 60 g for 30 s to separate the myocytes from the collagenase–trypsin solution.

7. Resuspend myocytes in sterile-filtered, Ca^{2+}-free Tyrode's buffer.
8. Wash the cells with Ca^{2+}-free Tyrode's buffer to remove residual enzyme.
9. Slowly replenish the cells with extracellular Ca^{2+} up to 1.25 mM (*see* **Notes 4** and **5**).

3.3. Mouse Ventricular Myocyte Isolation

1. A similar cannulation procedure is utilized in the mouse; however, a light magnifier is helpful when isolating the aorta.
2. Rapidly remove the heart from anesthetized (using the same protocol as listed in **Subheading 3.1.**) mice through cardiectomy and immediately mount it on a temperature-controlled (37 °C) perfusion system.
3. Perfuse the heart with modified Tyrode solution (Ca^{2+} free) for 2 min.
4. Digest the heart for 15–20 min with collagenase D in modified Tyrode solution.
5. Remove the digested heart from the cannula.
6. Cut the left ventricle into small pieces in the modified Tyrode solution.
7. Agitate these pieces gently and resuspend the pellet of cells in modified Tyrode solution.
8. Allow the cells to settle for another 20 min at room temperature.
9. During this period add extracellular Ca^{2+} incrementally, back up to 1.20 mM.
10. For your experiments, always use the isolated myocytes within 8 h after isolation. We have, however, had some success with culturing mouse myocytes for up to 48 h (*see* **Notes 4** and **5**).

3.4. Cell Shortening/Relengthening

Mechanical properties of ventricular myocytes are assessed using a video-based edge-detection system (IonOptix Corporation, Milton, MA, USA) *(4)*.

1. Place the cells in a Warner chamber mounted on the stage of an inverted microscope and superfuse ($\sim$1 ml/min at 25 °C) with the cell shortening/relengthening buffer. Only rod-shaped myocytes with clear edges are selected for recording of mechanical properties or intracellular Ca^{2+} transients *(4)*. Do not use the myocytes with obvious sarcolemmal blebs or spontaneous contractions.
2. Field stimulate the cells with a suprathreshold (50%) voltage at a frequency of 0.5 Hz, 3 ms duration, using a pair of platinum wires placed on opposite sides of the chamber connected to a FHC (Fred Haer Company, Inc., Bowdoing, ME, USA) stimulator. The polarity of the stimulatory electrodes is reversed frequently to avoid possible buildup of electrolyte by-products (*see* **Note 6**).
3. The myocyte being studied is displayed on the computer monitor using an IonOptix MyoCam camera that rapidly scans the image area every 8.3 ms such that the amplitude and velocity of shortening/relengthening can be recorded with good fidelity.
4. Use the soft-edge software to capture changes in cell length during shortening and relengthening (*see* **Fig. 1**).

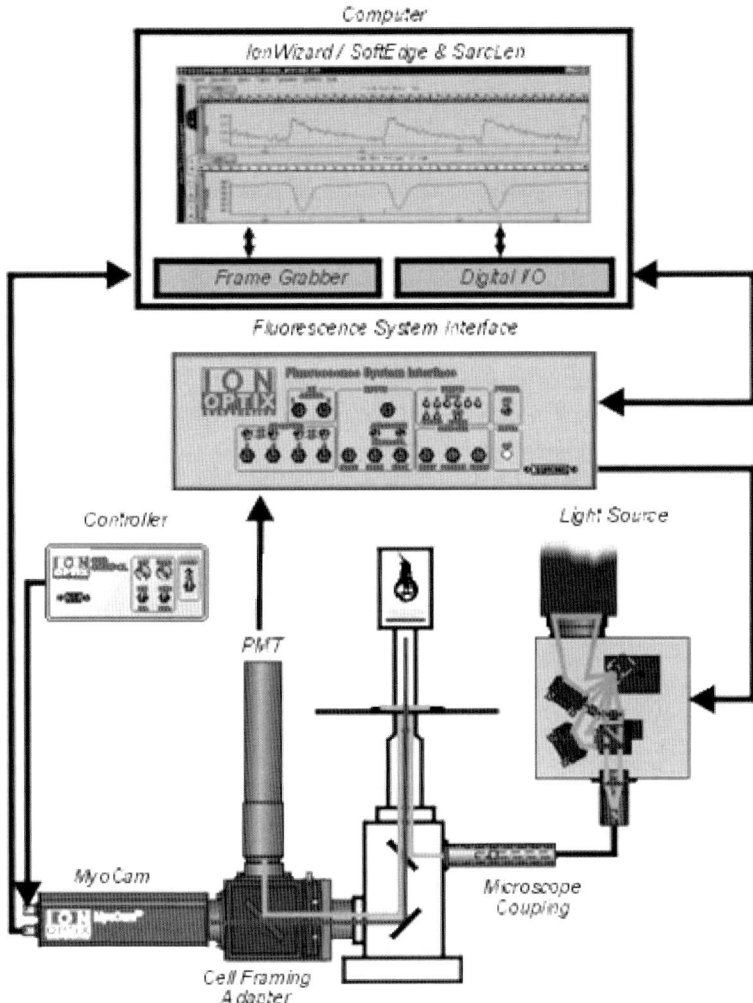

Fig. 1. Schematic of the IonOptix video-based SoftEdge MyoCam® system. This figure shows how edge-detection and intracellular Ca²⁺ can be monitored simultaneously in the same myocyte (kindly provided by Dr. Doug Tillotson from the IonOptix Corporation).

3.5. Intracellular Fluorescence Measurement

A separate cohort of myocytes is loaded with Fura-2/AM (0.5 μM) for 15 min, and fluorescence measurements are recorded with a dual-excitation fluorescence photomultiplier system (Ionoptix) as previously described *(4)*.

1. Place the myocytes in a chamber on the stage of an Olympus IX-70 inverted microscope and image through a Fluor 40× oil objective.
2. While the cells are being stimulated to contract at 0.5 Hz, expose them to light emitted by a 75-W lamp and pass through either a 360-nm or a 380-nm filter (bandwidths are ±15 nm).
3. Fluorescence emissions can be detected between 480 and 520 nm by a photomultiplier tube after first illuminating the cells at 360 nm for 0.5 s then at 380 nm for the duration of the recording protocol (333 Hz sampling rate).
4. Repeat the 360-nm excitation scan at the end of the protocol and infer the qualitative changes in intracellular Ca^{2+} concentration ($[Ca^{2+}]_i$) from the ratio of the fluorescence intensities at the two wavelengths.

Fluorescence and edge-detection can be monitored in the same myocyte (*see* **Fig. 1**); however, it is well known that many dyes can have a buffering effect, and therefore, it is not recommended to present functional data in dye-loaded cells (*see* **Note 7**).

4. Notes

1. Inadequate heparanization: Injection of heparin into the gut, gall bladder, or other organs in the abdominal cavity may result in inadequate heparanization and blood clotting in the coronary arteries. Enzymatic digestion and yield of myocytes will be significantly affected. If after cannulation and perfusion with buffer large blood clots are present in either the ventricle or the coronary arteries, increase the amount of heparin given in subsequent animals as well as take caution to inject medially to the hind limb in future studies. This will insure that heparin is given intraperitoneally.
2. Accumulation of air bubbles: Accumulation of air bubbles in the heart may buildup over the time of perfusion, causing an occlusion of perfusion. Therefore, make sure that the perfusion system is clear of all air bubbles at the start of the experimental protocol. Should any air bubbles accumulate in the perfusion system, an air trap device (most likely a valve in the perfusion apparatus or a 0.45-μm pore size cellulose acetate filter) may be placed within the perfusion path. Make sure to change the perfusion filter frequently as the air bubbles may accumulate over time.
3. Perfusion system maintenance: It is essential to wash the entire perfusion system with 70% ethanol and distilled water several times before and after each isolation. Enzyme may stick to the glassware tubing due to inappropriate washing after

perfusion or bacteria may grow within the system, contaminating the perfusion system.

4. Myocyte isolation and yield: The nature and quality of the myocyte isolation procedure may be the single most important factor in the success of mechanical assessment of myocytes and also for reliable comparison between experimental and control groups. Although the cell culture medium may provide a favorable environment for the cells to recover from enzymatic trauma, the yield, survival, and viability of the myocytes are largely dependent on the quality of isolation. Caution has to be taken when working with mice myocytes in such variables as temperature, type of enzyme, and difficulty of culturing. It is imperative that the temperature of the buffer leaving the perfusion cannula be $37 \pm 1\,°C$ and only collagenase D from Boehringer Mannheim (Ingelheim, Germany) is used in the isolation procedure for mice myocytes. A good isolation will yield approximately 70–80% viable cells or 4–5 million rod-shaped myocytes in both rats and mice.

5. Inadequate cell yield: The major problem in myocyte isolation is inadequate cell yield. If the animal under study is older than average or substantially overweight (such as in an obesity study), the heart may need to be digested for a longer period of time. Also, the perfusion pressure can be increased to enhance the digestion pressure within the heart. Make sure that when triturating the heart tissue, large clumps of tissue are not left out. These clumps will not be further digested and many cells will be lost. On the contrary, younger animals will require less perfusion pressure and shorter digestion times. All of this comes with practice in determining adequate perfusion and digestion.

6. Inability to stimulate myocytes to contract: A particular problem normally encountered in myocyte studies is the inability to get the cells to respond to electrical stimulation. Make sure to clean the platinum wires routinely to remove any electrolyte by-product buildup. Switching the polarity of the stimulating electrodes is sometimes helpful in "kick starting" the cells to contract. It is also helpful to allow the cells to recover for a short period of time ($\sim 15\,min$) once in the Warner chamber. It is essential that $CaCl_2$ be added to the HEPES buffer to facilitate contraction.

7. Inability to obtain a fluorescence signal: Fura-2/AM is cleaved into AM and Fura-2, which is able to penetrate the cell membrane and bind to intracellular Ca^{2+}. If the Fura-2-loaded cell is excited, it will fluoresce during contraction, and the signal will be displayed on the computer monitor. Make sure that the shutter is open, the emission filter is incorporated, and the room is as dark as possible. We have found that the addition of $1\,\mu l$ of Fura-2/AM stock to the Warner chamber will provide a beautiful signal after $15\,min$.

Acknowledgments

The work conducted in the laboratory of Dr. Ren has been supported by grants from the National Institute of Health, American Diabetes Association, and American Heart Association—Northland Affiliate, the Max Baer Fund, and the North Dakota Experimental Program to Stimulate Competitive Research. The authors are indebted to Dr. Doug Tillotson and colleagues at IonOptix Corporation for all of their hard work over the past several years in helping us to improve the video-based edge-detection system.

References

1. Wang, Z., Mukherjee, R., Lam, C.F., and Spinale, F.G. (1996) Spatial characterization of contracting cardiac myocytes by computer-assisted, video-based image processing. *Am. J. Physiol.* **39**, H769–H779.
2. Mukherjee, R., Crawford, F.A., Hewett, K.W., and Spinale, F.G. (1993) Cell and sarcomere contractile performance from the same cardiocyte using video microscopy. *J. Appl. Physiol.* **74**, 2023–2033.
3. Wold, L.E., Norby, F.L., Hintz, K.K., Colligan, P.B., Epstein, P.N., and Ren, J (2001) Prenatal ethanol exposure alters ventricular myocyte contractile function in the offspring of rats: influence of maternal Mg^{2+} supplementation. *Cardiovasc. Toxicol.* **1**, 215–224.
4. Wold, L.E., Relling, D.P., Duan, J., Norby, F.L., and Ren, J. (2001) Abrogated leptin-induced cardiac contractile response in ventricular myocytes under spontaneous hypertension: role of Jak/STAT pathway. *Hypertension* **39**, 69–74.

18

Studying Ischemia and Reperfusion in Isolated Neonatal Rat Ventricular Myocytes Using Coverslip Hypoxia

Kelly R. Pitts and Christopher F. Toombs

Summary

In vitro experimental models designed to study the effects of hypoxia and ischemia typically employ oxygen-depleted media and/or hypoxic chambers. These approaches, however, allow for metabolites to diffuse away into a large volume and may not replicate the local buildup of metabolic byproducts that occur in ischemic myocardium in vivo. Coverslip hypoxia (CSH) is a recently described method for studying hypoxia and ischemia derived from the byproducts and metabolites of contractile ventricular myocytes. Hence, this method is dependent on the purity and contractile activity of the isolated myocytes. We describe herein methods for isolating neonatal rat ventricular myocytes with these characteristics, as well as means for performing CSH, identifying viable and compromised myocytes after coverslipping, and tracking pH changes during CSH.

Key Words: Ischemia; Reperfusion; Neonatal rat; Ventricular myocyte; Fluorescence; pH; Annexin.

1. Introduction

Evaluation of hypoxic and ischemic insults on isolated cardiac myocytes typically involves atmospheric chambers or modified culture media. Replicating a hypoxic environment using a nitrogen chamber is a classical manner to deprive a population of myocytes of oxygen, and several metabolic *(1–3)*, genetic *(4–6)*, and functional changes *(3,7)* have been identified using this method. Reproducing the noxious metabolic environment associated with ischemia, however, is quite difficult. Glucose-depleted media *(1,8)*, ischemic solutions

From: *Methods in Molecular Medicine, Vol. 139: Vascular Biology Protocols*
Edited by: N. Sreejayan and J. Ren © Humana Press Inc., Totowa, NJ

(7,9–11), and low pH *(12)* have all been used to mimic environmental changes that occur during ischemia, but these techniques fail to concentrate myocyte metabolic wastes similar to ischemia in vivo.

Coverslip hypoxia (CSH) is a novel method for studying the effects of hypoxia and ischemia in isolated neonatal rat ventricular myocytes (NRVMs). This method involves placing a glass coverslip over an existing myocyte culture to create a diffusion barrier, resulting in rapid hypoxia and concentration of metabolic byproducts. As a result, intracellular pH of synchronously contracting myocytes under the coverslip decreases rapidly. This method is amenable to video, fluorescence, and electron microscopies, as well as protein and nucleic acid work, and can allow the end user to study effects induced by ischemia, including contractile dysfunction, pH changes, apoptotic and necrotic events, and gene modulation.

2. Materials

2.1. Cell Isolation and Culture

1. For convenience, several of the key reagents for isolation of NRVMs are available in a kit from Worthington Biochemical (Lakewood, NJ, USA) (no. LK003300).
2. Sprague–Dawley rat pups, 0–2 days old (Charles River, Wilmington, MA, USA).
3. Dulbecco's modified Eagle's medium (DMEM; JRH Biosciences, Lenexa, KS, USA) supplemented with 10% fetal bovine serum (FBS), 100 U/mL penicillin, 100 μg/mL streptomycin, and 0.3 mg/mL L-glutamine (Invitrogen, Carlsbad, CA, USA).
4. Hanks balanced salt solution (HBSS; Invitrogen).
5. Leibowitz L15 medium (L15; Worthington Biochemical), powder formulation, order no. LK003250, and make as directed.
6. Growth Factor Reduced MatriGel (BD Biosciences, Franklin Lakes, NJ, USA).
7. 1× Ads buffer: All reagents are from Sigma (St. Louis, MO, USA)—120 mM NaCl, 5.5 mM glucose, 11 mM NaH_2PO_4, 5.4 mM KCl, 0.44 mM $MgSO_4$, and 20 mM HEPES, pH 7.3 in distilled water. Make 1 L, filter sterilize, and store in 50 mL aliquots at 4°C.
8. 10× Ads buffer: Reagents are the same as 1× Ads—1.2 M NaCl, 55 mM glucose, 110 mM $NaH_2PO_4 \cdot H_2O$, 54 mM KCl, 4.4 mM $MgSO_4 \cdot 6H_2O$, and 200 mM HEPES, pH 7.3 in distilled water. Make 1 L, filter sterilize, and store in 50 mL aliquots at 4°C.
9. Trypsin (Worthington Biochemical), 1000-μg vial.
10. Soybean Trypsin Inhibitor (Worthington Biochemical), 2000-μg vial.
11. Collagenase (Worthington Biochemical), 1500-U vial.

12. LifterSlips 22×25 mm (Erie Scientific, Portsmouth, NH, USA), 100-mm tissue culture dishes (VWR, West Chester, PA, USA), sterile razor blades (VWR), 50-mL conical tubes (VWR), 10-mL and 25-mL serological pipettes (VWR), 70-μm disposable filters for 50-mL conical tubes (VWR), and 35-mm tissue culture dishes (VWR; *see* **Note 1**).

2.2. Fluorescence Microscopy

1. DMEM without Phenol Red (Invitrogen) supplemented with 10% FBS, 100 U/mL penicillin, 100 μg/mL streptomycin, and 0.3 mg/mL L-glutamine (Invitrogen).
2. Staining solution: Dilute tetramethylrhodamine ethyl ester (TMRE; Invitrogen) to 100 nM and AlexaFluor488-Annexin V (Invitrogen) to 50 μL/mL in DMEM without phenol red. Heat DMEM to 37 °C prior to making dilutions and adding to myocytes.
3. Carboxyseminaphthorhodafluor-4F (SNARF-4F; Invitrogen) loading solution: Dilute the 50 μL of SNARF-4F stock solution (50 μg in 50 μL DMSO) into a total of 3 mL DMEM without phenol red. The final concentration of SNARF-4F is 10 μM. Heat loading solution to 37 °C prior to adding to myocytes.

3. Methods

The key to CSH is in obtaining NRVMs that (i) are free of cardiac fibroblasts (i.e., >90% myocytes) and (ii) have the ability to electrically couple and synchronously contract with a beat frequency of approximately 2 Hz. In our hands, approximately one million myocytes per 35-mm dish routinely produces a monolayer with these characteristics, which may be a good starting point when developing CSH in your own laboratory. However, the end user would be well advised to spend the time necessary to (i) become proficient at isolating NRVMs with consistent yields and (ii) titrating plating densities to obtain reproducible monolayers with the appropriate contractile properties.

3.1. Isolation of NRVMs

The directions below assume that myocytes are prepared from 60 rat pups. This can be scaled upward without impacting yield but scaling down below 30 rat pups will dramatically decrease yield.

3.1.1. Tissue Harvest and Trypsin Digest

1. Harvest ventricles from day 0–2 Sprague–Dawley rat pups. First, cervically dislocate the head from the body and spray the chest and abdomen with 70% ethanol. Next, make a approximately 10-mm incision with surgical scissors just to the right of the sternum. Finally, apply a squeezing pressure to the body from behind, allowing the apex of the heart to protrude out of the incision. Cut the ventricles away from the

protruding heart and place in a 50-mL conical tubes containing 40 mL cold HBSS (up to 30 ventricles/tube).
2. Wash ventricles twice with cold HBSS to remove residual tissue and blood.
3. Pour off HBSS and place ventricles in 100-mm culture dish (up to 30 ventricles/dish).
4. Dice ventricles in dish using a sterile razor.
5. Make up 20 mL trypsin digestion solution by reconstituting one vial of trypsin with 2 mL HBSS and diluting into 18 mL HBSS for a total of 20 mL trypsin solution.
6. Place 10 mL trypsin solution into a dish containing up to 30 diced ventricles. About 20 mL of trypsin solution is enough to digest 60 ventricles—two dishes of 30 ventricles each with 10 mL trypsin solution.
7. Digest at 4°C overnight with rocking.

3.1.2. Collagenase Digest and Plate Coating

1. Transfer digested tissue to 50-mL conical tube (one dish to one conical tube).
2. Reconstitute one vial of trypsin inhibitor with 1 mL HBSS and add inhibitor solution to one conical tube.
3. Mix gently and place conical tube at 37 °C for 5 min. Plate coating should be done during this 5-min period (*see* **Subheading 3.2.**).
4. Remove conical tube from 37 °C and combine into one conical tube, that is, two conical tube into one (*see* **Note 2**).
5. Reconstitute 1 vial of collagenase with 5 mL L15 media and add collagenase solution to one conical tube (*see* Note **3**).
6. Digest at 37 °C for only 30 min with shaking, making sure to lay the conical tube on its side so that tissue shakes well (*see* **Note 4**).

3.1.3. Filtering and Gradient Separation

1. Place a 70-μm filter in a clean 50-mL conical tube and prewet with 4 mL L15. Collect and retain this volume, as it will catch the cells as they come over the filter into the conical tube.
2. Remove conical tube containing the collagenase digestion from the water bath and allow undigested tissue to settle to bottom of conical tube ($\sim$ 2 min).
3. Remove approximately 15 mL of supernatant from conical tube and pass over filter.
4. Triturate tissue in remaining volume until completely dissolved ($\sim$ 10 pass through a 10-mL pipette).
5. Pass remaining volume over filter, remove and discard filter, cap filtered cells, and set aside at room temperature.
6. Prepare the gradient as follows:
    a. Make stock solution of Percoll by mixing nine parts Percoll with one part 10× Ads buffer. Make at least 100 mL of this stock for two gradients—each gradient containing the material from 30 pups.

Table 1
Percoll Stock and Ads Buffer Volumes

Density (g/mL)	Percoll stock (mL)	1× Ads buffer (mL)	Total (mL)
1.050	18	32	50
1.060	22.2	27.8	50
1.082	31.6	18.4	50

The volumes of Percoll stock and 1× Ads buffer needed to make each of the three Percoll solutions that comprise the gradient are listed.

    b. Make 50 mL each of the Percoll solutions according to **Table 1**.
    c. Prepare Percoll gradient by *underlayment*—first, transfer 12 mL of 1.050 g/mL to a 50-mL conical tube using a 10-mL pipette; second, carefully plunge a 10-mL pipette containing 12 mL of 1.060 g/mL through the 1.050 layer to the bottom of the conical tube and slowly deliver the 1.060 layer; and third, plunge a 10-mL pipette containing isolated cells resuspended in 12 mL of 1.082 g/mL through the 1.050 and 1.060 layers to the bottom of the conical tube and deliver the 1.082 layer.

7. Pellet digested cells at approximately 300 *g* for 6 min at room temperature.
8. Aspirate and discard supernatant; resuspend cell pellet in a total of 24 mL of 1.082 Percoll solution.
9. Underlay 12 mL of cell-containing 1.082 Percoll solution in 1 gradient (*see* **Note 5**).
10. Centrifuge gradients at 1460 g (3000 rpm) in a Beckman Allegra 6R tabletop centrifuge fitted with a GH-3.8A swinging bucket rotor for 30 min at room temperature, making sure that deceleration is set to no brake (*see* **Note 6**).
11. After centrifugation, remove myocyte layer from the 1.082/1.060 interface (i.e., the lowest interface) and pool myocytes from both gradients in one conical tube.
12. Wash pooled myocytes once in an equal volume of DMEM; pellet cells at approximately 300 *g* for 6 min at room temperature and wash once again in 10 mL DMEM; pellet again and resuspend in 40 mL DMEM.
13. Count myocytes using hemocytometer and calculate final myocyte yield.

3.2. Preparing Dishes and Plating Myocytes

1. Prepare a 100 µg/mL MatriGel solution in cold HBSS (*see* **Note 7**). Coat each 35-mm dish with 3 mL of this solution. Plan on approximately one million myocytes/pup and approximately one million myocytes per plate, so make up the same number of dishes as the number of pups used (i.e., prepare 60 dishes when harvesting from 60 pups). Let dishes coat at room temperature in the culture hood until myocytes are ready to plate.

2. After obtaining the final myocyte yield, dilute myocytes to a concentration of one million myocytes/3 mL DMEM.
3. Before plating myocytes, thoroughly aspirate the MatriGel coating solution from each 35-mm dish.
4. Plate diluted myocyte solution into each dish. Use a 25-mL serological pipette having a volume capacity of at least 30 mL. Gently swirl the cells and immediately take 30 mL into the pipette. Dispense into dishes in 3 mL aliquots (approximately one million myocytes/dish).
5. Incubate plated myocytes at 37 °C in 5% CO_2 for 24–48 h prior to experimentation (*see* **Note 8**).

3.3. Coverslipping Myocyte Monolayers

The original description of CSH employed 18-mm round coverslips *(13)*. We have since found that LifterSlips (thick glass slips with printed edges that hold the slip above a substrate; especially useful for maintaining uniform distribution of reagent over a microarray chip) produce a more uniform insult and help standardize data from dish to dish. LifterSlips are also easier to place on and remove from a myocyte monolayer, preventing any accidental damage to myocytes during experiments.

1. Prepare several 22-mm × 25-mm LifterSlips for coverslipping by equilibrating them to 37 °C in a 10-cm dish containing approximately 30 mL DMEM. Place this 10-cm dish in a 37 °C, 5% CO_2 incubator (*see* **Note 9**).
2. Remove a 35-mm dish of synchronously contracting myocytes from the incubator and place on a benchtop (*see* **Note 10**).
3. Using a forceps remove a LifterSlip from the 10-cm dish in the incubator, briefly shake off excess DMEM, and angle the LifterSlip (printed edges down) into the 35-mm dish. Once the leading edge of the LifterSlip is at the bottom of the dish, release the forceps allowing the rest of the LifterSlip to slide under the DMEM surface and settle on top of the monolayer (*see* **Fig. 1**) (*see* **Note 11**).
4. Once the LifterSlip has settled onto the monolayer, gently place the 35-mm dish back in the incubator for the determined experimental time. Do not stack.
5. After the experimental time has elapsed, gently remove the 35-mm dish and, with a large-tipped lab marker, mark the location of each corner of the LifterSlip on the underside of the dish from beneath.
6. Place the dish on a benchtop and gently remove the LifterSlip. This is most easily done by grasping the unprinted edge of the LifterSlip with a forceps and slowly lifting this one edge out of the DMEM while pivoting on the opposite edge. Care must be taken to remove the LifterSlip slowly so as not to damage the myocytes underneath.

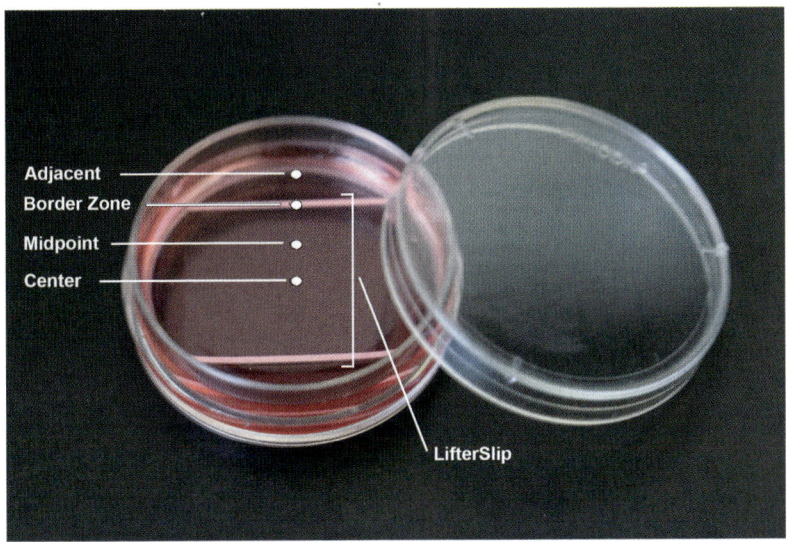

Fig. 1. Myocytes coverslipped with a LifterSlip and regions often imaged. A 22-mm × 25-mm LifterSlip covers approximately 66% of the myocyte monolayer in a 35-mm dish. The approximate adjacent, border zone, midpoint, and center regions are depicted.

3.4. Staining and Imaging Coverslipped Myocytes to Assess Survival

1. Aspirate media from dish and replace with 2 mL staining solution. Note: Always add solution slowly against the wall of the dish to avoid disturbing any dead or dying myocytes.
2. Incubate myocytes for 20 min at 37 °C in 5% CO_2.
3. Image myocytes using a confocal fluorescence microscope, preferably equipped with an environmental stage to maintain proper temperature and atmosphere. TMRE-positive myocytes are alive and appear red, whereas annexin V-positive myocytes are dead or dying and appear green. When imaging different regions of myocytes coverslipped for 3 h or less, the end user should see a gradient of viability progressing from most viable at the border zone to least viable at the center (*see* **Figs 1** and **2A**).

3.5. Staining and Imaging Coverslipped Myocytes to Assess pH Changes

The end user should note that assessing pH changes is performed real time during CSH rather than at the end of a predetermined coverslipping period.

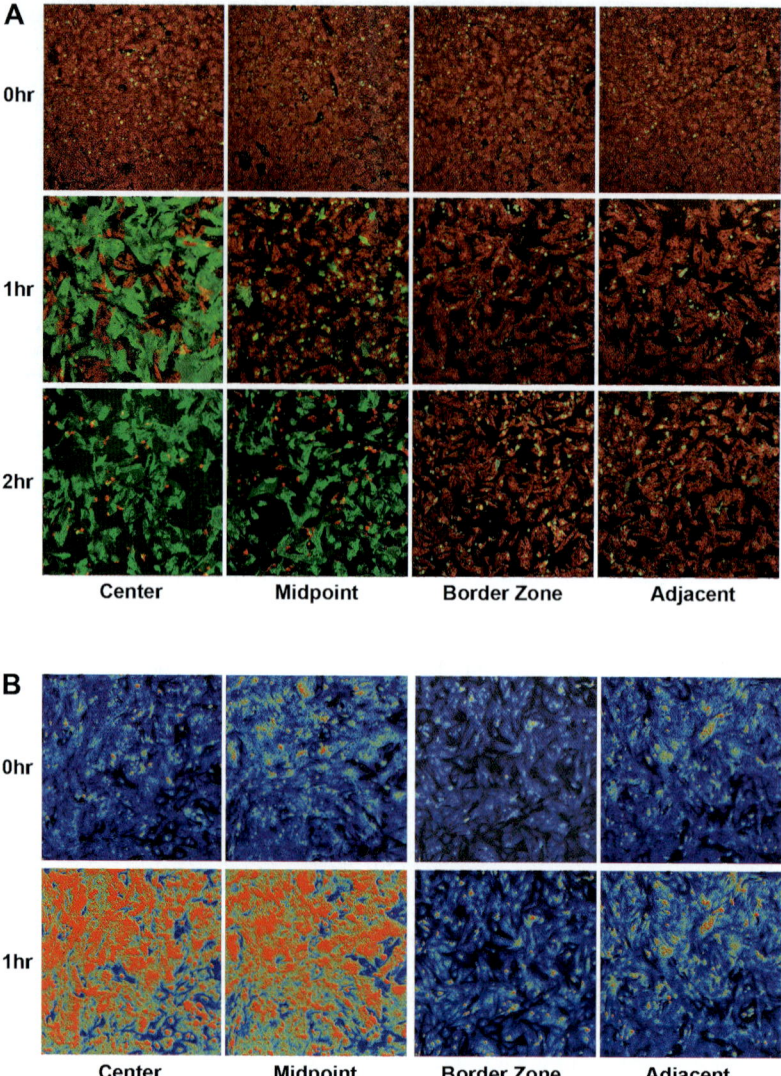

Fig. 2. Viability and pH can be assessed using the coverslip hypoxia model. Portions of synchronously contracting monolayers of neonatal rat ventricular myocytes were coverslipped for varying durations. Coverslips were removed and cells post-stained with tetramethylrhodamine ethyl ester (TMRE) (red) and AlexaFluor488-annexin V (green). (**A**) Images of myocytes under the center of the coverslip (center), at the coverslip edge (border zone), and halfway between (midpoint) taken after 5 min of coverslipping (0 h) show that all myocytes are TMRE(+)/annexin V(−), similar to

Therefore, myocytes are loaded with a pH-sensitive fluorescent dye, cover-slipped, and then imaged with the coverslip in place. Intensity is imaged using a 565-nm to 615-nm bandpass filter.

1. After identifying a 35-mm dish of myocytes with the proper contractile properties (*see* **Subheading 3.3.**), aspirate media from dish and replace with 3 mL loading solution.
2. Incubate myocytes for 40 min at 37 °C in 5% CO_2 to allow SNARF-4F to load into the cells.
3. Place 35-mm dish containing SNARF-4F-loaded myocytes on a confocal fluorescence microscope equipped with an environmental stage.
4. Place LifterSlip onto myocyte monolayer as described in **Subheading 3.3**.
5. Take $t = 0$ min image as a baseline, image thereafter based on experimental design. When imaging at 1 h or less, the end user should see intracellular and extracellular intensity increase, indicating that pH is dropping (*see* **Fig. 2B**).
6. Images are converted to pseudocolor intensity maps and can be qualitatively compared with a pH calibration curve to assess approximate pH.

3.6. Potential Uses for the CSH Model

1. Heterologous cell sandwich assay: To assess how a particular cell type responds to ischemic conditions, it can be cultured on a LifterSlip and then delivered directly to a myocyte monolayer to undergo ischemic insult for a defined period. The cell type of interest is kept segregated from the myocytes, allowing a complete and pure recovery without contaminating cells. The exposed cells can be studied to identify genes or proteins that are modulated during ischemia.
2. Local delivery of therapeutic compounds: The effect of a test compound to perform during ischemia can be assessed by adhering the compound to the LifterSlip in a

─────────────────────────

Fig. 2. (Continued) myocytes in images taken adjacent to the slip edge (adjacent). At 1 and 2 h, however, the number of TMRE(+) cells declines while the number of annexin V(+) cells increases. A progressive decrease in TMRE fluorescence intensity is observed at center regions beginning at 1 h. This progressive decrease is not seen at midpoint regions until 2 h. (**B**) Portions of synchronously contracting myocyte monolayers loaded with SNARF-4F were coverslipped. With the coverslip in place, images at center, midpoint, border zone, and adjacent regions were acquired at 0 and 1 h. Pseudocolored intensity maps (blue > green > yellow > red) of the corresponding fluorescence images allow for estimates of intracellular pH in time. These pseudocolored intensity maps reveal a rapid and substantial drop in pH under the coverslip in center and midpoint regions between 0 and 1 h, whereas the pH at border zone and adjacent regions does not change substantially.

formulation allowing its elution. Test compound can then be delivered directly to the myocytes monolayer, and its local concentration is kept constant particularly if its half-life under oxidative conditions is short.

4. Notes

1. The end user might be tempted to plate myocytes in six-well plates. However, not every well will have the qualities necessary to proceed with coverslipping. In our experience, plating myocytes in 35-mm dishes allows for the greatest flexibility during experimentation.
2. Keep this ratio the same if scaling up—four conical tubes into two, six conical tubes into three, and so on.
3. About 5 mL of collagenase solution is enough to digest 60 ventricles.
4. The collagenase digest has proven to be one of the most critical steps in the isolation procedure. Be vigilant regarding the 30-min time frame—both longer and shorter digestion times dramatically decrease myocyte yield and viability.
5. The 24 mL contains the cells from 60 hearts—separate only 30 hearts worth of material on one gradient.
6. Any braking during deceleration will disrupt the gradient.
7. MatriGel is a liquid at 4 °C but rapidly solidifies upon warming to room temperature. As a rule, we store MatriGel in 500 μL aliquots at 4 °C in its liquid form for fast and routine use. Transfer the required volume of MatriGel quickly to cold HBSS and dispense the solution immediately into the plates for a uniform coating. This process should take no longer than 5 min.
8. A dish of myocytes can be used in experiments once the monolayer is synchronously contracting with a beat frequency of approximately 2 Hz. The end user must determine what point in the 24-h to 48-h period yields the most dishes with this characteristic.
9. LifterSlips must be equilibrated for at least two reasons: (i) there should be no temperature differential in the glass that might affect myocyte contractility when the coverslip is initially placed and (ii) a wetted LifterSlip breaks any surface tension that would prevent it from settling properly in the myocyte dish.
10. Dishes can be quickly segregated into "contractors" and "non-contractors" by checking for the proper beat frequency using a phase-contract microscope. This should be done in batches to keep temperature variations in the dishes to a minimum. Perform this segregation at the beginning of the day the experiments are to be run. After segregating, allow the "contractors" to re-equilibrate in the incubator for 60 min before beginning experiments.
11. The end user should practice this method of LifterSlip placement on synchronously contracting monolayers at a microscope several times before performing any experiment to ensure that the monolayers are not damaged during coverslip placement.

References

1. Safran, N., Shneyvays, V., Balas, N., Jacobson, K.A., Nawrath, H., and Shainberg, A. (2001) Cardioprotective effects of adenosine A1 and A3 receptor activation during hypoxia in isolated rat cardiac myocytes. *Mol. Cell. Biochem.* **217**, 143–152.
2. Webster, K.A., Discher, D.J., Kaiser, S., Hernandez, O., Sato, B., and Bishopric, N.H. (1999) Hypoxia-activated apoptosis of cardiac myocytes requires reoxygenation or a pH shift and is independent of p53. *J. Clin. Invest.* **104**, 239–252.
3. Webster, K.A., Discher, D.J., and Bishopric, N.H. (1995) Cardioprotection in an in vitro model of hypoxic preconditioning. *J. Mol. Cell. Cardiol.* **27**, 453–458.
4. Kubasiak, L.A., Hernandez, O.M., Bishopric, N.H., and Webster, K.A. (2002) Hypoxia and acidosis activate cardiac myocyte death through the Bcl-2 family protein BNIP3. *Proc. Natl. Acad. Sci. U. S. A.* **99**, 12825–12830.
5. Long, X., Boluyt, M.O., Hipolito, M.L., Lundberg, M.S., Zheng, J.S., O'Neill, L., Cirielli, C., Lakatta, E.G., and Crow, M.T. (1997) p53 and the hypoxia-induced apoptosis of cultured neonatal rat cardiac myocytes. *J. Clin. Invest.* **99**, 2635–2643.
6. Todor, A., Sharov, V.G., Tanhehco, E.J., Silverman, N., Bernabei, A., and Sabbah, H.N. (2002) Hypoxia-induced cleavage of caspase-3 and DFF45/ICAD in human failed cardiomyocytes. *Am. J. Physiol. Heart Circ. Physiol.* **283**, 990–995.
7. Louch, W.E., Ferrier, G.R., and Howlett, S.E. (2002) Changes in excitation-contraction coupling in an isolated ventricular myocyte model of cardiac stunning. *Am. J. Physiol Heart Circ. Physiol.* **283**, 800–810.
8. Shiraishi, J., Tatsumi, T., Keira, N., Akashi, K., Mano, A., Yamanaka, S., Matoba, S., Asayama, J., Yaoi, T., Fushiki, S., Fliss, H., and Nakagawa, M. (2001) Important role of energy-dependent mitochondrial pathways in cultured rat cardiac myocyte apoptosis. *Am. J. Physiol. Heart Circ. Physiol.* **281**, 1637–1647.
9. Arutunyan, A., Webster, D.R., Swift, L.M., and Sarvazyan, N. (2001) Localized injury in cardiomyocyte network: a new experimental model of ischemia-reperfusion arrhythmias. *Am. J. Physiol. Heart Circ. Physiol.* **280**, 1905–1915.
10. Ferrier, G.R., Moffat, M.P., and Lukas, A. (1985) Possible mechanisms of ventricular arrhythmias elicited by ischemia followed by reperfusion. Studies on isolated canine ventricular tissues. *Circ. Res.* **56**, 184–194.
11. Vanden Hoek, T.L., Shao, Z., Li, C., Zak, R., Schumacker, P.T., and Becker, L.B. (1996) Reperfusion injury on cardiac myocytes after simulated ischemia. *Am. J. Physiol.* **270**, 1334–1341.
12. Hyatt, C.J., Lemasters, J.J., Muller-Borer, B.J., Johnson, T.A., and Cascio, W.E. (1998) A superfusion system to study border zones in confluent cultures of neonatal rat heart cells. *Am. J. Physiol.* **274**, 2001–2008.
13. Pitts, K.R., and Toombs, C.F. (2004) Coverslip hypoxia: a novel method for studying cardiac myocyte hypoxia and ischemia in vitro. *Am. J. Physiol. Heart Circ. Physiol.* **287**, H1801–H1812.

19

Isolation and Functional Studies of Rat Aortic Smooth Muscle Cells

Nair Sreejayan and Xiaoping Yang

Summary

Migration, proliferation, and collagen synthesis by vascular smooth muscle cells are thought to be key events involved in the pathogenesis of cardiovascular disease. Following endothelial injury, smooth muscle cells (SMCs) in the intima of the blood vessels assume a synthetic, promitogenic phenotype resulting in their migration, proliferation, and deposition of extracellular matrix within the neointimal tissue. This chapter describes a method of isolation of SMCs from rat aorta and in vitro assays to characterize these abnormal SMC functions.

Key Words: Smooth muscle cells; Proliferation; Migration; Collagen synthesis.

1. Introduction

The primary event in the development of atherosclerosis and restenosis following percutaneous transluminal coronary angioplasty is thought to involve injury to the endothelium. Following arterial injury, quiescent smooth muscle cells (SMCs) are exposed to growth factors such as platelet-derived growth factor (PDGF) that cause them to migrate from the media to the intima. Once they reach the intima, these cells proliferate and deposit extracellular matrix. Consequently, migration, proliferation, and collagen synthesis by SMCs are therefore thought to be key events in the pathogenesis of cardiovascular diseases.

The SMCs used for functional studies in vitro should be free of contamination with adventitial fibroblasts and endothelial cells. Isolating cells from pup rats (6–8 days old) would ensure easy removal of these extraneous cells. As repeated

From: *Methods in Molecular Medicine, Vol. 139: Vascular Biology Protocols*
Edited by: N. Sreejayan and J. Ren © Humana Press Inc., Totowa, NJ

subculturing of SMCs results in de-differentiation and phenotypic changes, it is important that freshly isolated, early passages of SMCs be used for functional assays described here.

In this chapter, we describe a simple method of isolation of aortic SMCs from rat pups. We also describe assays to assess the function of SMCs such as cell migration (chemotaxis and chemokinesis), depolymerization of actin filaments, cell proliferation, and collagen synthesis. These tools would be of use to study drug candidates for their potential utility in treating vascular disorders.

2. Materials

2.1. Isolation of Aortic SMCs

1. Sprague–Dawley rat pups, 6–9 days old (1 litter).
2. Dissecting board, scissors, forceps, Petri dish, and glass pipette.
3. Solution A: Suspension-Minimum Essential Media (S-MEM) (Invitrogen, Invitrogen, Carlsbad, CA) containing 0.2 mM calcium chloride, 100 U/mL penicillin, and 100 μg/mL streptomycin.
4. Digesting media: 10 mL of solution A containing 15 U/mL Elastase (Calbiochem, San Diego, CA), 200 U/mL collagenase Type I (Sigma), 0.4 mg/mL trypsin inhibitor (Soybean) (Sigma, St. Louis, MO), and 1.7 mg/mL of bovine serum albumin (BSA). Sterile filter this solution through a 0.22-μm low-protein retention filter (Millipore, Billerica, MA).
5. Culture media: 500 mL Dulbecco's modified Eagle's medium with Hams-F12 (DMEM/F12, 1:1) supplemented with 5 μg/mL insulin, 5 μg/mL transferrin, and 5 ng/mL selenous acid (ITS; Collaborative Research, Lexington, MA, USA), plus 50 U/mL penicillin and 50 μg/mL streptomycin.
6. Wash media: S-MEM supplemented with 0.1% BSA (about 25 mL). Filter this media through a 0.22-μm filter.
7. Fetal bovine serum (FBS).
8. Sodium pentobarbital 1 g/mL.
9. Ethanol 70%.

2.2. Cell Migration

1. SMCs (early passage).
2. Modified Boyden chambers (Costar, Corning, NY, USA): transwell filter inserts with 8 μm pore size.
3. Chemokines, such as PDGF.
4. Trypsin–EDTA solution.
5. Fixative.
6. Hematoxylin (0.4%).
7. Nunc chamber slides.

8. Fixing solution: Formaldehyde (3.7%) and Triton X-100 (2%) in phosphate-buffered saline (PBS).
9. Rhodamine–phalloidin (Molecular Probes).
10. Fluorescence microscopy (Nikon TE300 inverted microscope equipped with a Cascade 650 cooled CCD digital camera, ×100 magnification).

2.3. Cell Proliferation

1. [Methyl-^{3}H]thymidine (μCi/mmol) (NEN, Boston, MA, USA).
2. PBS supplemented with 1 mM calcium chloride.
3. Trichloroacetic acid: 10% solution in ice-cold water.
4. Ethanol: ether (2:1 vol/vol).
5. Extraction reagent: 0.1% sodium dodecyl sulfate and 0.1 N NaOH.
6. Scintillation cock tail.
7. Scintillation counter.
8. BCA protein assay kit (Pierce, Rockford, IL, USA).

2.4. Collagen Synthesis

1. Sodium ascorbate.
2. L-[^{3}H]proline (100 mCi/mmol).
3. Proline-free DMEM (NEN).
4. β-Aminopropionitrile.
5. Buffer 1: 0.65 M NaCl; 0.1 M Tris, pH 7.4; 4.7 mM $CaCl_2$ containing 2.5 mg/mL *N*-ethylmaleimide; and 50 μg/mL BSA.
6. Trichloroacetic acid (10% and 5%).
7. Sodium hydroxide (0.1 N).
8. Ethyl alcohol (90%).
9. Collagenase form VII (Sigma).

3. Method

3.1. Isolation of Aortic SMCs

1. Euthanize the rats (1 litter, about 6–7 rats) by injecting 0.25 mL intraperitoneal injection of sodium pentobarbital.
2. Place the rat in a supine position and pin the paws of the rat to a dissection board.
3. Wipe the thorax and the abdominal region of the rat with 70% ethanol.
4. Remove the skin from the thorax; open the thorax to expose the heart and lungs.
5. Dissect out the thoracic aorta from its origin just above the heart to the iliac bifurcation (*see* **Note 1**).
6. Place the aorta in a Petri dish with pre-warmed (37 °C) solution A.
7. Place the Petri dish under a dissection microscope, and using the forceps, gently remove the connective tissues and adventitia from the aorta (*see* **Note 1**). If done carefully, the adventitia will peel off the aorta as a single unit, like a snake skin.

When the adventitia is removed, the aorta will appear as a smooth tube clear of any surrounding tissues.

8. Wash aorta in solution A by repeatedly pipetting out with a wide-mouthed, glass pipet (eight washes) (*see* **Note 2**).
9. Chop aorta into small pieces (*see* **Note 1**).
10. Place chopped tissue in 7 mL of digesting solution taken in a sterile 15-mL centrifuge tube and incubate at 37 °C for about 60–75 min with gentle shaking (*see* **Note 3**).
11. Spin the tube at 1000 g for 5 min. Gently discard the supernatant.
12. Wash pellet by swirling gently (slight dissociation) with 5-ml glass pipette three times with solution A (5 mL).
13. Repeat steps 11 and 12 two more times.
14. Very gently dissociate to single cells, by pipetting the pellet up and down a few times using a glass pipette.
15. Collect supernatant containing single cells by spinning at 100 g for 30 s.
16. Repeat steps 14–16 four more times.
17. Pool the supernatant and count the cells using a cell counter. The yield is generally $3–3.5 \times 10^6$ single cells per litter.
18. Plate the cells into culture dishes containing cell culture media, at a density of about 2.5×10^4 cells/cm^2 (*see* **Note 4**).
19. Place the cell culture dish in a humidified atmosphere of 5% CO_2 and 95% air.
20. Most cells adhere to the plate within 2 h of seeding and begin to spread after overnight incubation.
21. Cells will be confluent in 4 or 5 days.
22. Confirm the purity of the cells by staining with alpha-smooth muscle-specific antiactin antibody. More than 99% of the cells isolated by this procedure would be stained with this antibody.

3.1.1. Cell Migration: Chemotaxis (see **Note 5**)

1. Trypsinize overnight serum-starved cells and wash three times with serum-free media containing 0.5% BSA.
2. Count the cells using a hemocytometer.
3. Plate the cells on the upper side of the transwell at 40,000 cells/well in 100 μL of serum-free medium. Add 500 μL of the same media to the lower chamber.
4. Incubate the transwell plate for 2–4 h to let the cells attach.
5. Add chemokine (we have used PDGF at a concentration 10 ng/mL) to the lower chamber. Add the experimental agent to the upper chamber at the same time.
6. Place the plate in the CO_2 incubator and allow the cells to migrate overnight.
7. Remove the filter inserts from wells with forceps. Tap the insert against the edge of the well to remove any additional buffer.
8. Remove the non-migrating cells from the upper chamber with a cotton swab.

9. Submerge the inserts in three fixing solutions for 10 min followed by 0.4% hematoxylin for 5 min. The cells embedded in the bottom membrane are stained by this procedure.
10. Count the number of stained nuclei per high-power field in a microscope ($\times$200) to determine the number of cells migrated.
11. Count each sample randomly in four separate locations in the center of the membrane.
12. Report SMC migration activity as number of cells migrated per field of view. Results of a representative experiment showing the migration of SMCs in response to stimulation with PDGF are shown in **Fig. 1**.

3.1.2. Cell Migration: Monolayer Wound-Healing Assay (see **Note 6**)

1. Culture SMCs in Petri dish till they are confluent. A good, uniform monolayer is essential for the success of this assay. Do not over grow either.
2. Optional step: To prevent proliferation, treat confluent cells with serum-free medium containing hydroxyurea (5 mmol/L) for 24 h prior to the start of the experiment (*see* **Note 7**).
3. Once confluent, remove the media and mark the plates by drawing a line along the diameter of the plate with a black marker (on the bottom side).
4. Wound the cell monolayer perpendicular to the marked line, using a gel-loading microtip (*see* **Note 8**).
5. Wash the plates three times with PBS.

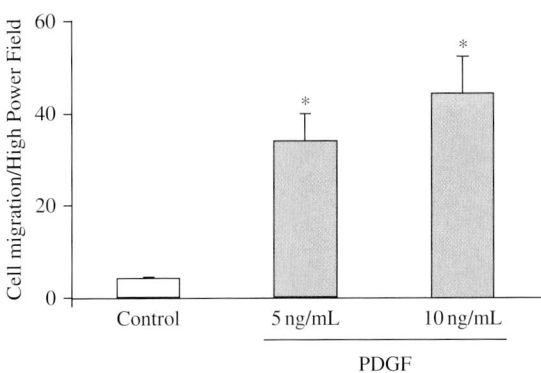

Fig. 1. Vascular smooth muscle cells were stimulated with or without platelet-derived growth factor (PDGF) (5 and 10 ng/mL), and cell migration was assessed using the transwell migration assay. Results are the mean $\pm$ SE from five independent experiments. *$p < 0.001$ compared to unstimulated cells.

6. Photograph the area with a video camera system using NIH Image software [at the intersection of the marked line and wound edge at 0 h (WW_0)].
7. Add media with or without test agents.
8. Incubate the plates for 24 h.
9. Wash the plates three times with PBS.
10. Photograph the area of migration at the intersection of the marked line and wound edge at 24 h (WW_{24}).
11. Calculate cell migration as wound width covered at time 24 (WW_0-WW_{24}) and express as percent control.
12. Repeat measurements at different areas about the wound.
13. Repeat experiments at least three times.
14. Each experiment should be conducted three times in duplicate, and each plate can be wounded twice. The number of experiments conducted would be considered to be six, although the results are the means of 12 observations. Photographs in **Fig. 2** are representative phase-contrast views showing cell migration following 0 and 24 h of treatment with PDGF in the injury model.

3.1.3. Procedure of Actin-Filament Staining (see **Note 9**)

1. Seed cells onto Nunc chamber slides (10,000 cells for each well) in the growth media (200 μL/chamber) and allow the cells to attain approximately 60% confluency.
2. Serum starve the cells for 24 h with the same growth medium (devoid of ITS but containing 0.1% serum).
3. Treat cells with the chemokine and/or experimental agent in serum-free media for 30 min.
4. Wash cells twice with pre-warmed PBS, pH 7.4 (250 μL each wash).

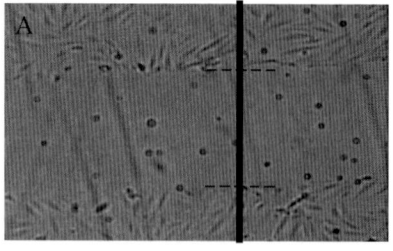 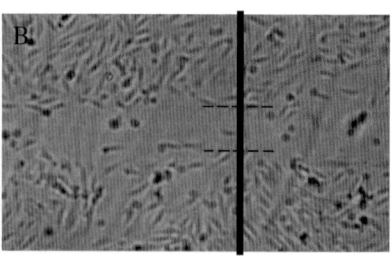

Fig. 2. Phase-contrast photographs of vascular smooth muscle cells stimulated with platelet-derived growth factor (10 ng/mL) at 0 min (**A**) and after 24 h (**B**). Confluent aortic smooth muscle cells were wounded with a gel-loading tip in the center of plates that were marked to localize the wound site (vertical line) The horizontal hatched-lines indicate the leading edges of the wound.

5. Fix the sample in the fixing solution for (200 μL) 10 min at room temperature (*see* **Note 10**).
6. Wash twice with PBS containing 0.1% BSA.
7. Incubate cells in rhodamine–phalloidin diluted 1:100 in PBS for 15 min.
8. Rinse three times in PBS, 5 min/wash.
9. Mount for fluorescence microscopy using mounting medium.
10. PDGF-induced disassembly of actin-filaments in SMCs is shown in **Fig. 3**.

3.2. Cell Proliferation

1. Plate SMCs in 24-well cell culture dishes at a density of 2×10^5 cells per well and grow to 90% confluency in DMEM-12 containing 10% serum and ITS.
2. Render confluent cells quiescent by placing it in the growth media containing 0.1% FBS (but lacking ITS) for 24 h prior to the experiment.
3. To investigate the effects of experimental agents on mitogenesis, incubate quiescent cells for 20 h in the absence or presence of experimental agents.
4. After 20 h of incubation with the experimental agent add 3 μCi of [^{3}H]thymidine to each well and incubate for an additional 2 h (*see* **Note 11**).
5. Stop the experiment by rapidly washing cells three times with ice-cold PBS supplemented with 1 mM calcium.
6. Wash the plates three times with ice-cold 10% trichloroacetic acid.
7. Wash the plates three times with ethanol : ether (2:1 vol/vol).

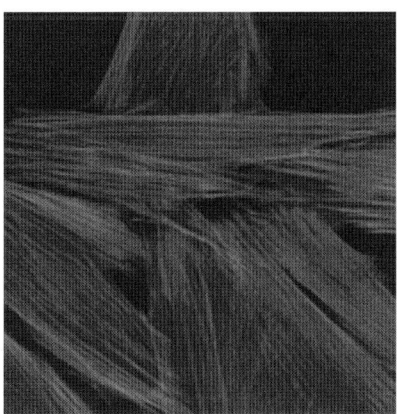 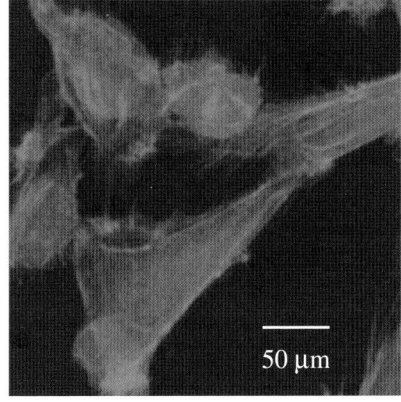

Control   PDGF (10 ng/mL)

Fig. 3. Platelet-derived growth factor (PDGF)-induced dissociation of actin microfilaments in vascular smooth muscle cells. Cells were grown on slides in the presence or absence of PDGF (10 ng/mL) for 30 min. Following fixing and permeabilization, cells were stained with rhodamine–phalloidin and examined by fluorescence microscopy.

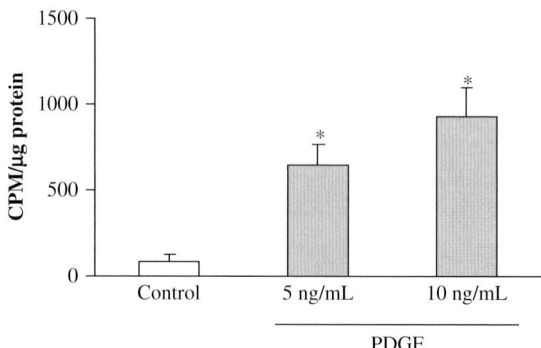

Fig. 4. Vascular smooth muscle cells were treated with or without platelet-derived growth factor (PDGF) (5 and 10 ng/mL) for 24 h, and incorporation of [³H]thymidine was monitored and normalized to protein. Results are the mean ± SE from three independent experiments. *$p < 0.005$ compared to control.

8. Air-dry the plates.
9. Add 100 μL extraction reagents to each well in the plates and leave it overnight.
10. Pipette out 25 μL of the extract into a scintillation vial containing 5 mL of scintillation fluid.
11. Determine the radioactivity using a scintillation counter.
12. With another aliquot of 25 μL, determine the protein concentration using the bicinchoninic acid method using BSA as standard.
13. Express proliferation as counts per minute per milligram of protein. **Figure 4** shows the results from a representative experiment.

3.3. Collagen Synthesis

1. Grow the SMCs to confluence and render them quiescent as described in **Subheading 3.2.**
2. Stimulate the cells with experimental agent in the presence of sodium ascorbate (10 μg/mL) for 28 h.
3. Incubate control cultures with ascorbate plus vehicle.
4. Change the medium to proline-free DMEM containing 1% dialyzed FBS, fresh sodium ascorbate (10 μg/ml), and β-aminopropionitrile (80 μg/ml).
5. Pulse the cells with 15 μCi/mL L-[³H]proline. Continue incubation for another 20 h (*see* **Note 11**).
6. Carefully pipette out the culture media and add vol/vol to cold buffer 1.
7. Remove an aliquot and add 10% TCA and allow to flocculate for 30 min at 4 °C.
8. Spin down the tubes at 1000 g for 10 min to pellet the TCA-precipitated material.
9. Wash twice with 5% trichloroacetic acid (TCA).

10. Wash twice with cold 95% ethanol.
11. Air-dry the precipitate and dissolve in 0.1 N sodium hydroxide and count as described in **Subheading 3.2**. Record the count as CPM_1.
12. Digest a second aliquot with collagenase Type VII (10 U/mL medium) for 90 min at 37 °C.
13. Treat identically as the non-digested aliquot. Record the count as CPM_2.
14. Determine the relative rate of collagen synthesis, assuming that the ratio of proline residues in collagen relative to non-collagen protein by the following equation (*see* **Note 12**):

$$\frac{CPM_2 - CPM_1}{[(CPM_2 - CPM_1) + CPM_1] \times 5.4}$$

4. Notes

1. Care should be exercised not to stretch the aorta during dissection and processing.
2. Take about 3 mL solution A in four or five Petri dishes. Place the aorta in one of these dishes. Pipette out the aorta with media to a fresh Petri dish (use a sterile glass pipette with a wide bore for pipetting). Repeating this procedure 8–10 times will ensure the removal of blood within the aorta. After this procedure, the aorta should appear as a smooth translucent tube.
3. It is important to check the digestion every 5 min after 60 min of incubation. The completion of digestion is indicated by homogenization of the aorta in the digestion media. Avoid over-digestion or under-digestion!
4. Growth media at this stage is devoid of fetal calf serum. It contains insulin, transferrin, and selenium.
5. The transwell migration assay represents chemotaxis, the directional movement of cells in response to a chemotactic agent.
6. The monolayer wound-healing assay represents chemokinesis or random cell migration.
7. Hydroxyurea is used to prevent artifacts due to the proliferation of cells (*1*). However, experiments performed in the absence of hydroxyurea have shown to produce qualitatively similar results (*2*).
8. This step requires practice. One should hold the pipette tip as perpendicular as possible and draw a line using a ruler as a guide. Uniform pressure should be applied to the tip during the scratching procedure so as to obtain a uniform wound on the monolayer.
9. Reorganization of actin cytoskeleton plays an important role in mediating cell migration (*3*). Dissociation of actin filament is associated with increased cell migration in vascular smooth muscle cells.
10. Phalloidin binding requires the F-actin to have a protein structure near native. Methanol or acetone used to fix and/or permeabilize essentially abolishes phalloidin binding.

11. Adequate precautions should be taken when using and disposing the radioactive chemical.
12. The relative rate of collagen synthesis in relation to the total proline-containing proteins is calculated assuming that the number of proline residues in collagen is 5.4-fold higher than that in the non-collagen proteins *(4)*.

References

1. Sarkar, R., Meinberg, E.G., Stanley, J.C., Gordon, D., Webb, R.C. (1996) Nitric oxide reversibly inhibits the migration of cultured vascular smooth muscle cells. *Circ. Res.* **78,** 225–230.
2. Brown, C., Pan, X., Hassid, A. (1999) Nitric Oxide and C-Type Atrial Natriuretic Peptide Stimulate Primary Aortic Smooth Muscle Cell Migration via a cGMP-Dependent Mechanism. *Circ. Res.* **84,** 655–667.
3. Hedberg, K.M., Bell, P.B. Jr. (1995) The effect of neomycin on PDGF-induced mitogenic response and actin organization in cultured human fibroblasts. *Ex. Cell Res.* **219,** 266–275.
4. Bühling, F., Wille, A., Röcken, C., Wiesner, O., Baier, A., Meinecke, I., Welte, T., Pap, T. (2005) Altered expression of membrane-bound and soluble CD95/Fas contributes to the resistance of fibrotic lung fibroblasts to FasL induced apoptosis. *Respir Res.* **6,** 37–45.

20

Detection of Reactive Oxygen Species and Nitric Oxide in Vascular Cells and Tissues
Comparison of Sensitivity and Specificity

Hua Cai, Sergey Dikalov, Kathy K. Griendling, and David G. Harrison

Summary

Reactive oxygen and nitrogen species are thought to contribute to pathogenesis of many cardiovascular diseases including hypertension, atherosclerosis, restenosis, heart failure, and diabetic vascular complications. Some of these reactive oxygen species also play an important role in vascular signaling. In this chapter, we describe various techniques that we have successfully employed to reliably measure superoxide and hydrogen peroxide. Because reactive oxygen species are capable of rapidly inactivating nitric oxide and because endothelial function characterized by nitric oxide bioavailability is an important indicator of vascular health, we have also included novel techniques capable of directly measuring nitric oxide radical from vascular cells and tissues.

Key Words: Reactive oxygen species; Nitric oxide; Vascular smooth muscle cells; Methods.

1. Introduction

Increased production of reactive oxygen species outstripping endogenous anti-oxidant defense systems has been referred to as oxidant stress, which in turn contributes to pathogenesis of many cardiovascular diseases including hypertension, atherosclerosis, restenosis, heart failure, and diabetic vascular complications *(1–4)*. In mammalian cells, potential enzymatic sources of reactive oxygen species include the mitochondrial electron transport chain, the arachidonic acid metabolizing enzymes lipoxygenase and cycloxygenase,

From: *Methods in Molecular Medicine, Vol. 139: Vascular Biology Protocols*
Edited by: N. Sreejayan and J. Ren © Humana Press Inc., Totowa, NJ

the cytochrome P450s, xanthine oxidase, NAD(P)H oxidases, uncoupled nitric oxide synthase (NOS), peroxidases, and other hemoproteins. Among biologically relevant and abundant reactive oxygen species, superoxide ($O_2^{\bullet-}$) and its dismutation product hydrogen peroxide (H_2O_2) appears most important in vascular signaling *(1–4)*. On the contrary, recent studies suggest that $O_2^{\bullet-}$ and H_2O_2 may have differential signaling roles in the vasculature under various conditions *(5–7)*. Both can be simultaneously produced by xanthine oxidase and some recently identified reduced nicotinamide dinucleotide (phosphate) (NAD(P)H) oxidases *(6)*. Thus, production of H_2O_2 may correlate with formation of $O_2^{\bullet-}$. It does not, however, always reflect rate of $O_2^{\bullet-}$ generation, because other systems such as mitochondria or uncoupled eNOS produce $O_2^{\bullet-}$ solely. The efficacy of $O_2^{\bullet-}$ dismutation into H_2O_2 is affected by the abundance of intracellular superoxide dismutase (SOD) and small reducing molecules such as glutathione. Herein, we describe various techniques that we have successfully employed to reliably measure $O_2^{\bullet-}$ or H_2O_2 differentially. Because reactive oxygen species are capable of rapidly inactivating nitric oxide (NO•) and endothelial function characterized by NO• bioavailability is an important indicator of vascular health *(1,2)*, this review also discusses novel techniques capable of directly measuring NO• from vascular cells and tissues.

Comprehensive reviews on methodologies detecting reactive oxygen species are available *(8,9)*. This chapter instead primarily focuses on simplified and user-friendly experimental protocols and technical notes that are practically applicable to vascular biological studies. Taking into consideration the complexity of the biology of reactive oxygen species and limitations of individual techniques, it is generally recommended that reactive oxygen species should be measured with at least two different assays.

1.1. Superoxide Analysis

Though superoxide ($O_2^{\bullet-}$) is very short lived, largely cell impermeable and less likely to serve as a dominant signaling intermediate compared to other reactive oxygen species including H_2O_2 and NO• it remains the precursor of many biologically relevant and important reactive oxygen species. More importantly, it rapidly reacts with NO• to inactivate NO• in a diffusion-limited fashion, which in turn leads to NO• deficiency and endothelial dysfunction, hallmarks of many vascular diseases, including hypertension, atherosclerosis, and diabetic vascular complications *(1–4)*. Many assays have been used to detect $O_2^{\bullet-}$ in test tubes, with cultured cells or intact blood vessels. Only the assays with which we are personally experienced are discussed in this chapter.

1.1.1. Dihydroethidium Staining

Dihydroethidium (DHE) or hydroethidine is a cell-permeable compound that, upon entering the cells, interacts with $O_2^{\bullet-}$ to form oxyethidium *(10)*, which in turn interacts with nucleic acids to emit a bright red color detectable qualitatively by fluorescent microscope *(8)*. Recent studies by Zhao et al. *(10)* demonstrated that $O_2^{\bullet-}$ oxidation of DHE yields oxyethidium, rather than the previously assumed ethidium bromide. The authors characterized this novel compound using HPLC. Fink et al. *(11)* adapted this high-performance liquid chromatography (HPLC)-based DHE assay for specific and quantitative detection of $O_2^{\bullet-}$ from biological samples, including cultured endothelial cells and isolated murine blood vessels.

1.1.2. Electron Spin Resonance with Superoxide-Specific Spin Traps

Free radicals such as $O_2^{\bullet-}$ and $NO^{\bullet}$ have unpaired electrons, and thus are paramagnetic and detectable by electron spin resonance (ESR). When conjugated to specific "spin traps"—compounds capable of selectively reacting with reactive oxygen species to prolong the half-lives of these molecules—free radicals-spin trap adducts generate characteristic signature spectrum that can be quantitatively analyzed using ESR *(12)*. **Figure 1** illustrates $O_2^{\bullet-}$ conjugates after reacting with the traditional nitrone spin traps DMPO and DEPMPO. The nitrone family of $O_2^{\bullet-}$ spin traps has been thoroughly reviewed by Zweier and colleagues *(13)*.

Recent studies however have demonstrated that cyclic hydroxylamines such as 1-hydroxy-3-carboxy-2,2,5,-tetramethyl-pyrrolidine hydrochloride (CPH) can react with $O_2^{\bullet-}$ to form a stable nitroxide radical with a much longer half-life (*see* **Fig. 2**). This is a distinct advantage over nitrone spin traps such as DEPMPO and DMPO, which form unstable $O_2^{\bullet-}$ adducts in biological samples *(14)*. The reaction of cyclic hydroxylamines with $O_2^{\bullet-}$ is also a 100-fold faster than those with nitrone spin traps *(15)*.

More importantly, cyclic hydroxylamines have been recently shown to specifically detect $O_2^{\bullet-}$ with high sensitivity in biological samples. They were

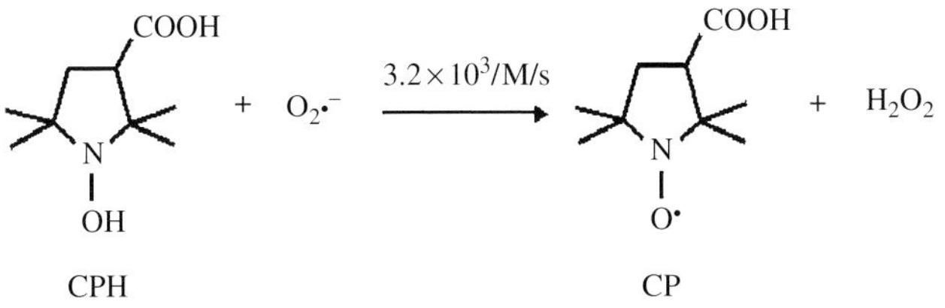

Fig. 1. Reaction of 1-hydroxy-3-carboxy-2,2,5,-tetramethyl-pyrrolidine hydrochloride (CPH) with $O_2^{\bullet-}$.

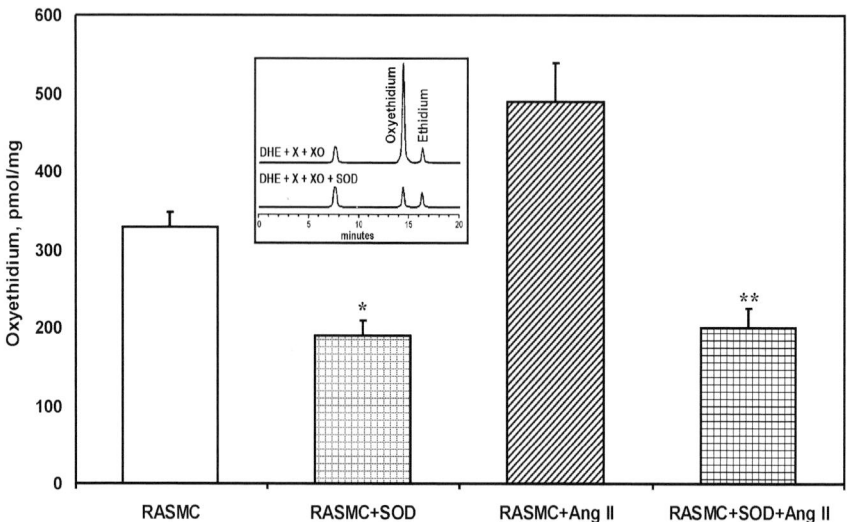

Fig. 2. Measurements of intracellular $O_2^{\bullet-}$ by dihydroethidium–HPLC. Inset shows typical HPLC diagram illustrating the formation $O_2^{\bullet-}$-specific adduct, oxyethidium. Treatment of rat aortic smooth muscle cells (RASMC) with polyethylene glycol (PEG)–superoxide dismutase (SOD) (18-h incubation with 25 U/ml) strongly inhibited oxyethidium formation in angiotensin II-stimulated (100 nmol/l, 4 h) cells. Data are presented as mean ± SEM. $^*p < 0.01$ versus RASMC, $^{**}p < 0.01$ versus angiotensin II.

used to detect $O_2^{\bullet-}$ formation from angiotensin II-stimulated or oscillatory shear stress-stimulated endothelial cells *(16,17)* or coronary artery homogenates of heart failure patients *(18)*. ESR has a detection limit of approximately 1 nmol/L $(10^{-9}\,M)$ for $O_2^{\bullet-}$, representing one of the most sensitive, quantitative, and characteristic means of measuring $O_2^{\bullet-}$.

1.1.3. Cytochrome C Reduction Assay

Traditionally, in the pre-ESR era, the cytochrome C reduction assay was considered the "gold standard" for detection of $O_2^{\bullet-}$. The rate constant for $O_2^{\bullet-}$ reduction of ferricytochrome C is approximately $1.5 \times 10^5/M/s$ *(8)*, allowing relatively sensitive detection of $O_2^{\bullet-}$.

1.1.4. Lucigenin Assay

Lucigenin at low concentrations ($< 5\,\mu M$) has been shown to detect $O_2^{\bullet-}$ without background noise derived from its own redox cycling *(9)*. The experimental procedure, precautions, advantages, and potential caveats of this assay have been comprehensively reviewed previously by Munzel et al. *(9)*. Lucigenin remains a valid method for $O_2^{\bullet -}$ production, when appropriate experimental procedures are used, as clearly described in the abovementioned review.

1.2. Detection of Hydrogen Peroxide

Emerging evidence has demonstrated a critical role of H_2O_2 in vascular signaling (*19–20*). Different from the very short-lived $O_2^{\bullet -}$, H_2O_2 is more stable and freely diffusible among adjacent cells. Its production can be rapidly increased through agonist-provoked activation of vascular NAD(P)H oxidases. Uniquely, H_2O_2 also amplifies its own production and modulates endothelial function through complex mechanisms (*19,20*). These properties of H_2O_2 make it an ideal signaling molecule. Hydrogen peroxide levels in vascular cells and tissues can be monitored by the following assays.

1.2.1. Dichlorofluorescein Fluorescent Assay

The chemical basis of the assay is that upon entry into cells, 2'-7'-dichlorodihydrofluorescein diacetate (DCFH-DA; Molecular Probes, Eugene, OR, USA) is cleaved by intracellular esterases to form DCFH, which is then oxidized by peroxides to highly fluorescent DCF.

1.2.2. Amplex Red Assay

Amplex Red (chemical name *N*-acetyl-3,7-dihydroxypgenoxazine) is a commercial compound from Molecular Probes. It reacts with extracellular H_2O_2 in the presence of peroxidase to form the highly fluorescent substance resorufin.

1.2.3. Hydrogen Peroxide Electrode

Liu and Zweier (*21*) previously reported that a *o*-phenylenediamine dihydrochloride (*o*-PD)-coated platinum microelectrode is capable of detecting H_2O_2 specifically and quantitatively from activated neutrophils. Although the specificity of the electrode is well characterized in the study and it detected 6–8 μmol/l H_2O_2 released from activated neutrophils, it remains unclear whether it is able to detect the small fraction of H_2O_2 released from vascular cells that have much lower production of H_2O_2.

1.3. Detection of Nitric Oxide Radical

It has been challenging to detect nitric oxide radical (NO$^{\bullet}$) directly from biological samples. Earlier studies mostly assessed NO$^{\bullet}$ "production" through indirect measurements of NO$^{\bullet}$ synthase activity using the L-arginine conversion assay or NO$^{\bullet}$ metabolites nitrite and nitrate using the Griess reagent. These assays, however, only reflect the enzyme activity or cumulative NO$^{\bullet}$ production, providing no information on actual bioavailable NO$^{\bullet}$. Studies in the past few years have shown that a NO•-selective electrode and ESR represent specific and quantitative assays for detection of functional NO•.

1.3.1. NO•-Specific Microelectrode

Friedemann and colleagues *(22)* previously showed that nafion and *o*-PD-coated carbon electrode can directly detect NO•. We adapted this method for direct NO• detection from porcine endocardium with and without atrial fibrillation *(23)*. These studies showed that the NO•-selective electrode is able to detect NO• in the low micromolar range.

1.3.2. ESR with NO•-Specific Spin Traps

Several iron compounds have been used to specifically trap NO•, including dithiocarbamate (DTC), N-methylglucamine dithiocarbamate (MGD), and diethyldithiocarbamate (DETC) *(24)*. Although MGD is useful for extracellular detection of NO• *(25–27)*, iron-DETC is particularly useful for detection of NO• in the cellular lipid membrane *(28,29)*. The following protocols are described for both iron-MGD and iron-DETC.

2. Materials

2.1. Detection of Superoxide

2.1.1. DHE Staining—Microscopic Method

1. Endothelial cells, vascular smooth muscle cells, or isolated vascular segments.
2. Tissue Freezing Medium (Triangle Biomedical Sciences, Durham, NC, USA).
3. Cover slips and slides (Fisher Scientific, Pittsburgh, PA).
4. DHE (Molecular Probes) stock solution: 2–5 mM dissolved in dimethylsulfoxide, prepare fresh in dark.
5. Phosphate-buffered saline (PBS).
6. Mounting media: Prolong anti-fade (Molecular Probes).
7. Fluorescent microscope.
8. Tissue Freezing Medium (Triangle Biomedical Sciences).

2.1.2. DHE Staining—HPLC Method

1. Endothelial cells, vascular smooth muscle cells, or isolated vascular segments.
2. Modified Kreb's/4-(2-hydroxyethyl)-1-piperazineethanesulfonic acid (HEPES) buffer: 99.0 mM NaCl, 4.69 mM KCl, 1.87 mM $CaCl_2$, 1.20 mM $MgSO_4$, 25 mM $NaHCO_3$, 1.03 mM K_2HPO_4, 20 mM sodium-HEPES, and 11.1 mM D-glucose (pH 7.35) containing 25 or 50 μM DHE Tissue Freezing Medium (Triangle Biomedical Sciences).
3. Acetonitrile (37–47%).
4. Trifluoroacetic acid (0.1%).

5. Xanthine oxidase (5 mU/ml).
6. Xanthine (10–100 μM).
7. Tissue Grinder (Fisher Scientific).
8. Syringe filter, 0.22 μm.
9. HPLC equipped with a sensitive fluorescent detector (Beckman Coulter, Fullerton, CA; Schimadzu by Fisher Scientific, Pittsburgh, PA).
10. C-18 reverse phase column (Necleosil 250, 4.5 mm; Sigma-Aldrich, St. Louis, MO, USA).

2.1.3. ESR

1. Endothelial or vascular smooth muscle cells.
2. Modified Krebs/HEPES buffer (*see* **Subheading 2.1.1.2.**).
3. Cyclic hydroxylamine CPH or 1-hydroxy-3-methoxycarbonyl-2,2,5,5-tetramethyl-pyrrolidine (CMH) stock solution (10 mM) (Alexis Biochemicals, San Diego, CA, USA) in modified Kreb's/HEPES buffer containing metal chelator, 25–50 μM deferoximine, and 3.5 μM DETC. This stock solution should be de-oxygenated by nitrogen gas continuously to maintain low background oxidation of the spin traps.
4. Lysis buffer containing protease inhibitors: 50 mM Tris–HCl buffer, pH 7.4, containing 0.1 mM ethylenediamine tetraacetic acid (EDTA), 0.1 mM ethylene glycol tetraacetic acid (EGTA), 1 mM phenylmethyl sulfonyl fluoride, 2 μM bestatin, 1 μM pepstatin, and 2 μM leupeptin.
5. NADPH (0.2 mM).
6. Xanthine (0.1 mM).

2.1.4. Cytochrome C Reduction Assay

1. Endothelial cells or vascular cells.
2. Acetylated cytochrome C: 50 μM.
3. Potassium phosphate buffer pH 7.4 (KPi buffer): 50 mM K_2HPO_4 and 50 mM KH_2PO_4, pH 7.8.
4. Catalase 1 U/μl.
5. SOD (1 U/μl).
6. Fluorescent plate reader.

2.2. Detection of Hydrogen Peroxide

2.2.1. Dichlorofluorescein Fluorescent Assay

1. Cultured cells.
2. Dichlorofluorescein (DCF): 30 μM.
3. Fluorescent plate reader.

2.2.2. Amplex Red Assay

1. Amplex Red Assay kit from Molecular Probes.

2.3. Detection of Nitric Oxide Radical

2.3.1. NO•-Specific Microelectrode

1. Carbon fiber electrodes (100 μm length and 30 μm outer diameter; Word Precision Instruments, Sarasota, FL, USA).
2. *o*-PD solution (in 0.1 M PBS with 100 μM ascorbic acid).
3. Nafion (5% in aliphatic alcohols; Sigma-Aldrich).
4. Modified Kreb's/HEPES buffer.
5. Axopatch 200B amplifier (Axon Instruments, Union City, CA, USA).
6. Silver/silver chloride reference electrode.

2.3.2. Iron-DETC for Trapping of NO•

1. Culture endothelial cells.
2. Saline (0.9% NaCl).
3. $Fe^{2+}(DETC)_2$: $FeSO_4 \cdot 7H_2O$, 4.45 mg/10 ml for 1.6 mmol/l stock and DETC, 7.21 mg/10 ml for 3.2 mmol/l stock.
4. PBS.
5. Modified Kreb's/HEPES buffer.
6. Ferrous sulfate (4 mM).
7. *N*-methyl-D-glucamine dithiocarbamate MGD (20 mM).

3. Methods

3.1. Detection of Superoxide Anion

3.1.1. DHE Assay—Microscopic Method (See **Notes 1 and 2**)

1. Culture endothelial cells or vascular smooth muscle cells on glass cover slips in six-well plates. Alternatively, embed freshly isolated vascular segments (2 mm) in tissue freezing medium, section to 30 μm, and mount on cover slips.
2. Dilute DHE in PBS to final concentration of 2–5 μmol/l. Add 200 μl drops of DHE–PBS solution on cell monolayer or tissue section and incubate at 37 °C in dark for 30 min.
3. Rinse off excess DHE with PBS twice, drip off excess liquid, and mount cover slips to microscopic slides using mounting media.
4. Capture images immediately with a fluorescent microscope at excitation and emission wavelengths of 520 and 610 nm, respectively.
5. If drug treatment is desired, pre-incubate with cells or tissue sections prior to addition of DHE.

3.1.2. DHE Assay—HPLC Method (See **Notes 3 and 4**)

1. Treat endothelial cells or 2-mm vessel segments (*see* **Note 4**) with desired agonists and incubate in modified Kreb's/HEPEs buffer at 37 °C in dark for 15 min–1 h.

2. At end of the incubation, collect cells by gentle scraping, homogenize cells or vessel segments in chilled methanol using a Tissue Grinder, and filter suspensions through a 0.22-μm syringe filter.

3. Analyze the supernatant using a HPLC equipped with a C-18 reverse phase column and a sensitive fluorescent detector, at excitation and emission wavelengths of 480 and 580 nm, respectively. A gradient of acetonitrile from 37 to 47% over 23 min at a flow rate of 0.5 ml/min mixed with 0.1% trifluoroacetic acid is used as solvent to separate DHE, ethidium, and oxyethidium.

4. Calibration of oxyethidium formation from DHE is obtained by incubation of DHE with xanthine oxidase (5 mU/ml) and increasing concentrations of xanthine (10–100 μM).

5. A representative HPLC–DHE assay of $O_2^{\bullet-}$ production from angiotensin II-stimulated vascular smooth muscle cells is presented in **Fig. 3**.

3.1.3. ESR with Superoxide-Specific Spin Traps *(See* **Notes 5 and 6***)*

1. Culture endothelial or vascular smooth muscle cells to confluence on 100-mm Petri dishes. Wash off media and collect cells in modified Kreb's/HEPES buffer.

2. Gently scrape cells off culture dish and collect in ice-cold Kreb's/HEPEs buffer. After centrifuging at 500 *g* for 10 min at 4 °C, re-suspend cell pellets in freshly prepared deferoxamine-containing Kreb's/HEPEs buffer. Usually a density of 1×10^6 cells/10 μl is desired.

3. Mix in an Eppendorf tube 10 μl cell suspension, 80 μl of Kreb's/HEPEs buffer, and 10 μl CPH or CMH stock solution (final concentration 1 mM).

4. Immediately transfer sample into glass capillaries and load capillaries into appropriate ESR cavity for 10-min time scan.

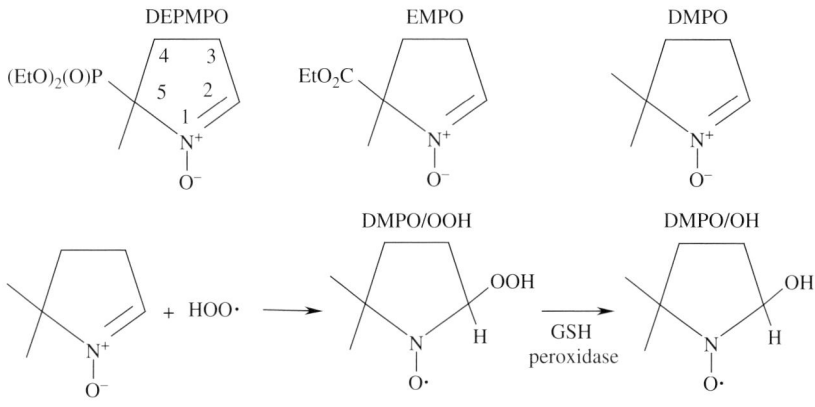

Fig. 3. Nitrone spin traps and formation of $O_2^{\bullet-}$-derived radical adducts.

3.1.3.1. ANALYSIS OF $O_2^{\bullet-}$ IN MEMBRANE PREPARATIONS OF CELLS OR TISSUES TO ASSESS NADPH OXIDASE AND XANTHINE OXIDASE ACTIVITY

1. Homogenize cultured vascular endothelial cells, smooth muscle cells, or vessel segments in lysis buffer using a Tissue Grinder.
2. Centrifuge samples at $750\,g$ at $4\,°C$ for $10\,min$. Remove supernatant and centrifuge at $30,000\,g$ at $4\,°C$ for $30\,min$.
3. Resuspend membrane pellet in $150\,\mu l$ lysis buffer and measure protein concentration using the Bradford method.
4. Prepare cyclic hydroxylamine CPH or CMH stock solution as described above.
5. Mix in an Eppendorf tube with $10\,\mu l$ membrane preparation: $80\,\mu l$ of Kreb's/HEPEs buffer and $10\,\mu l$ of the CPH stock solution ($1\,mM$). Activity of NADPH oxidase can be determined by adding NADPH ($0.2\,mM$) as the substrate, whereas activity of xanthine oxidase is determined by adding xanthine ($0.1\,mM$).
6. Immediately transfer membrane-spin trap mix into glass capillaries and load capillaries into appropriate ESR cavity for 10-min time scan. Use the following ESR settings described below:

Bruker EMX. Field sweep, $50\,G$; microwave frequency, $9.78\,GHz$; microwave power, $20\,mW$; modulation amplitude, $2\,G$; conversion time, $1312\,ms$; time constant, $656\,ms$; 512 points resolution; and receiver gain, 1×10^5.

Miniscope 200. Biofield, $3350\,G$; field sweep, $40\,G$; microwave frequency, $9.78\,GHz$; microwave power, $20\,mW$; modulation amplitude, $3\,G$; 4096 points resolution; and receiver gain, 500.

3.1.4. Cytochrome C Reduction Assay (See **Notes 7 and 8**)

1. Culture endothelial cells or vascular smooth cells in 6-well or 12-well plates or prepare vessel segments ($3 \times 2\,mm$ rings). Wash off media or blood residues using warm PBS.
2. Incubate cells or tissues with acetylated cytochrome C ($50\,\mu M$) in KPI buffer containing $1\,U/\mu l$ catalase (to prevent re-oxidation of reduced cytochrome C by H_2O_2) in the dark at $37\,°C$ for $1\,h$. An identical set of samples is incubated in the presence of SOD ($1\,U/\mu l$) for subtraction of the SOD-inhibitable signal.
3. Transfer $200\,\mu l$ post-incubation supernatant to a fresh 96-well plate and read the plate with a fluorescent plate reader at 540, 550, and $560\,nm$.
4. Calculate the optical density (OD) of reduced cytochrome C using the path-length-corrected values: $OD_{550nm} - (OD_{540nm} + OD_{560nm})/2$.
5. Subtract OD from that of identical samples containing SOD. Covert to $O_2^{\bullet-}$ concentration using the extinction coefficient equation of $Em_{550nm} = 2.1 \times 10^4\,M/cm$.

3.2. Detection of Hydrogen Peroxide

3.2.1. DCF Fluorescent Assay (See **Note 9**)

1. Culture endothelial cells or vascular smooth muscle cells on 100-mm Petri dishes.
2. Aspirate media, rinse cells with PBS, and load cells or vessel segments with freshly prepared DCFH-DA (30 μM) for 15 min to allow intracellular conversion of DHCF prior to stimulation with desired agonists.
3. At the end of incubation, gently scrape cells off dish in 1 ml ice-cold PBS, load 200 μl cell suspensions ($\sim 2 \times 10^5$ cells) into 96-well plate, and read with a fluorescent plate reader at excitation and emission wavelengths of 475 and 525 nm, respectively.

3.2.2. Amplex Red Assay (See **Note 10**)

1. Culture endothelial cells or vascular smooth muscle cells on P100 Petri dishes or prepare three vessel segments of 2 mm size.
2. Incubate cells or vessel segments in Krebs Ringer phosphate-glucose buffer (see manufacturer's manual) containing 100 μM Amplex Red and 1 U/ml horseradish peroxidase at 37 °C for 30 min, with or without desired agonists.
3. At end of incubation, transfer 200 μl post-incubation buffers into 96-well plate and read with a fluorescent plate reader at excitation and emission wavelengths of 530 and 580 nm, respectively.
4. Generate calibration curve using resorufin standard supplied by the manufacturer.

3.3. Detection of Nitric Oxide Radical

3.3.1. NO•-Specific Microelectrode (See **Notes 11 and 12**)

1. Coat bare carbon fiber electrodes (100 μm length and 30 μm outer diameter; Word Precision Instruments) with nafion and o-PD. Coat with freshly made o-PD solution at constant potential (+0.9 V vs. Ag/AgCl reference electrode) for 45 min. Dip in nafion solution for 3 s and dry for 5 min at 85 °C. The nafion-coating cycle should be repeated 10–15 times.
2. Culture endothelial cells on 35-mm dishes or prepare fresh tissue samples in freshly made modified Kreb's/HEPES buffer.
3. Place the electrode tip at the surface of an individual cell, endocardium, or lumen of blood vessels, and then withdraw precisely 5 μm.
4. Record NO•-dependent oxidation currents (voltage clamp mode, hold at 0.65 V, approximately the voltage for peak NO• oxidation) immediately after addition of agonists using an Axopatch 200B amplifier (Axon Instruments). A silver/silver chloride reference electrode is used. Use pCLAMP 7.0 program (Axon Instruments) for delivery of voltage protocols and data acquisition and analysis.

5. Calculate NO• concentrations from a standard curve obtained using dilutions of de-oxygenated, saturated NO• gas solutions.

3.3.2. Iron-DETC Protocol for Intracellular Trapping of NO•
*(See **Notes 13 and 14**)*

1. Culture endothelial cells on 100-mm Petri dishes or prepare vascular segments (6–12 2-mm vessel segments).
2. Bubble freshly prepared saline (0.9% NaCl) with nitrogen gas to remove oxygen.
3. Aspirate media and rinse cells with warm PBS once, add 1.5 ml modified Kreb's/HEPES buffer with or without desired agonists, then mix $Fe^{2+}(DETC)_2$, and immediately add to culture dish (500 μl of each solution, final volume 2.0 ml).
4. Incubate in cell culture incubator for desired period for cumulative trapping of NO•.
5. Aspirate buffer, gently collect cells into a 1-ml insulin syringe, snap freeze in liquid nitrogen, then transfer sample column into a finger dewer, and capture $Fe^{2+}(DETC)_2$–NO• signal using ESR at the following settings:

Bruker EMX: Field sweep, 160 G; microwave frequency, 9.39 GHz; microwave power, 10 MW; modulation amplitude, 3 G; conversion time, 2621 ms; time constant, 328 ms; modulation amplitude, 3 G; receiver gain, 1×10^4; and four scans.

Miniscope 200: Biofield, 3267; field sweep, 100 G; microwave frequency, 9.78 GHz; microwave power, 40 mW; modulation amplitude, 10 G; 4096 points resolution; and receiver gain, 900.

3.3.2.1. IRON-MGD PROTOCOL FOR EXTRACELLULAR TRAPPING OF NO•
*(See **Notes 15 and 16**)*

1. Culture endothelial cells on 100-mm Petri dishes or prepare vascular segments (6–12 2-mm vessel segments).
2. Bubble freshly prepared saline (0.9% NaCl) with nitrogen gas to remove oxygen, and then make stock solutions of $FeSO_4 \cdot 7H_2O$, 4 mM, and MGD, 20 mM.
3. Prepare stock solutions of $Fe^{2+}MGD$ by mixing $FeSO_4$ and MGD at the ratio of 1:5 or 1:10 (final Fe concentration: 0.5 mM).
4. Aspirate media and rinse cells with warm PBS once, add 1.5 ml modified Kreb's/HEPES buffer with or without desired agonists, then mix $Fe^{2+}(MGD)_2$, and immediately add to culture dish.
5. Incubate in cell culture incubator for desired period for cumulative trapping of NO•.
6. Collect 1 ml of post-incubation supernatant into a 1-ml insulin syringe and snap freeze in liquid nitrogen, then transfer sample column into a finger dewer, and capture $Fe^{2+}MGD$–NO• signal using ESR settings as described above for $Fe^{2+}(DETC)_2$.

4. Notes

1. DHE is cell permeable, which allows intracellular detection of $O_2^{\bullet-}$. It is relatively specific with minimal potential of being oxidized by H_2O_2 or other reactive oxygen species *(30)*. It is very sensitive (fluorescent detection with HPLC identifies as low as 1 pmol/mg protein $O_2^{\bullet-}$, see protocol below) *(11)* and generally reproducible. The procedure is convenient to use as an initial screening for $O_2^{\bullet-}$ production from both cells and tissues. Results from this widely used $O_2^{\bullet-}$-detection method have been confirmed by recent ESR studies that unequivocally identify and quantify individual reactive oxygen species. For example, aortic $O_2^{\bullet-}$ production was found to be increased in angiotensin II-induced hypertension using Lucigenin *(31)* or DHE assays *(32)*, and this was recently confirmed by ESR $O_2^{\bullet-}$ trapping with the $O_2^{\bullet-}$-specific spin trap cyclic hydroxylamine *(16)*.

2. DHE staining is semi-quantitative. It was demonstrated that DHE can be oxidized by excessive cytochrome C (when the molar ratio of cytochrome C to DHE is > 10) *(30,33)* or other heme-containing proteins such as hemoglobin *(33)*. Thus, caution should be used when interpreting DHE data in cases where apoptosis or mitochondrial damage is involved or post-hemorrhage endothelium is studied. To minimize influences of this non-specific reaction on data analysis, it is recommended to use cell permeable SOD such as polyethylene glycol-conjugated SOD (Sigma-Aldrich) in parallel and only compare responses that are SOD inhibitable. Another caution is *not* to use lower wavelengths of 490–495/580–600 nm *(34)*. Recent studies demonstrated that peroxidase-catalyzed oxidation of DHE by H_2O_2 produced fluorescent products detectable at the abovementioned wavelengths *(34)*. Using 520/610 nm could potentially avoid these non-specific overlapping fluorescent signals caused by H_2O_2. In many cases, researchers have to use the wavelengths with which the microscope is equipped, and this may have caused variations in data reporting.

3. DHE–HPLC provides quantitative, sensitive, and specific detection of $O_2^{\bullet-}$ from vascular cells and tissues. Fink and colleagues *(11)* have shown that the specific HPLC peak for oxyethidium is completely preventable by SOD but is not formed in response to H_2O_2 or peroxynitrite.

4. Modestly larger amounts of biological materials are needed compared to fluorescent imaging of DHE (i.e., three 2-mm aortic rings are required for HPLC assay of DHE compared to only one 2-mm ring for fluorescent microscope). This is, however, a minimal disadvantage considering the quantitative nature of the HPLC-based assay.

5. ESR spin trapping of $O_2^{\bullet-}$ has been proven to selectively detect $O_2^{\bullet-}$ from cultured vascular cells or tissue homogenates in the low nanomolar range *(16,18)*. Our unpublished data suggest that this method is also applicable to intact, isolated blood vessels, but a slightly larger capillary and fitting holder is required to accommodate vessel segments inside the ESR cavity. In addition, cyclic hydroxylamine compounds can be injected into mice for in vivo detection of $O_2^{\bullet-}$, but the preferential tissue distribution patterns of these compounds remains unclear *(35,36)*.

6. CMH is cell permeable and more suitable for detection of intracellular $O_2^{\bullet-}$. CPH, however, is used for extracellular detection of $O_2^{\bullet-}$. The advantage of the spin traps is that the ESR spectrum of the radical spin traps works as a "finger print" of the trapped radical. The problem, however, deals with the limited stability of the radical adducts. Even relatively stable $O_2^{\bullet-}$ adducts of DMPO and DEPMPO can be decomposed by intracellular peroxidases into the corresponding hydroxyl adducts (**Fig. 3**). In case of cyclic hydroxylamines, though very stable when forming adducts with $O_2^{\bullet-}$, they can also trap peroxynitrite to form identical nitroxides *(15)*, limiting specificity. Thus, it is highly recommended that only the SOD-inhibitable fraction of the ESR signals is compared between experimental groups.
7. Because cytochorome C is cell impermeant, this assay provides quantitative detection of *extracellular* $O_2^{\bullet-}$.
8. Reduced cytochrome C can be oxidized by peroxidases or oxidants such as H_2O_2 and peroxynitrite *(8,37)*. This will underestimate the rate of production of $O_2^{\bullet-}$. Our own experience suggests that the reproducibility is not great, and the assay is not sensitive enough to detect the low levels of $O_2^{\bullet-}$ found in vascular cells. As cytochrome C is also strictly cell *im*permeable, it is more applicable to activated neutrophils where $O_2^{\bullet-}$ is produced in large quantities extracellularly. It is likely that some intracellular $O_2^{\bullet-}$ can be transported through membrane pores to extracellular space. However, as regulation of this process is poorly understood, it is difficult to predict the intracellular $O_2^{\bullet-}$ production rate based on cytochrome C data. Nevertheless, the cytochrome C assay can be used as a secondary method for $O_2^{\bullet-}$ detection in conjunction with another more sensitive method for intracellular $O_2^{\bullet-}$ to estimate overall bioactive $O_2^{\bullet-}$.
9. This assay can be used for intracellular detection of H_2O_2 and is relatively specific. DCF assay is only semi-quantitative, because it is difficult to generate a standard curve that mimics the intracellular situation. In the past, we attempted to add esterases to ex vivo incubation of DCFH-DA, but this was not sufficient to produce optimal concentration-absorbance curves when serial dilutions of standard were used. It is unclear whether other enzymes besides esterases are involved in converting DCFH-DA intracellularly. Earlier work by LeBel et al. *(38)* demonstrated that DCF is not oxidized by hydroxyl radical or $O_2^{\bullet-}$ directly, although it can also be oxidized by peroxynitrite *(39)*. Similar to $O_2^{\bullet-}$ assays, it is always recommended to use cell-permeable catalase (i.e., polyethylene glycol-conjugated catalase from Sigma-Aldrich) and compare only the catalase-inhibitable signal, because DCF reacts with other cellular peroxides and can redox cycle *(40)*. Nevertheless, the DCF assay can be used for relative comparison among different experimental groups.
10. The Amplex Red assay is good for specific, extracellular detection of H_2O_2 and is also quantitative. Recent studies using this method have demonstrated basal endothelial H_2O_2 production at the rate of 30 pmol/mg protein/min *(17,27)*. However, this assay may not accurately reflect intracellular formation of H_2O_2

although H_2O_2 is expected to diffuse to the outside of the cell to reach equilibrium. It is unclear how much H_2O_2 is able to diffuse out of the cell, considering presence of intracellular catalase and small thiols that are capable of "trapping" H_2O_2 intracellularly. The activities of these H_2O_2-consuming enzymes or small molecules may change with agonist treatment to compound the capability of the assay to reflect actual H_2O_2 production rate.

11. Control experiments showed that nafion and *o*-PD coatings effectively eliminated electrode responsiveness to other oxidizable species, including nitrate, nitrite, and H_2O_2. This method is not only specific and quantitative, it also offers "real-time" monitoring of $NO^{\bullet}$ production over a time course that has particular utility for signal transduction studies examining temporal and spatial activation of $NO^{\bullet}$-producing pathways.

12. This assay requires an entire patch-clamp electrophysiology setup. It only detects $NO^{\bullet}$ from released from single endothelial cell or the most surrounding few endothelial cells. This is because $NO^{\bullet}$ has limited diffusion ability to reach the electrode in the in vitro system. It may not reflect overall $NO^{\bullet}$ production very accurately.

13. Several advantages exist for $NO^{\bullet}$ detection using ESR and $Fe^{2+}(DETC)_2$. $Fe^{2+}(DETC)_2$ is specific for bioactive $NO^{\bullet}$ and does not detect nitrite or nitrate. In addition, the stability of the $Fe^{2+}(DETC)_2–NO^{\bullet}$ complex allows for measurement of the cumulative amount of bioactive $NO^{\bullet}$ produced over time.

14. Limitations of this technique include the special handling required for $Fe^{2+}(DETC)_2$ colloid to prevent oxidation. In addition, $Fe^{2+}(DETC)_2–NO^{\bullet}$ can be oxidized by extracellular H_2O_2 or $O_2^{\bullet-}$ to form the ESR silent, $Fe^{3+}(DETC)_2–NO^{\bullet}$. However, our data showed that bolus addition of $100\,\mu mol/l\ H_2O_2$ or the same amount of $O_2^{\bullet-}$ generated by xanthine oxidase decreased the $Fe^{2+}(DETC)_2–NO^{\bullet}$ signal by 20%. Both H_2O_2 and $O_2^{\bullet-}$ caused line broadening of the ESR spectra due to accumulation of Fe^{3+} in the samples. The effect of the line broadening on the quantification of $NO^{\bullet}$ can be avoided by double integration of the ESR signal of $Fe^{2+}(DETC)_2–NO^{\bullet}$.

15. $Fe^{2+}MGD$ stays in the extracellular compartment and thus allows detection of $NO^{\bullet}$ that is diffused out the of endothelial cells. It may thus reflect amount of $NO^{\bullet}$ that is available to travel within intracellular space. However, how much $NO^{\bullet}$ eventually reaches adjacent vascular smooth muscle cells also depends on extracellular SOD activity in vivo.

16. It has been previously reported that autoxidation of MGD may cause formation of $O_2^{\bullet-}$ and H_2O_2, which may interfere with quantitative detection of $NO^{\bullet}$ *(41)*. It was also shown that anaerobic solutions of $Fe^{2+}MGD$ can reduce nitrite to $NO^{\bullet}$ *(42)*. However, this reaction is negligible under normoxic conditions.

Acknowledgments

This work is supported by NIH/NHLBI grants HL077440 and HL081571 American Heart Association Scientist Development Grant (no. 0435189N to H.C.), American Diabetes Association Research Award (H.C.), Career Development Award from the Schweppe Foundation (H.C.), and Start-up Fund from the University of Chicago (H.C.).

References

1. Cai, H., and Harrison, D. G. (2000) Endothelial dysfunction in cardiovascular diseases: the role of oxidant stress, *Circ. Res.* **87**, 840–844.
2. Cai, H., Griendling, K. K., and Harrison, D. G. (2003) The vascular NAD(P)H oxidases as therapeutic targets in cardiovascular diseases, *Trends Pharmacol. Sci.* **24**, 471–478.
3. Griendling, K. K., and FitzGerald, G. A. (2003) Oxidative stress and cardiovascular injury: part I: basic mechanisms and in vivo monitoring of ROS, *Circulation* **108**, 1912–1916.
4. Griendling, K. K. (2004) Novel NAD(P)H oxidases in the cardiovascular system, *Heart* **90**, 491–493.
5. Jung, O., Marklund, S. L., Geiger, H., Pedrazzini, T., Busse, R., and Brandes, R. P. (2003) Extracellular superoxide dismutase is a major determinant of nitric oxide bioavailability: in vivo and ex vivo evidence from ecSOD-deficient mice, *Circ. Res.* **93**, 622–629.
6. Khatri, J. J., Johnson, C., Magid, R., Lessner, S. M., Laude, K. M., Dikalov, S. I., Harrison, D. G., Sung, H. J., Rong, Y., and Galis, Z. S. (2004) Vascular oxidant stress enhances progression and angiogenesis of experimental atheroma, *Circulation* **109**, 520–525.
7. Yang, H., Roberts, L. J., Shi, M. J., Zhou, L. C., Ballard, B. R., Richardson, A., and Guo, Z. M. (2004) Retardation of atherosclerosis by overexpression of catalase or both Cu/Zn-superoxide dismutase and catalase in mice lacking apolipoprotein E, *Circ. Res.* **95**, 1075–1081.
8. Tarpey, M. M., Wink, D. A., and Grisham, M. B. (2004) Methods for detection of reactive metabolites of oxygen and nitrogen: in vitro and in vivo considerations, *Am. J. Physiol. Regul. Integr. Comp. Physiol.* **286**, R431–R444.
9. Munzel, T., Afanas'ev, I. B., Kleschyov, A. L., and Harrison, D. G. (2002) Detection of superoxide in vascular tissue, *Arterioscler. Thromb. Vasc. Biol.* **22**, 1761–1768.
10. Zhao, H., Kalivendi, S., Zhang, H., Joseph, J., Nithipatikom, K., Vasquez-Vivar, J., and Kalyanaraman, B. (2003) Superoxide reacts with hydroethidine but forms a fluorescent product that is distinctly different from ethidium: potential implications in intracellular fluorescence detection of superoxide, *Free Radic. Biol. Med.* **34**, 1359–1368.

11. Fink, B., Laude, K., McCann, L., Doughan, A., Harrison, D. G., and Dikalov, S. (2004) Detection of intracellular superoxide formation in endothelial cells and intact tissues using dihydroethidium and an HPLC-based, *Am. J. Physiol. Cell Physiol.* **287**, C895–902.

12. Buettner, G. R., and Mason, R. P. (1990) Spin-trapping methods for detecting superoxide and hydroxyl free radicals in vitro and in vivo, *Methods Enzymol.* **186**, 127–133.

13. Villamena, F. A., and Zweier, J. L. (2004) Detection of reactive oxygen and nitrogen species by EPR spin trapping, *Antioxid. Redox Signal.* **6**, 619–629.

14. Dikalov, S. I., Dikalova, A. E., and Mason, R. P. (2002) Noninvasive diagnostic tool for inflammation-induced oxidative stress using electron spin resonance spectroscopy and an extracellular cyclic hydroxylamine, *Arch. Biochem. Biophys.* **402**, 218–226.

15. Dikalov, S., Grigor'ev, I. A., Voinov, M., and Bassenge, E. (1998) Detection of superoxide radicals and peroxynitrite by 1-hydroxy-4-phosphonooxy-2,2,6,6-tetramethylpiperidine: quantification of extracellular superoxide radicals formation, *Biochem. Biophys. Res. Commun.* **248**, 211–215.

16. Landmesser, U., Cai, H., Dikalov, S., McCann, L., Hwang, J., Jo, H., Holland, S. M., and Harrison, D. G. (2002) Role of p47(phox) in vascular oxidative stress and hypertension caused by angiotensin II, *Hypertension* **40**, 511–515.

17. McNally, J. S., Davis, M. E., Giddens, D. P., Saha, A., Hwang, J., Dikalov, S., Jo, H., and Harrison, D. G. (2003) Role of xanthine oxidoreductase and NAD(P)H oxidase in endothelial superoxide production in response to oscillatory shear stress, *Am. J. Physiol. Heart Circ. Physiol.* **285**, H2290–H2297.

18. Spiekermann, S., Landmesser, U., Dikalov, S., Bredt, M., Gamez, G., Tatge, H., Reepschlager, N., Hornig, B., Drexler, H., and Harrison, D. G. (2003) Electron spin resonance characterization of vascular xanthine and NAD(P)H oxidase activity in patients with coronary artery disease: relation to endothelium-dependent vasodilation, *Circulation* **107**, 1383–1389.

19. Cai, H. (2005) Hydrogen peroxide regulation of endothelial function: mechanisms, consequences and origins, *Cardiovasc. Res.* **68**, 26–36.

20. Cai, H. (2005) NAD(P)H oxidase-dependent self-propagation of hydrogen peroxide and vascular disease, *Circ. Res.* **96**, 818–822.

21. Liu, X., and Zweier, J. L. (2001) A real-time electrochemical technique for measurement of cellular hydrogen peroxide generation and consumption: evaluation in human polymorphonuclear leukocytes, *Free Radic. Biol. Med.* **31**, 894–901.

22. Friedemann, M. N., Robinson, S. W., and Gerhardt, G. A. (1996) o-Phenylenediamine-modified carbon fiber electrodes for the detection of nitric oxide, *Anal. Chem.* **68**, 2621–2628.

23. Cai, H., Li, Z., Goette, A., Mera, F., Honeycutt, C., Feterik, K., Wilcox, J. N., Dudley, S. C., Jr., Harrison, D. G., and Langberg, J. J. (2002) Downregulation of

endocardial nitric oxide synthase expression and nitric oxide production in atrial fibrillation: potential mechanisms for atrial thrombosis and stroke, *Circulation* **106**, 2854–2858.

24. Vanin, A. F., Huisman, A., and van Faassen, E. E. (2002) Iron dithiocarbamate as spin trap for nitric oxide detection: pitfalls and successes, *Methods Enzymol.* **359**, 27–42.

25. Komarov, A. M., Wink, D. A., Feelisch, M., and Schmidt, H. H. (2000) Electron-paramagnetic resonance spectroscopy using N-methyl-D-glucamine dithiocarbamate iron cannot discriminate between nitric oxide and nitroxyl: implications for the detection of reaction products for nitric oxide synthase, *Free Radic. Biol. Med.* **28**, 739–742.

26. Xia, Y., Cardounel, A. J., Vanin, A. F., and Zweier, J. L. (2000) Electron paramagnetic resonance spectroscopy with N-methyl-D-glucamine dithiocarbamate iron complexes distinguishes nitric oxide and nitroxyl anion in a redox-dependent manner: applications in identifying nitrogen monoxide products from nitric oxide synthase, *Free Radic. Biol. Med.* **29**, 793–797.

27. Cai, H., McNally, J. S., Weber, M., and Harrison, D. G. (2004) Oscillatory shear stress upregulation of endothelial nitric oxide synthase requires intracellular hydrogen peroxide and CaMKII, *J. Mol. Cell Cardiol.* **37**, 121–125.

28. Kleschyov, A. L., and Munzel, T. (2002) Advanced spin trapping of vascular nitric oxide using colloid iron diethyldithiocarbamate, *Methods Enzymol.* **359**, 42–51.

29. Cai, H., Li, Z., Dikalov, S., Holland, S. M., Hwang, J., Jo, H., Dudley, S. C., Jr., and Harrison, D. G. (2002) NAD(P)H oxidase-derived hydrogen peroxide mediates endothelial nitric oxide production in response to angiotensin II, *J. Biol. Chem.* **277**, 48311–48317.

30. Benov, L., Sztejnberg, L., and Fridovich, I. (1998) Critical evaluation of the use of hydroethidine as a measure of superoxide anion radical, *Free Radic. Biol. Med.* **25**, 826–831.

31. Rajagopalan, S., Kurz, S., Munzel, T., Tarpey, M., Freeman, B. A., Griendling, K. K., and Harrison, D. G. (1996) Angiotensin II-mediated hypertension in the rat increases vascular superoxide production via membrane NADH/NADPH oxidase activation. Contribution to alterations of vasomotor tone, *J. Clin. Invest.* **97**, 1916–1923.

32. Rey, F. E., Cifuentes, M. E., Kiarash, A., Quinn, M. T., and Pagano, P. J. (2001) Novel competitive inhibitor of NAD(P)H oxidase assembly attenuates vascular O(2)(−) and systolic blood pressure in mice, *Circ. Res.* **89**, 408–414.

33. Papapostolou, I., Patsoukis, N., and Georgiou, C. D. (2004) The fluorescence detection of superoxide radical using hydroethidine could be complicated by the presence of heme proteins, *Anal. Biochem.* **332**, 290–298.

34. Patsoukis, N., Papapostolou, I., and Georgiou, C. D. (2005) Interference of non-specific peroxidases in the fluorescence detection of superoxide radical by hydroethidine oxidation: a new assay for H(2)O(2), *Anal. Bioanal. Chem.* **81**, 1065–1072.

35. Fink, B., Dikalov, S., and Bassenge, E. (2000) A new approach for extracellular spin trapping of nitroglycerin-induced superoxide radicals both in vitro and in vivo, *Free Radic. Biol. Med.* **28**, 121–128.
36. Kozlov, A. V., Szalay, L., Umar, F., Fink, B., Kropik, K., Nohl, H., Redl, H., and Bahrami, S. (2003) Epr analysis reveals three tissues responding to endotoxin by increased formation of reactive oxygen and nitrogen species, *Free Radic. Biol. Med.* **34**, 1555–1562.
37. Thomson, L., Trujillo, M., Telleri, R., and Radi, R. (1995) Kinetics of cytochrome c2+ oxidation by peroxynitrite: implications for superoxide measurements in nitric oxide-producing biological systems, *Arch. Biochem. Biophys.* **319**, 491–497.
38. LeBel, C. P., Ischiropoulos, H., and Bondy, S. C. (1992) Evaluation of the probe 2′, 7′-dichlorofluorescin as an indicator of reactive oxygen species formation and oxidative stress, *Chem. Res. Toxicol.* **5**, 227–231.
39. Kooy, N. W., Royall, J. A., and Ischiropoulos, H. (1997) Oxidation of 2′, 7′-dichlorofluorescin by peroxynitrite, *Free Radic. Res.* **27**, 245–254.
40. Rota, C., Chignell, C. F., and Mason, R. P. (1999) Evidence for free radical formation during the oxidation of 2′-7′-dichlorofluorescin to the fluorescent dye 2′-7′-dichlorofluorescein by horseradish peroxidase: possible implications for oxidative stress measurements, *Free Radic. Biol. Med.* **27**, 873–881.
41. Tsuchiya, K., Jiang, J. J., Yoshizumi, M., Tamaki, T., Houchi, H., Minakuchi, K., Fukuzawa, K., and Mason, R. P. (1999) Nitric oxide-forming reactions of the water-soluble nitric oxide spin-trapping agent, MGD, *Free Radic. Biol. Med.* **27**, 347–355.
42. Tsuchiya, K., Yoshizumi, M., Houchi, H., and Mason, R. P. (2000) Nitric oxide-forming reaction between the iron-N-methyl-D-glucamine dithiocarbamate complex and nitrite, *J. Biol. Chem.* **275**, 1551–1556.

21

Assessment of Protein Glycoxidation in Ventricular Tissues

Shi-Yan Li and Jun Ren

Summary

Advanced glycation end products are permanently modified protein derivatives formed in the presence of reducing sugars, such as glucose, fructose, hexose-phosphates, trioses, and triose-phosphates by non-enzymatic glycation and oxidation ("glycoxidation") reactions and further irreversible rearrangements. Numerous studies have revealed the pivotal role of protein glycoxidation in the pathogeneses of diabetes-related and age-related diseases. Protein glycoxidation is generally recognized both as a hallmark and as a promoter for progression of diabetes-related and age-related ailments, particularly in cardiovascular system such as increased vascular and myocardial stiffness, endothelial dysfunction, altered vascular injury responses, and atherosclerotic plaque formation. An appropriate surveillance on abnormal protein glycoxidation at an early stage of disease progression is of clinical and practical importance to handle diabetes-related and age-related cardiovascular complications especially those leading to ventricular dysfunction.

Key Words: Advanced glycation end products; Diabetes; Glycoxidation.

1. Introduction

Advanced glycation end products (AGEs) are usually formed over weeks to months by non-enzymatic glycation and oxidation ("glycoxidation") reactions between carbohydrate-derived carbonyl groups and protein amino groups, commonly known as the Maillard reaction *(1,2)*. Glycoxidation of proteins has generated great clinical and research interest, as it may be generally recognized as both a marker and a promoter in diabetes-related and aging-related diseases *(3,4)*. It has long been realized that these diseases are associated with increased post-translational chemical modification of proteins in the form of AGEs. AGEs

From: *Methods in Molecular Medicine, Vol. 139: Vascular Biology Protocols*
Edited by: N. Sreejayan and J. Ren © Humana Press Inc., Totowa, NJ

could be viewed as the "tip of the iceberg" and be regarded as a useful index of the progression in diabetes-related and aging-related disease *(5–8)*. AGEs formation starts with the reaction of the amino groups of proteins, particularly the side chains of lysine, arginine, and histidine, with reducing sugars, such as glucose, fructose, hexose-phosphates, trioses, and triose phosphates. Formation of early glycation end products, such as Schiff bases and Amadori products, is reversible. Further molecular rearrangements, often involving oxidation, eventuate in the formation of AGEs *(9,10)*. Importantly, in contrast to Amadori product precursors, AGE–protein adducts, such as carboxymethyl-lysine and pentosidine linked to polypeptides, are quite stable and virtually irreversible once formed. Long-lived structural proteins such as collagen are particularly vulnerable to AGE formation by nature of their slow turnover rate *(11)*. As formation of AGEs seems to be virtually irreversible, their accumulation in tissues leads to a chronic, long-term "memory" for previous stress and could cause the host response to future challenges quite different from that observed in normal tissues *(12,13)*. Three considerations make protein glycoxidation measurement of special practical interest. First, protein glycoxidation by glucose, fructose, and more highly reactive, metabolic pathway-derived intermediates in target tissues is viewed as a characteristic hallmark in development or progression of diabetes-related and aging-related diseases *(14–16)*. Second, an appropriate surveillance of the abnormal accumulation of glycoxidized protein at an early stage of diabetes and aging is of clinical and practical importance *(17,18)*, as protein glycoxidation induces permanent abnormalities in structure and function of molecules and increases oxidative stress, which consequently promote the pathogenesis of diabetic complications and changes associated with aging. Third, practical techniques applied in the assay of protein glycoxidation may be utilized as possible diagnostic and therapeutic tools in diabetes-related and age-related cardiovascular diseases *(19,20)*.

2. Materials

2.1. Ventricular Tissue Lysate

1. Fresh or frozen ventricular tissues from aged or diabetic mice (*see* **Note 1**).
2. Radio immunoprecipitation (RIPA) lysis buffer for total tissue protein lysate preparation: 150 mM NaCl, 0.5% deoxycholic acid, 1% Triton-X 100, 0.1% sodium dodecyl sulfate (SDS), 1 mM ethylene glycol-bis(2-aminoethylether)-N, N, N', N'-tetraacetic acid, 1 mM ethylene diamine tetra acetic acid, 50 mM dithiothreitol (DTT), and 20 mM Tris–HCl, pH 7.4. Immediately prior to use, add 1% (v/v) protease inhibitor cocktail (Sigma P-8340, Sigma-Aldrich Corporation, St. Louis, Missouri, USA).
3. PRO250 Homogenizer (PRO Scientific Inc., Oxford, Connecticut, USA).
4. Sonic Dismembrator Model 100 (Fisher Scientific, Pittsburgh, Pennsylvana, USA).

2.2. Quantification of Protein Oxidation

1. 1% Streptomycin sulfate.
2. 2.5 M HCl.
3. 20% (w/v) trichloroacetic acid (TCA) solution.
4. 10 mM 2,4-dinitrophenylhydrazine (2,4-DNPH) solution.
5. Ethyl alcohol, absolute.
6. 2-(2-butoxyethoxy) ethyl acetate.
7. 6 M Guanidine solution.

2.3. Immunoblot Detection of Oxidized Protein by SDS–PAGE

1. Derivatization solution: 10 mM 2,4-DNPH in 2 M HCl.
2. Derivatization-control solution: 2 M HCl.
3. Neutralization solution: 2 M Tris base and 30% (v/v) glycerol.
4. Mixture of standard proteins with attached dinitrophenyl (DNP) residues (OxyBlot™ Protein Standard; CHEMICON Internal Inc., Temecula, California, USA) (*see* **Note 2**).
5. Primary antibody: Rabbit anti-DNP-antibody (DakoCytomation, Glostrup, Denmark).
6. Secondary antibody: Goat anti-rabbit IgG (H + L), horseradish peroxidase (HRP) conjugated (CHEMICON Internal Inc.).
7. Separating buffer (4×): 1.5 M Tris–HCl, pH 8.7 and 0.4% SDS. Store at room temperature.
8. Stacking buffer (4×): 0.5 M Tris–HCl, pH 6.8 and 0.4% SDS. Store at room temperature.
9. Acrylamide (30%)/bis solution (37.5:1) and *N, N, N, N′*-teramethyl-etylenediamine (TEMED; Bio-Rad, Hercules, CA, USA) (*see* **Note 3**).
10. Ammonium persulfate: Prepare 10% solution in water and immediately freeze in single-use aliquots (200 μl) at −20 °C (*see* **Note 4**).
11. Water-saturated isobutanol: Shake equal volumes of water and isobutanol in a glass bottle and allow to separate at room temperature. Use the top layer. Store at room temperature.
12. Running buffer (5×): 125 mM Tris, 960 mM glycine, and 0.5% (w/v) SDS. Store at room temperature.
13. Gel loading buffer (1×): 65 mM Tris–HCl (pH 6.8), 10% (v/v) glycerol, 180 mM 2-mercaptoethanol, 2% (w/v) SDS, and 0.002% (w/v) bromophenol blue.
14. Transfer buffer: 12 mM Tris, 96 mM glycine, and 20% (v/v) methanol (*see* **Note 5**).
15. Phosphate-buffered saline with Tween (PBS-T) (10×): 1.36 M NaCl, 15 mM KCl, 100 mM Na_2HPO_4, 15 mM KH_2PO_4, pH 7.4, and 0.05% (v/v) Tween 20.
16. Blocking/dilution buffer: 1% BSA/PBS-T.
17. SDS (10%).
18. Enhanced chemiluminescent (ECL) reagents from Pierce (Rockford, IL, USA) and Bio-Max ML film from Kodak (Rochester, NY, USA) (*see* **Note 6**).

2.4. Quantification of Protein Advanced Glycosylation

1. Carbonate-coating buffer (10×): 1.5 M sodium carbonate, 3.5 M sodium bicarbonate, and 0.3 M sodium azide, pH 9.6.
2. Tris-buffered saline with Tween (10×) washing solution: 1 M NaCl, 100 mM Tris–HCl, pH 7.4, and 1% Tween 20.
3. Blocking buffer: 1× Tris-buffered Saline Tween-20 (TBS-T) with 1% BSA.
4. Primary antibody: 6D12 monoclonal antibody (Trans Genic Inc., Kumamoto, Japan, Cod No. KH001) (*see* **Note 7**).
5. Secondary antibody: Goat anti-mouse IgG (L + H), HRP conjugated (cat no. AP308P, CHEMICON Internal Inc.).
6. NeA-Blue tetramethylbenzidine (TMB) substrate (aqueous) for HRP (Clinical Science Products Inc., Mansfield, Massachusetts, USA).
7. Stopping solution: 1 N H_2SO_4.
8. Ninety-six-well Assay Plate (Nunc-Immuno™ Plate and Maxi Sorp™ Surface; Naige Nunc International, Roskilde, Denmark).
9. Incubator.
10. SpectraMax 190 Microplate Reader System.

2.5. Assay of Protein Advanced Glycosylation by Two-Dimensional Gel Electrophoresis

1. ReadyPrepTM 2D Cleanup Kit (cat. no. 163-2130, Bio-Rad): Precipitation agent 1, precipitation agent 2, wash reagent 1, wash reagent 2, and wash 2 additive.
2. Rehydration buffer: 8 M urea, 2% (w/v) 3[(3-Cholamidopropyl)dimethylammonio]-propane sulfonic acid buffer, 50 mM DTT, 0.5% (w/v) Bio-Lyte3/10 ampholytes, and bromophenol blue (trace).
3. Equilibration buffer I: 6 M urea, 2% (w/v) SDS, 0.375 M Tris–HCl (pH 8.8), 20% (w/v) glycerol, and 2% (w/v) DTT.
4. Equilibration buffer II: 6 M urea, 2% SDS, 0.375 M Tris–HCl (pH 8.8), and 20% glycerol. Immediately prior to use, add 3% (w/v) iodoacetamide.
5. Running buffer (5×): 125 mM Tris, 960 mM glycine, and 0.5% (w/v) SDS. Store at room temperature.
6. Overlay agarose: 0.5% (w/v) low-melting point agarose in 1× running buffer.
7. Silver Stain Plus (cat no. 161-0449, Bio-Rad).
8. Transfer buffer: 12 mM Tris, 96 mM glycine, and 20% (v/v) methanol.
9. Nitrocellulose membrane and 3MM chromatography paper (Whatman, Maidstone, UK).
10. TBS-T washing solution: 1× TBS and 0.1% (v/v) Tween 20.
11. Blocking buffer: 5% (w/v) non-fat milk in 1× TBS-T.
11. Primary antibody: 6D12 monoclonal antibody (Trans Genic Inc.).
12. Secondary antibody dilution: Goat anti-mouse IgG (L + H), HRP conjugated (CHEMICON Internal Inc).

3. Methods

3.1. Quantification of Protein Oxidation

1. Finely mince a 200-mg sample of cardiac tissue.
2. Homogenize the minced tissue in 3 ml of RIPA lysis buffer.
3. Transfer the protein lysates to a microcentrifuge tube and centrifuge at $6000\,g$ for 10 min. Discard the pellet.
4. Aliquot 250 µl of the lysate in two eppendorf tubes.
5. Add 250 µl 2.5 M HCl to one tube (serves as blank for that sample) and add 250 µl of 10 mM 2,4-DNPH (dissolved 2.5 M HCl) to the second tube.
6. Incubate the tubes in the dark for 15 min at room temperature, votexing every 5 min (*see* **Note 8**).
7. At the end of 15 min, add 500 µl 20% TCA to each tube and centrifuge at $11,000\,g$ for 3 min at room temperature.
8. Remove supernatant, wash pellet with ethyl alcohol/2-(2-butoxyethoxy) ethyl acetate (1:1, v/v), and incubate for 10 min. Following washing, centrifuge the samples at $11,000\,g$ for 3 min at room temperature. Repeat this washing procedure two more times.
9. After the last wash, add 500 µl 6 M guanidine HCl to pellet and incubate at $37\,°C$ for 30–60 min to dissolve the pellet.
10. Centrifuge at $11,000\,g$ for 1 min to remove insoluble debris.
11. Read the absorbance of the clear supernatant at 360 nm.
12. Determine the protein contents of the final sample blanks (usually 1:200 dilution) by Bradford method. Use the 6 M Guanidine HCl for standard with BSA solution.
13. The final protein carbonyl content = absorption at $360\,m \times 45.45$ nmol/protein content (mg).

3.2. Immunoblot Detection of Protein Oxidation

3.2.1. Sample Preparation

1. These instructions assume the use of Bio-Rad minigel system for two pieces of gels. They are easily adaptable to other formats.
2. Transfer 10 µl ($\sim$ 20–50 µg) of protein-lysed extracts into each of two 1.5 ml-microcentrifuge tubes (*see* **Note 9**).
3. Denature each 10 µl aliquot of protein by adding 10 µl of 10% SDS for a final concentration of 5% SDS.
4. Derivatize the sample by adding 20 µl of $1 \times$ 2,4-DNPH solution to one of the tubes. To the aliquot designated as the negative control, add 20 µl of $1 \times$ derivation-control solution instead of the 2,4-DNPH solution.
5. Incubate both tubes at room temperature for 15 min (*see* **Note 10**).
6. Add 20 µl of neutralization solution to both the treated sample and the negative control.

7. Add 30 µl gel loading buffer to each preparation and centrifuge the tube at maximum speed for 2–5 min at room temperature to clarify the protein sample. The supernatant can be used directly for gel loading or stored at 4 °C for 1 week (*see* **Note 11**).

3.2.2. SDS–PAGE

1. Prepare a 0.75-mm thick, 10% gel by mixing 5.2 ml 4× separating buffer with 6.6 ml acrylamide/bis solution, 8.2 ml water, 100 µl 10% ammonium persulfate solution, and 10 µl TEMED. Pour the gel, leaving space for a stacking gel, and overlay with water-saturated isobutanol. The gel should polymerize in about 30 min.
2. Pour off the isobutanol and rinse the top of the gel twice with water.
3. Prepare the stacking gel by mixing 2.66 ml of 4× stacking buffer with 1.38 ml acrylamide/bis solution, 2.6 ml water, 60 µl ammonium persulfate solution, and 6 µl TEMED. Use about 0.5 ml of this to quickly rinse the top of the gel, and then pour the stack and insert the comb. Polymerize the stack gel within 30 min.
4. Prepare the running buffer by diluting 200 ml of the 5× running buffer with 800 ml of water in a measuring cylinder. Cover with Para-Film and invert to mix.
5. Once the stacking gel has set, carefully remove the comb and use a 3-mL syringe fitted with 22-G needle to wash the wells with running buffer.
6. Add the running buffer to the upper and lower chambers of the gel unit and load equal amount of 25 µg (5–20 µl) of each sample in a well. Include one well for prestained molecular weight markers and one well for mixture of standard proteins with attached DNP residues as positive control.
7. Complete the assembly of the gel unit and connect to a power supply. The gel can be run for about 1.5 h at 90 V through the stacking gel and 120 V through the separating gel. The dye fronts (blue and pink) can be run off the gel.

3.2.3. Western Blotting

Electrophoretically transfer the samples that have been separated by SDS–PAGE onto nitrocellulose membranes as follows:

1. Prepare a tray of setup buffer that is large enough to lay out a transfer cassette with its pieces of foam and with two sheets of 3MM paper submerged on one side. Cut a sheet of the nitrocellulose just larger than the size of separating gel and lay it on the surface of separate tray of distilled water to allow the membrane to wet by capillary action. Submerge the membrane in the setup buffer on top of the 3MM paper.
2. Wet two more sheets of 3MM paper in the setup buffer and carefully lay them on top of the gel, ensuring that no bubbles are trapped in the resulting sandwich. Lay the second wet foam sheet on the top and bottom of the transfer cassette closed.
3. Place the cassette in the transfer tank such that the nitrocellulose membrane is between the gel and the anode. It is vitally important to ensure this orientation, or the proteins will be lost from the gel into the buffer rather than transferred to the nitrocellulose.

4. Place the lid on the tank and activate the power supply. Transfer can be accomplished at 100 V for 1 h.
5. Once the transfer is complete, remove the cassette from the tank and carefully disassemble and remove the top sponge and sheets of 3MM paper. Leave the gel in place on top of the nitrocellulose so that the shape of the gel (including the cut corner for orientation) can be cut into the membrane using a razor blade. Discard the gel and excess nitrocellulose. The colored molecular weight markers should be clearly visible on the membrane.
6. Incubate the nitrocellulose in 30 ml blocking buffer for 1 h at room temperature on a rocking platform.
7. Discard the blocking buffer and quickly rinse the membrane once with TBS-T.
8. Incubate the membrane with primary antibody diluted 1:250 in TBS-T/3% BSA for 1 h at room temperature on a rocking platform (*see* **Note 12**).
9. Remove the primary antibody (*see* **Note 13**) and wash the membrane three times for 5 min each with 30 ml TBS-T.
10. Freshly prepare secondary antibody as 1:500-fold dilution in blocking buffer and add to the membrane. Incubate for 30 min at room temperature on a rocking platform (*see* **Note 14**).
11. Discard the secondary antibody and wash the membrane six times for 10 min each with TBS-T.
12. During the final wash, mix 2 ml aliquots of each portion of the ECL reagent. The subsequent steps are done in a dark room under safe light conditions.
13. Immediately following the final wash add the ECL reagent to the membrane and rotate by hand for 1 min to ensure even coverage.
14. Remove the membrane from the ECL reagent, drain off excess reagent with Kim-Wipes, and place the membrane between two acetate sheet protectors, cut to the size of an X-ray film cassette (*see* **Note 15**).
15. Place the acetate sheet containing membrane in an X-ray film cassette with X-ray film for a suitable exposure time, typically a few minutes. An example of the results produced is shown in **Fig. 1**.

3.3. Quantification of Protein Advanced Glycosylation by Enzyme-Linked Immunosorbent Assay

1. Predilute 1 ml coating buffer with 9 ml redistilled water. Mix 10 μl sample with 90 μl diluted coating solution into each well and incubate at 4 °C overnight.
2. Rinse the wells five times with TBS-T washing solution. Add 200 μl per well of blocking buffer (1% BSA in TBS-T) to the wells. Incubate for 30 min at room temperature or overnight at 4 °C.
3. Rinse the plates again five times with TBS-T washing solution. Add 50 μl per well of primary antibody (6D12) dilution (*see* **Note 16**). Incubate for 1.5 h at 37 °C.
4. Rinse plates five times with 1× TBS-T washing solution. Add 50 μl per well of HRP-conjugated secondary antibody (1:2000). Incubate for 1.5 h at 37 °C.

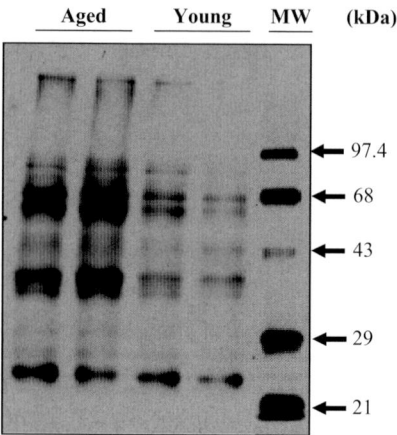

Fig. 1. Immunoblot detection of protein oxidation by protein carbonyl content in cardiac tissues from young and aged mice.

5. Rinse plates five times with TBS-T washing buffer. Add 100 μl NeA-Blue TMB substrate per well (*see* **Note 17**). Incubate for 10–30 min at 37 °C for adequate color development to occur.
6. Stop the reaction by adding 50 μl of 1 N H_2SO_4 or HCl. Agitate the plates to generate a uniform color before measuring the optical density at 450 nm.

3.4. Assay of Protein Advanced Glycosylation by Two-Dimensional Gel Electrophoresis

3.4.1. Sample Preparation

1. These instructions assume the use of Bio-Rad 7 cm ReadyStrip immobilized pH gradient (IPG) strips and the initial protein concentration larger than 5 mg/ml. They are easily adaptable to other formats.
2. Transfer 1–500 μg of protein-lysed extracts in a final volume of 100 μl into a 1.5-ml microcentrifuge tube (*see* **Note 18**).
3. Add 300 μl precipitation agent 1 to the protein sample and mix well by vortexing. Incubate on ice for 15 min (*see* **Note 19**).
4. Add 300 μl precipitating agent 2 to the mixture of protein and precipitating agent 1. Mix well by vortexing (*see* **Note 20**).
5. Centrifuge the tube(s) at maximum speed (> 12, 000 g) for 5 min to form a tight pellet. Remove the tube promptly once the centrifuge stops so that the pellet does not disperse.
6. Without disturbing the pellet, remove and discard the supernatant using a pipette.

7. Position the tube in the centrifuge as before (i.e., cap hinge and protein pellet facing outward) and centrifuge for 15–30 s to collect any residual liquid at the bottom of the tube. Use a pipette to carefully remove the remaining supernatant.

8. Add 40 µl of wash reagent 1 on top of the pellet. Position the tube in the centrifuge as before and centrifuge at maximum speed ($> 12,000\,g$) for 5 min (*see* **Note 21**).

9. With a pipette remove and discard the wash.

10. Add 25 µl of water on top of pellet. Vortex the tube for 10–20 s. Protein pellets may disperse but will not dissolve in the water.

11. Add 1 ml of wash reagent 2 (prechilled at $-20\,°C$ for at least 1 h) and 5 µl of wash 2 additive. Vortex the tube for 1 min (*see* **Note 22**).

12. Incubate the tube at $-20\,°C$ for 30 min. Vertex the tube for 30 s every 10 min during the incubation period.

13. After the incubation period, centrifuge the tube at top speed for 5 min to form a tight pellet. Remove and discard the supernatant. Centrifuge the tube briefly (15–30 s) and remove and discard any remaining wash. The pellet will appear white at this stage. Air-dry the pellet at room temperature for no more than 5 min (*see* **Note 23**).

14. Reuse each pellet by adding an appropriate volume of 2D dehydration/sample buffer to the pellet. Vortex the tube for at least 30 s. Incubate the tube at room temperature for 3–5 min. Vortex the tube again for approximately 1 min or pipette the solution up and down to fully resuspend (*see* **Note 24**).

15. Centrifuge the tube at maximum speed for 2–5 min at room temperature to clarify the protein sample. The supernatant can be used directly for first dimensional isoelectric focusing (IEF) in an IPG strip. Store any unused or remaining protein sample in a clean tube at $-80\,°C$ for later analysis.

3.4.2. Two-Dimensional Gel Electrophoresis

1. Pipette 125 µl rehydrated sample into adjacent channels (*see* **Note 25**).

2. When all the protein samples have been loaded into the rehydration/equilibration tray using forceps, peel the coversheet from one of the pH 3–10 ReadyStrip IPG strips. Gently place the trip gel side down onto the sample. The "+" and "pH 3–10" should be legible and positioned at the left side of the tray. Take care not to get the sample onto the plastic backing of the strips as this portion of the sample will not be absorbed by the gel material. Pour off the isobutanol and rinse the top of the gel twice with water.

3. Overlay each of the strips with 1–2 ml of mineral oil to prevent evaporation during the rehydration process. Add the mineral oil slowly, by carefully dripping the oil onto the plastic backing of the strips while moving the pipette along the length of the strip.

4. Cover the rehydration/equilibration tray with the plastic lid provided and leave the tray sitting on a level bench overnight (11–16 h) to rehydrate the IPG strips and load the protein sample.

5. Place a clean, dry PROTEAN IEF focusing tray, the same size as the rehydrating IPG strips, onto the laboratory bench.

6. Using forceps, place a paper wick at both ends of the channels covering the wire electrodes. Use channels with the same numbers as those used during rehydration.

7. Pipette 8 μl of water onto each wick to wet them.

8. Remove the cover from the rehydration/equilibration tray containing the IPG strips. Using forceps, carefully hold the strip vertically for about 7–8 s to allow the mineral oil to drain, then transfer the IPG strip to the corresponding channel in the focusing tray (maintain the gel side down). Repeat for all the strips. (*see* **Note 26**).

9. Cover each IPG strip with 2–3 ml fresh mineral oil. Check for, and if applicable remove, any trapped air bubbles beneath the strips. Place the lid onto the tray (positive "+" to the left when the inclined portion of the tray is on the right).

10. Place the focusing tray into the PROTEAN IEF cell and close the cover.

11. Program the PROTEAN IEF cell according to PROTEAN® IEF Cell Instruction Manual. Use the default cell temperature of 20 °C, with a maximum current of 50 μA/strip. Press START to initiate the electrophoresis run.

12. When the electrophoresis run has been completed, remove the IPG strips from the focusing tray and transfer them gel side up into a clean, dry disposable rehydration/equilibration tray that matches the length of the IPG (*see* **Note 27**).

13. Add 1.5–2 ml of equilibration buffer I to each channel containing an IPG strip with gel side up.

14. At the end of the 20-min incubation, discard the used equilibration buffer I by carefully decanting the liquid from the tray (*see* **Note 28**).

15. Add 1.5–2 ml of complete equilibration buffer II (containing iodoacetamide) to each strip.

16. Return the tray to the orbital shaker for 20 min. During the incubation, melt the overlay agarose solution in a microwave oven.

17. Remove an IPG strip from the disposable rehydration/equilibration tray and dip briefly into a tube, the same length or longer than the IPG strip length, containing the 1× Tris/glycine/SDS running buffer. Lay the strip gel side up and onto the back plate of the SDS–PAGE gel above the IPG well. Repeat this process for any remaining IPG strips.

18. Take the first SDS–PAGE gel with the IPG strip resting on the back plate and stand vertically with the short plate facing toward you. Use a Pasteur pipette and pipette overlay agarose solution into the IPG well of the gel.

19. Using the forceps, carefully push the strip into the well, taking care not to trap any air bubbles beneath the strip. When pushing the IPG strips with forceps be certain the forceps are pushing on the plastic backing to the strip and not the gel matrix.

20. Stand the gel(s) vertically by placing them in the gel box. Allow the agarose to solidify for 5 min before proceeding.

21. Complete the assembly of the gel unit followed by connection to a power supply. The electrophoresis can be run for approximately 1.5 h at 90 V through the stacking gel and 120 V through the separating gel. The dye fronts (blue and pink) may be run off the gel. The gel can be either stained with Bio-Safe Sliver Stain (cat. no. 161-0449, Bio-Rad), then imaged (*see* **Fig. 2A**), or proceeded to Western blot as the following.

3.4.3. Western Blotting for Protein-Advanced Glycosylation

1. Process the same steps 1–7 as described in **Subheading 3.2.3**.
2. Discard the blocking buffer and quickly rinse the membrane.
3. Add primary antibody (6D12) dilution (*see* **Note 16**) overnight at 4 °C on a rocking platform.
4. Remove the primary antibody and wash the membrane three times for 5 min each with 30 ml TBS-T.
5. Incubate the membrane with freshly prepared secondary antibody (1:1000-fold dilution in blocking buffer) for 1 h at room temperature on a rocking platform.
6. Discard the secondary antibody and wash the membrane three times for 10 min each with TBS-T.
7. The subsequent steps are similar to those described in **Subheading 3.2.3**.
8. An example of the results produced is shown in **Fig. 2B and C**.

3.4.4. Potential Experimental Problems and Alternative Approaches

Assessment of protein glycoxidation in ventricular tissues is of great clinical and practical importance in understanding the pathogenesis of cardiovascular diseases. However, in this protocol, we only provide a general tool to screen a panorama of protein glycoxidation in crude extracts of ventricular tissue. It is recommended that further protein fractionation (i.e., cytosolic, membrane, and nuclear and cytoskeletal fractions) will be conducted prior to performing the 1D or 2D gel electrophoresis if a protein in a specific cellular compartment is speculated to be glycoxidized. In addition, MALDI-MS analysis may be employed when combined with gel electrophoresis to identify the proteins of interest. For identification of membrane-associated protein glycoxidation, a potential experimental problem is the low extraction of integral membrane proteins especially in case of sacro(endo)plasmic reticulum Ca^{2+}-ATPase, a membrane protein with a critical role in the regulation of cardiac function. Low abundance proteins may be beneath the level of detection in a single gel. The dynamic range of the protein stain can limit the comparison of protein levels between gels. As an alternative approach, integral membrane proteins may be extracted with more vigorous separation techniques containing potent reducing agents such as tributyl phosphine or zwitterionic detergents *(21)*.

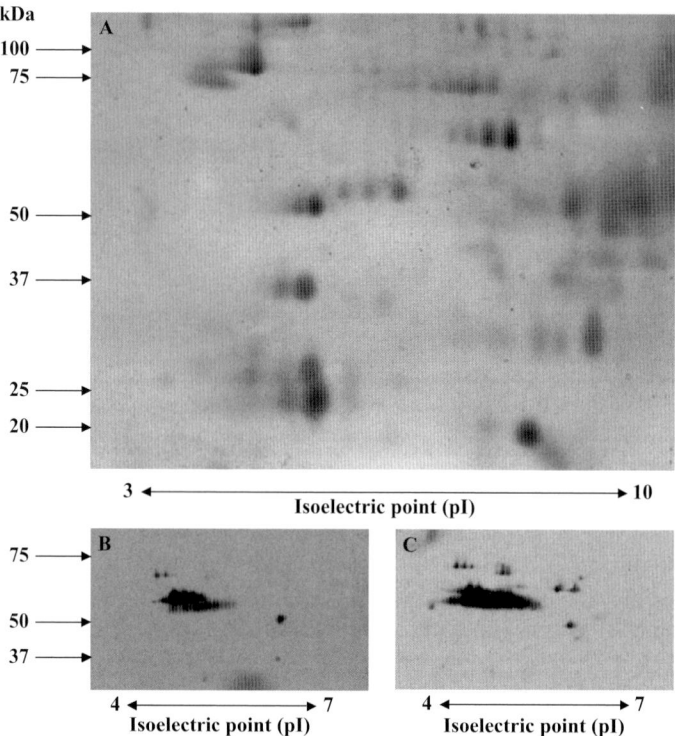

Fig. 2. Advanced glycation end products (AGEs) formation in cardiac tissue illustrated by two-dimensional (2D) gel electrophoresis. (**A**) Total protein imaged by 2D gel electrophoresis and silver stained. (**B**) AGEs immunoblot in cardiac tissue of young group. (**C**) AGEs immunoblot in cardiac tissue of aged group.

Overlapped spots and highly expressed housekeeping proteins can be avoided to a great extent through telescoping the first dimension gel isoelectric point range. The limitations in protein quantification in 2D PAGE have been successfully addressed through the use of ruthenium-based fluorescent stains, such as SYPRO Ruby. SYPRO Ruby has an extended dynamic range thereby permitting superior quantification of protein expression compared to silver stains *(22)*.

4. Notes

1. Aged mice refer to those 24 months of age or older, and diabetic mice refer to streptozotocin-induced (200 mg/kg, i.p.) experimental diabetes in mice.
2. Upon first use, warm the standard protein mixture to room temperature. Transfer the amount necessary for the first experiment into one tube and divide the remaining

mixture among several tubes. Store the aliquots at −15 to −25 °C. As the mixture contains SDS that might precipitate during storage, it may be necessary to warm the stock solution to 50 °C to redissolve the SDS. Aliquoting the mixture protects the protein standard from excessive freeze-thaw and heating cycles. Combine 2.5 μl of the molecular weight standard with attached DNP residues and 20 μl of 1× gel loading buffer prior to loading onto the gel.

3. Acrylamide/bis (30%) solution is a neurotoxin when unpolymerized, and so care should be taken to avoid exposure. TEMED is best stored at room temperature in a desiccator. Buy small bottles as it may decline in quality after opening, that is, gels will take longer to polymerize.

4. Unless stated otherwise, all solutions are to be prepared in water that has a resistivity of more than 18 MΩ-cm and total organic content of less than five parts per billion. This standard is referred to as "water" in this text.

5. Transfer buffer can be used for up to several times within 1 week so long as the voltage is maintained constant for each successive run (the current will increase each time). Adequate cooling to keep the buffer no warmer than room temperature by use of a refrigerated/icing bath is essential to prevent heat-induced damage to the apparatus and the experiment.

6. Quantification of data may be desired, and this can be done by scanning densitometry of the films, provided that care is taken to ensure that the signal has not saturated.

7. Upon first use, the 6D12 monoclonal antibody should be aliquoted to avoid repeated freezing and thawing.

8. Do not allow the reaction to proceed for more than 30 min, as side reactions other than hydrazone linkage may occur.

9. Avoid high protein concentration ($> 10 \mu g/\mu L$) to assure solubility during the derivatization reaction.

10. Do not allow the reaction to proceed longer than 30 min, as side reactions other than hydrazone linkage may occur.

11. It is not necessary to add 1× gel loading buffer to samples prior to loading the gel, as the addition of the neutralization solution makes the sample dense enough to sink to the bottom of the well. However, the addition of loading buffer will not adversely affect the electrophoresis of the sample.

12. Dilute rabbit anti-DNP antibody stock 1:500 to 1:750 in PBS-T/1% BSA just before use. Use enough solution so that the membrane is completely immersed in the solution (15 ml for a $10 \times 10 \text{cm}^2$ membrane). An alternative method would be to place the membrane in a heat-sealable plastic bag. In this case, the amount of solution required will be less than that used for a membrane treated in an open container.

13. The primary antibody can be saved for subsequent experiments (up to 3–5 blots over 2 weeks with the only adjustment required being increasing length of exposure to film at the ECL step) by addition of 0.02% final concentration sodium azide (conveniently done by dilution from a 10% stock solution; exercise caution as azide is highly toxic) and storage at 4 °C.

14. Dilute the secondary antibody goat anti-rabbit IgG (H+L), HRP conjugated, 1:2000 to 1:2500 in 1% BSA/1X PBS-T.
15. To identify the aligned signals on the subsequent film with the nitrocellulose, a square of luminescent tape (Sigma) to the edge of the acetate sheet can be applied to provide an alignment mark for the film and membrane.
16. Dilute 6D12 monoclonal antibody 1:250 to 1:500 in TBS-T/1% BSA.
17. Bring NeA-Blue TMB substrate (aqueous) to room temperature before use.
18. Sample quantities larger that 500 μg of protein may reduce the efficiency of the cleanup leading to poor-quality IEF.
19. When adding agent 1, do not touch protein sample with the pipette tip. The protein may precipitate on the tip causing sample loss.
20. When adding agent 2, do not touch protein and precipitating agent 1 with the pipette tip. The protein may precipitate on the tip causing sample loss.
21. A precipitate may dorm along the tube wall. In these cases, vortex and/or pipette wash solution over the pellet several times to ensure entire pellet is thoroughly washed.
22. Protein pellets will not dissolve in wash reagent 2. If wash reagent 2 is not completely chilled, quantitative recovery may be affected.
23. Once sufficiently dry, the pellet looks translucent. Do not over-dry pellets, otherwise pellets will be difficult to resuspend.
24. Sonication may be used to speed the process of resuspension.
25. To avoid confusing the samples, it is suggested to place samples on both sides of the tray and to maintain the same numbering system through the 2D PAGE process.
26. Drain the oil to allow removal of unabsorbed protein from surface of the gel. This should reduce the incidence of horizontal streaking.
27. Hold the strips vertically with the forceps and let the mineral oil drain from the strips for approximately 5 s before transfer.
28. Decanting is best carried out by pouring the liquid from the square side of the rehydration/equilibration tray, until the tray is positioned vertically. When most of the liquid has been descanted, "flick" the tray a couple of times to remove the last few drops of equilibration buffer I.

References

1. Sensi, M., Pricci, F., Andreani, D., and Di Mario, U. (1991) Advanced nonenzymatic glycation endproducts (AGE): their relevance to aging and the pathogenesis of late diabetic complications. *Diabetes Res.* **16,** 1–9.
2. Brownlee, M. (1995) Advanced protein glycosylation in diabetes and aging. *Annu. Rev. Med.* **46,** 223–234.
3. Vlassara, H., and Palace, M.R. (2003) Glycoxidation: the menace of diabetes and aging. *Mt. Sinai. J. Med.* **70,** 232–241.

4. Kass, D.A. (2003) Getting better without AGE: new insights into the diabetic heart. *Circ. Res.* **92**, 704–706.

5. Dawnay, A., and Millar, D.J. (1998) The pathogenesis and consequences of AGE formation in uraemia and its treatment. *Cell Mol. Biol.* **44**, 1081–1094.

6. Brownlee, M. (2000) Negative consequences of glycation. *Metabolism* **49**, 9–13.

7. Peppa, M., Uribarri, J., and Vlassara H. (2002) Advanced glycoxidation. A new risk factor for cardiovascular disease? *Cardiovasc. Toxicol.* **2**, 275–287.

8. Meli, M., Frey, J., and Perier C. (2003) Native protein glycoxidation and aging. *J. Nutr. Health Aging.* **7**, 263–266.

9. Tsukahara, H., Ohta, N., Sato, S., Hiraoka, M., Shukunami, K., Uchiyama, M., Kawakami, H., Sekine, K., and Mayumi, M. (2004) Concentrations of pentosidine, an advanced glycation end-product, in umbilical cord blood. *Free Radic. Res.* **38**, 691–695.

10. Knott, H.M., Brown, B.E., Davies, M.J., and Dean, R.T. (2003) Glycation and glycoxidation of low-density lipoproteins by glucose and low-molecular mass aldehydes. Formation of modified and oxidized particles. *Eur. J. Biochem.* **270**, 3572–3582.

11. Aronson, D. (2003) Cross-linking of glycation collagen in the pathogenesis of arterial and myocardial stiffening of aging and diabetes. *J. Hypertens.* **21**, 3–12.

12. Morgan, P.E., Dean, R.T., and Davies, M.J. (2002) Inactivation of cellular enzymes by carbonyls and protein-bound glycation/glycoxidation products. *Arch. Biochem. Biophys.* **403**, 259–269.

13. Cai, W., He, J.C., Zhu, L., Peppa, M., Lu, C., Uribarri J., and Vlassara H. (2004) High levels of dietary advanced glycation end products transform low-density lipoprotein into a potent redox-sensitive mitogen-activated protein kinase stimulant in diabetic patients. *Circulation* **110**, 285–291.

14. Zieman, S.J. and Kass, D.A. (2004) Advanced glycation endproduct crosslinking in the cardiovascular system: potential therapeutic target for cardiovascular disease. *Drugs* **64**, 459–470.

15. Forbes, J.M., Yee, L.T., Thallas, V., Lassila, M., Candido, R., Jandeleit-Dahm, K.A., Thomas, M.C., Burns, W.C., Deemer, E.K., Thorpe, S.M., Cooper, M.E., and Allen, T.J. (2004) Advanced glycation end product interventions reduce diabetes-accelerated atherosclerosis. *Diabetes* **53**, 1813–1823.

16. Li, S.Y., Du, M., Dolence, E.K., Fang, C.X., Mayer, G.E., Ceylan-Isik, A.F., LaCour, K.H., Yang, X., Wilbert, C.J., Sreejayan, N., and Ren J. (2005) Aging induces cardiac diastolic dysfunction, oxidative stress, accumulation of advanced glycation endproducts and protein modification. *Aging Cell* **4**, 57–64.

17. Luth, H.J., Ogunlade, V., Kuhla, B., Kientsch-Engel, R., Stahl, P., Webster, J., Arendt, T., and Munch, G. (2005) Age- and stage-dependent accumulation of advanced glycation end products in intracellular deposits in normal and Alzheimer's disease brains. *Cereb. Cortex.* **15**, 211–220.

18. Simm, A., Casselmann, C., Schubert, A., Hofmann, S., Reimann, A., and Silber, R.E. (2004) Age associated changes of AGE-receptor expression: RAGE upregulation is associated with human heart dysfunction. *Exp. Gerontol.* **39,** 407–413.

19. Dukic-Stefanovic, S., Schinzel, R., Riederer, P., and Munch, G. AGEs in brain ageing: AGE-inhibitors as neuroprotective and anti-dementia drugs? (2001) *Biogerontology* **2,** 19–34.

20. Bakris, G.L., Bank, A.J., Kass, D.A., Neutel, J.M., Preston, R.A., and Oparil, S. (2004) Advanced glycation end-product cross-link breakers: a novel approach to cardiovascular pathologies related to the aging process. *Am. J. Hypertens.* **17,** S23–S30.

21. Molloy, M.P., Herbert, B.R., William, K.L., and Gooley, A.A. (1999) Extraction of Escherichia coli proteins with organic solvents prior to two-dimensional electrophoresis. *Electrophoresis* **20,** 701–714.

22. Lopez, M.F., Berggren, K., Chernokalskaya, E., Lazarev A., Robinson, M., and Patton W.F. (2000) A comparsion of silver stain and SYPRO Ruby Protein Gel Stain with respect to protein detection in two-dimensional gel and identification by peptide mass profiling. *Electrophoresis* **21,** 3673–3683.

22

Assessment of PI-3 Kinase and Akt in Ischemic Heart Diseases in Diabetes

Takashi Matsui and Amy J. Davidoff

Summary

Diabetes mellitus is the most common disease in Westernized countries in large part because of the rising prevalence of obesity and physical inactivity. In addition, diabetes mellitus is an important risk factor for both heart failure and ischemic heart disease. As insulin resistance is known as an important pathophysiological feature in the cardiac diseases, understanding the mechanisms responsible for altered metabolism and insulin signaling in the diabetic heart may help identify novel targets in these conditions. Phosphatidylinositol (PI)-3 kinase (PI3K) and Akt are key signaling molecules in insulin and insulin-like growth factor-1 (IGF-1), which induce multiple biological effects in the heart such as cell survival and hypertrophy. Here, we have shown several fundamental techniques to study the role of PI3K and Akt in heart diseases.

Key Words: Phosphatidylinositol-3 kinase (PI3K); Akt; Phosphorylation; Kinase assay; Apoptosis; Cardiomyocyte.

1. Introduction

Human studies and animal experiments have shown that insulin-stimulated glucose uptake is significantly decreased in diabetic hearts, suggesting that insulin signaling pathway is impaired *(1)* (review in **ref. 2**). Insulin provides the dominant stimulus to enhance glucose uptake, largely through the activation of the downstream signaling molecules: insulin receptor substrate-1 (IRS-1), phosphatidylinositol-3 kinase (PI3K), and the serine/threonine kinase Akt. It is also known that PI3K and Akt provide potent stimuli for cell proliferation, growth, and survival in most tissues (*see* **Fig. 1**), including hearts *(3–9)*.

From: *Methods in Molecular Medicine, Vol. 139: Vascular Biology Protocols*
Edited by: N. Sreejayan and J. Ren © Humana Press Inc., Totowa, NJ

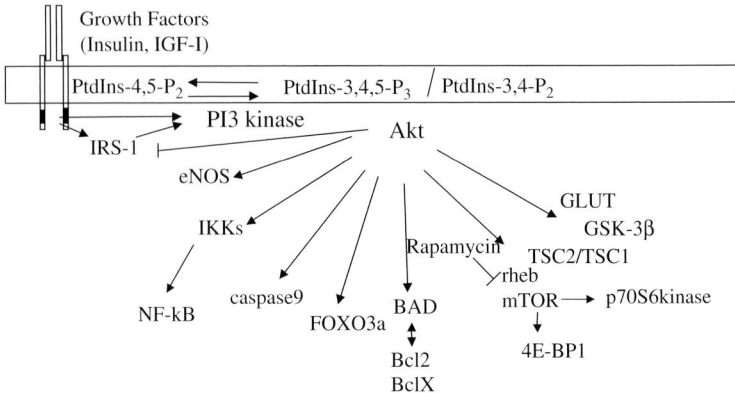

Fig. 1. Signaling pathway of Akt. PtdIns(3,4)P$_2$ and PtdIns(3,4,5)P$_3$ are generated mainly by D3 phosphorylation on PtdIns(4)P and PtdIns(4,5)P$_2$, respectively, with class I$_A$ phosphatidylinositol 3-kinase (PI3K) activated by tyrosine kinase receptors. Activated, phosphorylated Akt stimulates multiple downstream pathways, including endothelial nitric oxide synthase (eNOS), nuclear factor-κB (NF-κB), caspase 9, FOXO3a (FKHRL1), BAD, GSK-3β, mTOR, and glucose transporter (GLUT).

Although many papers have reported that cardiomyocytes regenerate from stem cells *(10)*, cardiomyocyte regeneration as therapeutic strategy still appears inadequate to repair injured myocardium in cardiac diseases *(11)*. For these reasons, there are significant clinical implications to increase our understanding of the mechanisms that control cardiomyocyte death and identify possible opportunities for intervention. In response to specific stimuli, cells can activate suicide pathways and undergo apoptosis (programmed cell death). Markers of apoptosis have been identified in a wide variety of cardiac conditions in patients *(12–16)*, including diabetes *(17)*. The biochemical hallmark of apoptosis is fragmentation of genomic DNA (DNA laddering) into multimers of approximately 200 bp fragments forming a characteristic ladder on electrophoresis (*see* **Fig. 2A**). We will describe the method of DNA laddering for samples from heart.

Hyperinsulinemia, seen in type 2 diabetes, chronically activates PI3K/Akt signaling pathway *(2)*. Chronic activation of Akt causes adverse effects on cardiac function in animal models *(18,19)*. The adverse effects of chronic Akt activation in heart may result through a negative-feedback system involving augmented IRS-1 degradation *(18)* and a change in the expression level of

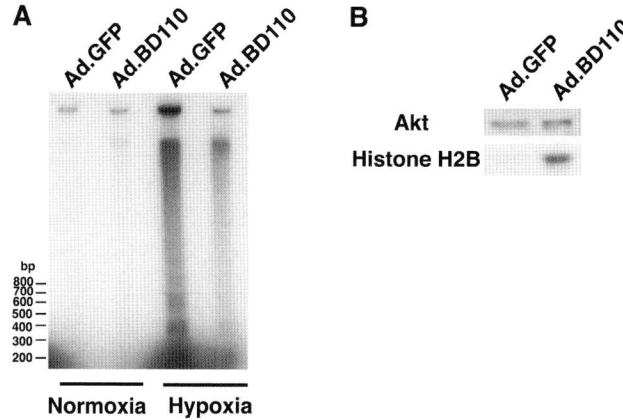

Fig. 2. (**A**) DNA laddering in rat neonatal cardiomyocytes. Twenty-four-hour hypoxic culture, set with 95% N_2, induced an increase on DNA laddering in rat neonatal cardiomyocytes infected with control virus carrying green fluorescence protein (GFP). An active mutant of PI3K (Ad.BD110) significantly decreased DNA laddering. BD110 is a constitutively active mutant of PI3K previously constructed by fusing the binding domain of the p85 regulatory subunit in-frame with the p110 catalytic domain *(7,22)*. (**B**) Akt activity in rat neonatal cardiomyocytes. Forty-eight hours after infection, cells were placed in serum-free media and examined for Akt kinase activity. Cell lysates were immunoprecipitated with anti-Akt antibodies and subjected to Western blotting (upper panel) and kinase assays to detect phosphorylation of the Akt substrate, histone H2B (lower panel). Ad.BD110 infection increased the endogenous Akt activity.

plasma membrane glucose transporter (GLUT) *(20)*. These biological effects are decided by "level" and "time" of kinase activation. Therefore, analyzing not only phosphorylation but also kinase activity is necessary for the study of PI3K/Akt receptor signaling pathway.

2. Materials

2.1. Cardiomyocyte Culture

The following protocols are described for assays involving neonatal rat cardiomyocytes but can be used for most cells/tissues. The procedures for isolating and culturing these cells have been described extensively; therefore, details are not included in this chapter. Typically, we culture neonatal rat cardiomyocytes in 60-mm dishes and allow them to become confluent before initiating our experiments. The following procedures are based on the amount of DNA, and Akt and PI3K activities that can be derived from each plate of confluent cells.

2.2. DNA Laddering

1. Lysis buffer: 100 mM Tris–HCl, pH 8.5, 5 mM ethylenediaminetetraacetic acid (EDTA), 0.2% sodium dodecyl sulfate (SDS), 200 mM NaCl, 100 μg/mL Proteinase K (Sigma, St. Louis, MO).
2. TE buffer: 10 mM Tris–HCl, pH 8.0, 1 mM EDTA, pH 8.0.
3. 10 μg/mL RNase.
4. EDTA 10 mM (dissolved in dH_2O).
5. Klenow DNA polymerase (New England BioLabs, Beverly, MA)

2.3. Akt Kinase Assay

1. SDS gel (10 or 12%).
2. Akt kinase assay (commercially available kits, e.g., from Cell Signaling).
3. Anti-Akt (Cell Signaling, Beverly, MA).

2.4. PI3K Assay

1. Lysis buffer: 20 mM Tris–HCl, pH 7.5, 150 mM NaCl, 1 mM EDTA, 1 mM ethyleneglycol-bis-N,N,N′, N′-tetraacetic acid (EGTA), 1% Triton, 2.5 mM sodium pyrophosphate, 1 mM β-glycerolphosphate, 1 mM Na_3VO_4, 1 μg/mL Leupeptin, and 1 mM phenylmethylsulfonyl fluoride (PMSF).
2. Protein G Sepharose beads (commercially available from many manufacturers).
3. L-a-phosphatidylinositol (liver salt solution) (Avanti Polar Lipids, Alabaster, AL).
4. Reaction mixture: 0.1 mM ATP, 50 mM Hepes, pH 7.1, 1 mM EGTA, 20 mM $MgCl_2$, 20 mM NaCl, 1 mM sodium phosphate, 0.2 M NaH_2CO_3, 0.2 M Na_2HCO_3, and 5 μCi of $^{32}\tilde{P}$-γ-ATP.
5. Thin-layer chromatography (TLC) aluminum sheets 20×20 cm—Silica gel 60 (Merck KGaA, Germany).
6. Buffers of TLC

   a. Pretreatment buffer; 1.95 g potassium oxalate, 0.6 mL 500 mM EDTA, 130 mL dH_2O, and 60 mL methanol.
   b. Separating buffer sufficient for two experiments; 141 mL methanol, 6 mL NH_4OH, 33.9 mL dH_2O, and 60 mL chloroform. Solution should be stored in glass.

7. Anti-phosphotyrosine (PY20 Upstate; for endogenous PI kinase activity).
8. Buffer A: 1% Nonidet P40 in phosphate-buffered saline (PBS).
9. Buffer B: 100 mM Tris–HCl, pH 7.6, 0.5 M lithium chloride.
10. Buffer C: 10 mM Tris–HCl, pH 7.6, 100 mM NaCl.

3. Methods

3.1. DNA Laddering

Using radiolabeled DNA, very little laddering is detected in untreated cultures but is significantly increased by exposure in hypoxia overnight (set with 95%N$_2$ in an incubation chamber) *(7)*. For other methods to detect apoptosis, please *see* **Notes 1** and **2**.

1. Expose neonatal cardiomyocytes to specific conditions such as ischemia or hypoxia.
2. Trypsinize cells and resuspend in 0.7 mL lysis buffer. This amount of lysis buffer is based on confluent cells in 60-mm culture dishes.
3. Incubate cell lysate at 37 °C for 4 h with agitation.
4. Subject cell lysates to isopropanol precipitation, phenol/chloroform extraction, and ethanol precipitation (see below):

    a. Apply 700 μL lysis buffer in the tube and incubate at 37 °C with agitation overnight (as previously described).
    b. Centrifuge (1000 g) for 10 min and remove the clear sup into a new 1.5-ml tube (discard the pellet containing debris).
    c. Apply 700 μL isopropanol in the tube with the lysate and mix the solution with vortex. This can be kept at −20 °C overnight.
    d. Centrifuge at max speed (20, 000 g) for 10 min, discard the supernatant, and re-suspend the pellet with 400 μL of TE.
    e. Add 400 μL Phe/Chl and vortex. Centrifuge at 20, 000 g for 10 min at room temperature (RT).
    f. Transfer the upper phase (almost 400 μL) to a new 1.5-ml tube.
    g. Add 40 μL 3M sodium acetate (pH 5.2) and 1 ml of EtOH.
    h. Keep it at −80 °C for at least 15 min (more time is fine). Centrifuge at 20, 000 g for 10 min at 4 °C.
    i. Discard the supernatant and add 0.7–0.8 ml 70% EtOH. Centrifuge at 20, 000 g for 10 min at 4 °C.
    j. Discard the supernatant carefully and dry the pellet.
    k. Resuspend the pellet with 30–50 μL of TE.
    l. Incubate the sample with 0.5 μL RNase for 30 min.
    m. Determine concentration of genomic DNA spectrophotometrically (5 μL sample+95 μL TE, read dissolved extract at OD260).

5. Label 1 μg of each DNA sample in total 20 μL of solution with 0.5 μCi [α −^{32}P]dCTP in the presence of 5 U of Klenow DNA polymerase for 10 min.
6. Terminate the reaction by adding EDTA (final concentration 10 mM) and incubating for 10 min at 75 °C.
7. Perform electrophoresis of the entire sample using a 1.8% agarose gel. Detect DNA damage by autoradiography.

8. Non-labeled genomic DNA demonstrates a laddering pattern in some settings. However, in most of the cases on cardiomyocytes, labeled genomic DNA is required because of low level of apoptosis *(7,9)*.

3.2. Akt Assays

Akt kinase activity can be measured by the extent of GSK3 phosphorylation. As phospho-GSK3 antibodies are available, this method is frequently used as a non-radioactive approach (kits are commercially available from several companies). Histone H2B can also be used as a substrate in Akt kinase assay (*see* **Fig. 2B**). A kinase assay is performed with immunoprecipitation of Akt from total cell lysate as described below (*see* **Note 3**).

1. Immunoprecipitate Akt from cell lysates in a reaction mixture containing Histone H2B and ATP.
2. Perform an SDS–polyacrylamide gel electrophoresis (PAGE) gel electrophoresis using a 15% gel.
3. Separate the lower molecular weight area in the gel [including radiolabeled Histone H2B (16 kDa)] for direct autoradiography. Immunoblot the higher molecular weight area in the gel (including Akt about 60 kDa) with antibody for Akt.
4. Confirm the amount of immunoprecipitated Akt by immunoblotting with antibody for Akt using the same membrane and qualitatively evaluating it with densitometry (or NIH image).

3.3. PI3K Activity

As PI3K is a lipid kinase, TLC is required to measure the lipid substrate (phosphatidylinositol) (*see* **Note 4**).

1. TLC plate: Pretreat TLC plate with the pretreatment buffer. After applying the buffer at the bottom of the glass chamber with a lid, place the TLC plate in the chamber. Wait until the TLC plate is covered with the buffer from the bottom to the top. After completely covered (wet), dry the plate in an oven for 10 min at 100 °C. At that time, note the direction (bottom to top) on the plate with a pencil (the plate should be placed with the same direction in the running buffer). The dried TLC plate can be stored several months at room temperature.
2. Immunoprecipitaion (IP): Prepare 500–700 μg whole cell lysate in lysis buffer (total 400–500 μL). Incubate lysates at 4 °C for 1 h after adding 2–3 μg anti-phosphotyrosine. After the incubation, add 40 μL Protein G-Sepharose (1:1 with lysis buffer) into each tube. As Protein G-Sepharose is heavy, mix well before applying to lysate. Incubate the lysate with Protein G-Sepharose at 4 °C for 2 h with agitation (rotation). Wash the Protein G-Sepharose beads twice with buffers A, B, and C; spin for approximately 15 s at top speed of a microfuge and aspirate the supernatant.

3. Reaction: L-α-phosphatidylinositol [10 μg/reaction (spot)] is used as a substrate for the PI3K reaction. Dilute L-α-phosphatidylinositol (0.5 μl/sample) in 50 mM Hepes buffer (pH 7.1), 1 mM EGTA, and 1 mM sodium phosphate, then briefly sonicated (for 1 min, three times on ice). Apply L-α-phosphatidylinositol solution (20 μL) to the IP beads made in step 2. Add the reaction mixture (15 μL in each sample to the beads) and incubate at 30 °C for 10 min. Terminate the reaction by adding 15 μL 4N HCl, followed by the addition of 130 μL 1:1 chloroform: methanol. Mix vigorously, centrifuge at 20, 000 g for 1 min.
4. The reacted substrate is in the bottom phase in the tube. Take 20–30 μL of this solution to spot on the TLC plate. The lipid substrate is separated on the TLC plate with the separating buffer in the glass chamber.
5. Dry the TLC plate at RT and evaluate autoradiographically.
6. PI3K activity is detected by the appearance of a specific radioactive spot corresponding to ^{32}P-PIP3 (*see* **Fig. 3**).

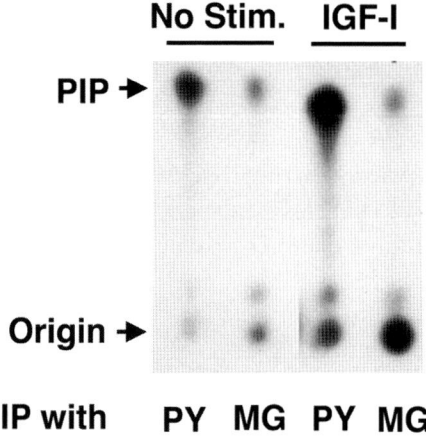

Fig. 3. Phosphatidylinositol-3 kinase (PI3K) activity in rat neonatal cardiomyocytes. Rat neonatal cardiomyocytes cultured in serum-free medium were untreated (No Stim.) or treated [insulin-like growth factor-1 (IGF-1)] with 100 nM of IGF-1. Eight hours later, cell lysates were immunoprecipitated (IP) with normal mouse globin (MG) or monoclonal anti-phospho-tyrosine PY20 (PY) in each group and PI3K activity assessed. Thin-layer chromatography separated PtdInsP (PIP). While no difference was seen on MG, immunocomplexes with PY significantly enhanced PI3K activity in IGF-1-stimulated cells.

Notes

1. Morphologically, apoptosis induces condensation of nuclear chromatin, which can be visualized with Hoechst33258 staining. However, it is not easy to distinguish cardiomyocytes from fibroblasts or other cells both in vitro and in vivo. In addition, the morphological change in nuclei is not specific for apoptosis.
2. Terminal deoxynucleotidyltransferase-mediated dUTP Nick End Labeling (TUNEL) is used for histological analysis. This method specifically labels the 3′-OH ends of fragmented nucleosomal DNA in situ in the presence of exogenously added TdT and digoxigenin-labeled dUTP. Double staining with both TUNEL and myocyte-specific staining such as α-actinin (to identify myocytes) is frequently used for this method.
3. We recommend using the non-radioactive method (i.e., generation of phospho-GSKβ production) for evaluating Akt activity rather than the histone $H2B/^{32}P$ method because it is safer and there are no apparent disadvantages of using the antibody approach.
4. PI3K is heterodimeric complex. Tyrosine kinase receptors such as insulin and IGF-1 activate (i.e., phosphorylate) class I_A PI3K through phosphorylation of IRS-1/2. Class I_A PI3Ks consist of a regulatory subunit (either p85α, p85β, or splice variants p50 or p55) and a catalytic subunit (either p110α, p110β, or p110δ) *(21)*. Therefore, anti-phosphotyrosine, anti-p85α, or anti-p110α antibodies are used for generating immunocomplexes for PI3K assays.

Acknowledgments

This work was supported in part by NIH HL04250 to T. Matsui and NIH R01 HL66895 and American Diabetes Association Research Award 7-04-RA-23 to A. J. Davidoff. The authors thank Dr. Kenta Hara for his helpful advice.

References

1. Kessler, A., Uphues, I., Ouwens, D.M., Till, M., and Eckel, J. (2001) Diversification of cardiac insulin signaling involves the p85 alpha/beta subunits of phosphatidylinositol 3-kinase. *Am. J. Physiol. Endocrinol. Metab.* **280**, E65–74.
2. Poornima, I.G., Parikh, P., and Shannon, R.P. (2006) Diabetic cardiomyopathy: the search for a unifying hypothesis. *Circ. Res.* **98**, 596–605.
3. Shioi, T., Kang, P.M., Douglas, P.S., Hampe, J., Yballe, C.M., Lawitts, J., Cantley, L.C., and Izumo, S. (2000). The conserved phosphoinositide 3-kinase pathway determines heart size in mice. *EMBO J.* **19**, 2537–2548.
4. Shioi, T., McMullen, J.R., Kang, P.M., Douglas, P.S., Obata, T., Franke, T.F., Cantley, L.C., and Izumo, S. (2002) Akt/Protein kinase B promotes organ growth in transgenic mice. *Mol. Cell. Biol.* **22**, 2799–2809.

5. Matsui, T., Li, L., Wu, J.C., Cook, S.A., Nagoshi, T., Picard, M.H., Liao, R., and Rosenzweig, A. (2002) Phenotypic spectrum caused by transgenic overexpression of activated Akt in the heart. *J. Biol. Chem.* **277**, 22896–22901.

6. Condorelli, G., Drusco, A., Stassi, G., Bellacosa, A., Roncarati, R., Iaccarino, G., Russo, M.A., Gu, Y., Dalton, N., Chung, C., et al. (2002) Akt induces enhanced myocardial contractility and cell size in vivo in transgenic mice. *Proc. Natl. Acad. Sci. U. S. A.* **99**, 12333–12338.

7. Matsui, T., Li, L., del Monte, F., Fukui, Y., Franke, T.F., Hajjar, R.J., and Rosenzweig, A. (1999) Adenoviral gene transfer of activated phosphatidylinositol 3′-kinase and Akt inhibits apoptosis of hypoxic cardiomyocytes in vitro. *Circulation* **100**, 2373–2379.

8. Fujio, Y., Nguyen, T., Wencker, D., Kitsis, R.N., and Walsh, K. (2000) Akt promotes survival of cardiomyocytes in vitro and protects against ischemia-reperfusion injury in mouse heart. *Circulation* **101**, 660–667.

9. Matsui, T., Tao, J., del Monte, F., Lee, K.H., Li, L., Picard, M., Force, T.L., Franke, T.F., Hajjar, R.J., and Rosenzweig, A. (2001) Akt activation preserves cardiac function and prevents injury after transient cardiac ischemia in vivo. *Circulation* **104**, 330–335.

10. Yoon, Y.S., Lee, N., and Scadova, H. (2005) Myocardial regeneration with bone-marrow-derived stem cells. *Biol. Cell* **97**, 253–263.

11. Rosenzweig, A. (2006) Cardiac cell therapy–mixed results from mixed cells. *N. Engl. J. Med.* **355**, 1274–1277.

12. Narula, J., Haider, N., Virmani, R., DiSalvo, T.G., Kolodgie, F.D., Hajjar, R.J., Schmidt, U., Semigran, M.J., Dec, G.W., and Khaw, B.A. (1996) Apoptosis in myocytes in end-stage heart failure. *N. Engl. J. Med.* **335**, 1182–1189.

13. Mallat, Z., Tedgui, A., Fontaliran, F., Frank, R., Durigon, M., and Fontaine, G. (1996) Evidence of apoptosis in arrhythmogenic right ventricular dysplasia. *N. Engl. J. Med.* **335**, 1190–1196.

14. Vazquez-Jimenez, J.F., Qing, M., Hermanns, B., Klosterhalfen, B., Woltje, M., Chakupurakal, R., Schumacher, K., Messmer, B.J., von Bernuth, G., and Seghaye, M.C. (2001) Moderate hypothermia during cardiopulmonary bypass reduces myocardial cell damage and myocardial cell death related to cardiac surgery. *J. Am. Coll. Cardiol.* **38**, 1216–1223.

15. Narula, J., Pandey, P., Arbustini, E., Haider, N., Narula, N., Kolodgie, F.D., Dal Bello, B., Semigran, M.J., Bielsa-Masdeu, A., Dec, G.W., et al. 1999. Apoptosis in heart failure: release of cytochrome c from mitochondria and activation of caspase-3 in human cardiomyopathy. *Proc. Natl. Acad. Sci. U. S. A.* **96**, 8144–8149.

16. Abbate, A., Bussani, R., Biondi-Zoccai, G.G., Rossiello, R., Silvestri, F., Baldi, F., Biasucci, L.M., and Baldi, A. (2002) Persistent infarct-related artery occlusion is associated with an increased myocardial apoptosis at postmortem examination in humans late after an acute myocardial infarction. *Circulation* **106**, 1051–1054.

17. Frustaci, A., Kajstura, J., Chimenti, C., Jakoniuk, I., Leri, A., Maseri, A., Nadal-Ginard, B., and Anversa, P. (2000) Myocardial cell death in human diabetes. *Circ. Res.* **87**, 1123–1132.

18. Nagoshi, T., Matsui, T., Aoyama, T., Leri, A., Anversa, P., Li, L., Ogawa, W., del Monte, F., Gwathmey, J.K., Grazette, L., et al. (2005) PI3K rescues the detrimental effects of chronic Akt activation in the heart during ischemia/reperfusion injury. *J. Clin. Invest.* **115**, 2128–2138.

19. Shiojima, I., Sato, K., Izumiya, Y., Schiekofer, S., Ito, M., Liao, R., Colucci, W.S., and Walsh, K. (2005) Disruption of coordinated cardiac hypertrophy and angiogenesis contributes to the transition to heart failure. *J. Clin. Invest.* **115**, 2108–2118.

20. Matsui, T., Nagoshi, T., Hong, E.G., Luptak, I., Hartil, K., Li, L., Gorovits, N., Charron, M.J., Kim, J.K., Tian, R., et al. (2006) Effects of chronic Akt activation on glucose uptake in the heart. *Am. J. Physiol. Endocrinol. Metab.* **290**, E789–797.

21. Fruman, D.A., Meyers, R.E., and Cantley, L.C. (1998) Phosphoinositide kinases. *Annu. Rev. Biochem.* **67**, 481–507.

22. Kobayashi, M., Nagata, S., Kita, Y., Nakatsu, N., Ihara, S., Kaibuchi, K., Kuroda, S., Ui, M., Iba, H., Konishi, H., et al. (1997) Expression of a constitutively active phosphatidylinositol 3-kinase induces process formation in rat PC12 cells. Use of Cre/loxP recombination system. *J. Biol. Chem.* **272**, 16089–16092.

23

Two-Dimensional Gel Electrophoresis in Platelet Proteomics Research

Ángel García

Summary

Proteomics technology allows a comprehensive and efficient analysis of the proteome and has become an indispensable tool in biomedical research. Since the late 80s, advances on mass spectrometry (MS) instrumentation and techniques have revolutionized the way proteins can be analyzed. Such analysis would only be possible with a proper sample preparation and separation ahead of the MS step. Different gel and nongel-based methods are available for protein separation. This chapter will focus on the use of two-dimensional gel electrophoresis (2-DE) in proteomics and its application to platelet research. 2-DE separates proteins according to their isoelectric point (pI) and size (molecular weight) and allows the detection of thousands of proteins at a time. Platelets are enucleated cells that play a critical function in the control of bleeding and wound healing. As platelets do not have a nucleus, proteomics offers a powerful alternative approach to provide data on protein expression in these cells, helping to address their biology. This chapter presents a protocol for an efficient sample preparation, protein separation by 2-DE, and protein digestion ahead of the MS analysis. The experimental approach, already successfully applied to the study of the platelet proteome, includes recommendations for an efficient platelet preparation for proteomics studies.

Key Words: Proteomics; Two-dimensional gel electrophoresis; Protein digestion; Liquid chromatography-tandem mass spectrometry; Platelets.

1. Introduction

After the sequencing of the human genome, it has become more evident that the complexity of an organism is primarily given by a complex proteome. The proteome can be defined as the whole set of proteins present in a cell,

From: *Methods in Molecular Medicine, Vol. 139: Vascular Biology Protocols*
Edited by: N. Sreejayan and J. Ren © Humana Press Inc., Totowa, NJ

tissue, body fluid, or organism at one time, including all the isoforms and post-translational variants *(1)*. Taking into account the numerous post-translational modifications (PTMs), the estimated 20,000–25,000 human genes may generate around 1 million distinguishable functional entities at the protein level. The analysis of the complete protein set is an analytical challenge that is made more difficult by the broad dynamic range of protein quantities expressed and by the fact that proteins, unlike DNA sequences, cannot be amplified.

Proteomics technology allows a comprehensive and efficient analysis of the proteome (*see* **Fig. 1**). Mass spectrometry (MS) is currently the most important tool in proteomics. During the recent years, changes in MS instrumentation and techniques revolutionized the analysis of proteins. Two ionization techniques developed in the late 1980s were the reason for this step forward and contributed to the development of proteomics research: electrospray ionization (ESI) and matrix-assisted laser desorption ionization (MALDI) *(2,3)*. Those methods solved the problem of generating ions from large, nonvolatile analytes, such as proteins and peptides, without significant analyte fragmentation, which is the reason why they are called "soft" ionization methods *(4,5)*. MS-based

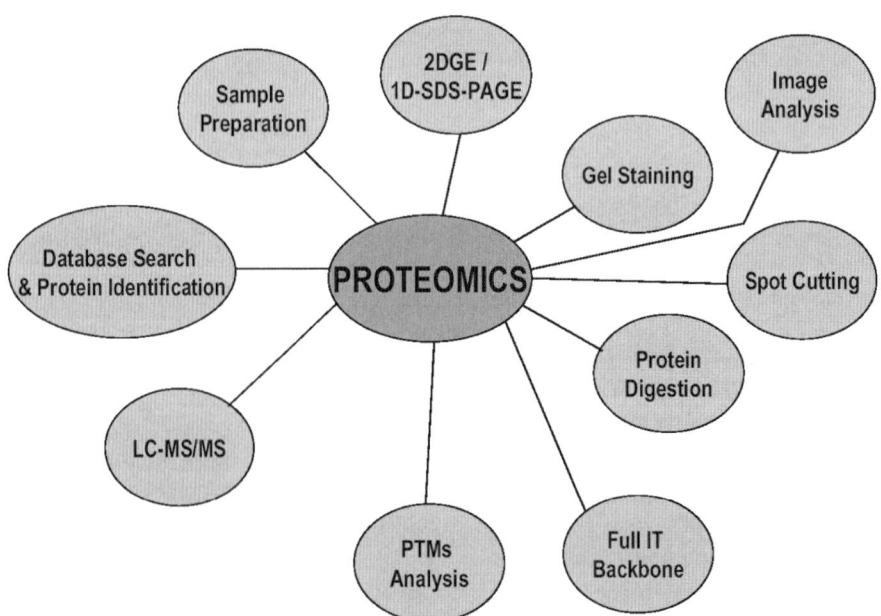

Fig. 1. Schematic representation of the basic components of a modern proteomics technology platform.

proteomics, which routinely achieves femtomole sensitivity, has a growing role in biomedical research where limited sample material is available.

Prior to MS analysis, proteins can be separated by different gel and nongel-based techniques, with sample preparation varying depending on the chosen approach. Two-dimensional gel electrophoresis (2-DE) is the only technique that allows the separation of up to several thousands of proteins at a time *(6,7)*. 2-DE separates proteins according to their isoelectric point (pI) and size (molecular weight). Following the electrophoresis, gels are stained, and the information on the presence of proteins, their up-regulation or down-regulation, and changes in the isoelectric point due to potential PTMs are provided by extensive image analysis. Proteins of interest can be excised, in-gel trypsinized, and identified by the application of powerful liquid chromatography-ESI-tandem MS (LC-EI-MS/MS), which provides data on the protein amino acid sequence (*see* **Fig. 2**). Powerful databases, such as SWISS-PROT and TrEMBL, can be used for protein identification. Recent technical improvements in 2-DE-based proteomics include a better solubilization of membrane proteins, enrichment of low-abundance proteins by sample pre-fractionation, novel highly sensitive water-soluble fluorescent dyes, and narrow pI range 2-D gels (*zoom* gels) for a detailed analysis of the region of interest *(8–10)*. Indeed, the use of sub-cellular pre-fractionation techniques in combination with 1-D sodium dodecyl sulfate–polyacrylamide gel electrophoresis (SDS–PAGE) and LC-MS/MS or multidimensional nanoscale capillary LC-MS/MS overcomes problems that arise from the fact that 2-DE cannot properly resolve many high molecular weight proteins or very basic and hydrophobic ones *(11–13)*.

Platelets are small enucleated cells that play a fundamental, life-saving role in hemostasis, and blood clotting at sites of vascular injury. Under normal conditions, platelets circulate in the blood as quiescent discs. However, when the endothelium of a blood vessel is damaged, platelets adhere to various extra-cellular matrix components, such as collagen, and undergo powerful activation. Under pathologic circumstances, platelets are involved in thrombosis and heart disease. Because platelets are enucleated, the analysis of the proteome is the best way to approach their biochemistry. In the recent years, 2-DE-based proteomics has been successfully applied to the study of the general proteome and signaling cascades in human platelets *(14–17)*.

This chapter will take the analysis of the platelet proteome as an example to provide some guidelines for an efficient sample preparation for 2-DE and help achieving optimal conditions for micro-preparative 2-DE and protein digestion prior to MS analysis.

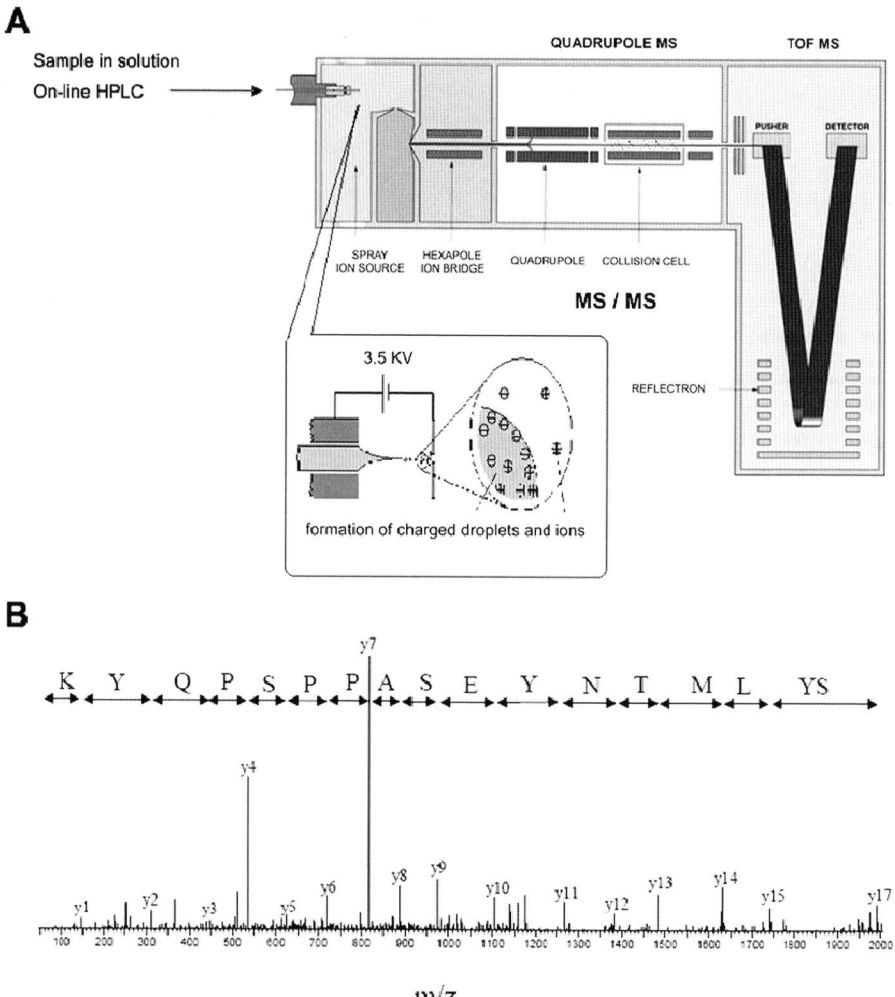

Fig. 2. Peptide sequencing using tandem mass spectrometry (MS/MS). (**A**) Representation of a typical quadrupole-time-of-flight (Q-TOF) mass spectrometer (modified with permission from Waters Corporation, Milford, MA, USA). Peptides are separated by high-pressure liquid chromatography (HPLC) in fine micro-capillaries and eluted into an electrospray ion source, where they are nebulized in small, highly charged droplets. After evaporation, multiply protonated peptides $(M + nH)^{n+}$ enter the mass spectrometer. A mass spectrum of all the peptides eluted at this time point is generated. A given peptide can be selected by the computer for MS/MS analysis. The peptide is first isolated at the quadrupole section of the mass spectrometer and fragmented by collision with gas. The corresponding peptide fragments enter the TOF section and are

2. Materials

2.1. Platelet Preparation and Lysis

1. Sterile sodium citrate 4% (w/v) solution, prostacyclin, ethylene glycol tetraacetic acid (EGTA), and all the chemicals mentioned below, unless specifically stated, are from Sigma (St. Louis, MO, USA).
2. Modified Tyrode's buffer (MTB): 134 mM NaCl, 0.34 mM Na_2HPO_4, 2.9 mM KCl, 12 mM $NaHCO_3$, 20 mM N-2-hydroxyethylpiperazine-N'-2-ethanesulfonic acid (HEPES), 5 mM glucose, 1 mM $MgCl_2$, pH 7.3. Glucose is added just before the experiment.
3. Acid–citrate–dextrose solution (ACD): 117 mM sodium citrate, 111 mM glucose, 78 mM citric acid.
4. Phosphatase inhibitors (Sigma). Prepare the following stocks: 100 mM sodium fluoride, 1 M sodium orthovanadate, and 1 M benzamidine. Aliquots can be kept at 4 °C for 2 weeks; sodium orthovanadate can be kept for longer at −20 °C. A phosphatase inhibitor cocktail would be the result of mixing equal volumes of the stocks shown above.
5. Protease inhibitors: cocktail from Sigma (keep at −20 °C).
6. Trichloroacetic acid (TCA) stock: 60% TCA (Sigma) in acetone (VWR International, Poole, UK). TCA should be kept at 4 °C and protected from light. TCA stock solution should be used within a day.
7. Leukocyte removal filters are from Pall (Portsmouth, Hampshire, UK).

2.2. 2-DE: Reagents and Isoelectric Focusing (First Dimension)

1. All the solutions, unless stated, are in milliQ water (*see* **Note 1**).
2. Reagents: 3-[(3-cholamidopropyl)dimethylammonio]-1-propane-sulfonic acid (CHAPS), tributyl phosphine, SDS, glycine, agarose, bromophenol blue, sodium fluoride, benzamidine, and sodium orthovanadate are from Sigma; glycerol and isopropanol from VWR International; 2-amino-2-hydroxymethyl-1,3-propanediol (Tris) and agarose from Roche Diagnostics (Hertfordshire, UK); dimethyl-benzyl-ammonium propane sulfonate-256 (NDSB-256) from Calbiochem-Novabiochem

---

Fig. 2. pulsed onton the detector, where they are recorded to produce the MS/MS spectrum. (**B**) Sequence analysis of peptides by collision-induced dissociation (CID). CID causes cleavage at the various amide bonds of the peptide to generate a series of fragments that differ by a single amino acid residue. The main ions generated are ions of types y and b, which contain the C terminus or the N terminus, respectively. As an example, an MS/MS profile that corresponds to a peptide that allowed the identification of *protein C7orf24* (Swiss-Prot accession number O75223) in human platelets is shown. The y ions that resulted from the peptide fragmentation are indicated *(16)*.

(Nottingham, UK); thiourea from Fluka (Steinheim, Switzerland); urea (ultrapure) and 1,2-dithiothreitol (DTT) from USB Corporation (Cleveland, OH, USA); carrier ampholytes (Servalytes) from Serva Electrophoresis GmbH (Heidelberg, Germany); DryStrip Cover Fluid and isoelectric focusing (IEF) electrode paper strips are from Amersham Biosciences currently GE Healthcare Ltd. (Buckinghamshire, UK).

3. Materials: immobilized pH gradient (IPG) strips and the electrophoresis equipment for IEF (Multiphor II, EPS 3500 XL power supply, reswelling cassette and reswelling tray) are from Amersham Biosciences. There are different ranges of IPG strips available depending on the region of interest. In platelet studies, pI 4–5, 5–6, 4–7, 6–11, and 3–10 have been the strips more commonly used so far.

4. Sample buffer for IEF: 5 M urea, 2 M thiourea, 2 mM tributyl-phosphine, 65 mM DTT, 65 mM CHAPS, 0.15 M NDSB-256, 1 mM sodium orthovanadate, 0.1 mM sodium fluoride, 1 mM benzamidine, trace of bromophenol blue. For pI 6–11 gels, isopropanol is added to the buffer at a final concentration of 10% (v/v). Store at −20 °C (*see* **Note 2**).

5. Protein quantification: Coomassie Plus protein assay reagent is from Pierce Biotechnology (Rockford, IL, USA); bovine serum albumin (BSA) is from Sigma.

2.3. 2-DE: Equilibration and SDS–PAGE (Second Dimension)

1. All equipment for SDS electrophoresis (Ettan DALT II gel system, power packs, Multitemp II thermostatic circulator) is from Amersham Biosciences.

2. Equilibration buffer: 4 M urea, 2 M thiourea, 2% (w/v) DTT, 50 mM Tris–HCl, pH 6.8 (pH 8 for pI 6–11 gels), 2% w/v SDS, 30% v/v glycerol, 50 μl of bromophenol blue stock (stock: 0.03 g in 10 ml milliQ water). Prepare fresh (*see* **Note 3**).

3. SDS–PAGE (second dimension) running buffer (stock 10×): for 3 l, dissolve 90.9 g Tris, 434 g glycine, and 30 g SDS in milliQ water. Take up to 30 l with milliRo water to have 1× SDS–PAGE running buffer.

4. Acrylamide 4× solution (40% w/v) is from Serva Electrophoresis GmbH (store at 4 °C); N, N′-methylenebisacrylamide (Bis), N, N, N′, N′-tetramethyl-ethylenediamine (TEMED) (*see* **Note 4**), and ammonium persulfate (APS) are from Bio-Rad (Hercules, CA, USA); Butan-2-ol and *n*-butanol are from VWR International.

5. Acrylamide stock solution (30.8% T, 2.6% C): 30% (w/v) acrylamide, 0.8% (w/v) Bis (this is a neurotoxin when un-polymerized and so care should be taken not to receive exposure). Filter and store at 4 °C.

6. Other solutions: 1.5 M Tris–HCl, pH 8.8 (store at 4 °C for 2 weeks); 10% (w/v) SDS (store at room temperature); 10% (w/v) APS (prepare fresh); 10% (v/v) TEMED (prepare fresh); displacing solution: 0.375 M Tris–HCl, pH 8.8, 50% (v/v) glycerol, bromophenol blue (prepare fresh); water-saturated butanol: 50 ml butan-2-ol or *n*-butanol plus 5 ml milliQ water (combine in a bottle and shake; use the top phase to overlay gels; store at room temperature); bagging buffer: 0.375 M Tris–HCl, pH 8.8 solution (store at 4 °C for 2 weeks).

2.4. Fixing and Staining

1. Reagents: ethanol, methanol, and acetic acid are from VWR International.
2. Fixing solution: 40% ethanol, 10% acetic acid.
3. Staining: SYPRO Ruby Protein gel stain is from Molecular Probes™ and Invitrogen (Paisley, UK). Dedicated polypropylene or polycarbonate containers are recommended. Protect from light (*see* **Note 5**).
4. Washing solution: 10% methanol, 7% acetic acid.

2.5. Protein Digestion

1. DTT and ammonium bicarbonate are from USB Corporation; acetonitrile, high-pressure LC (HPLC) grade water, formic acid, and iodoacetamide from Sigma; and trypsin (sequencing grade) from Roche Diagnostics.

3. Methods
3.1. Platelet Preparation

1. Fresh blood should be collected from normal healthy volunteers who were not on medication for the previous 10 days (assuming the study would be on normal human platelets). Only trained personnel should deal with human blood. Hepatitis B vaccine is needed. Ethical approval for the study should be granted.
2. Warm up 20 ml ACD and 50 ml MTB containing 0.045 g glucose (*see* **Subheading 2.1.**) to 30 °C (*see* **Note 6**). Make available several polypropylene centrifuge tubes and human blood disposal bags.
3. Obtain 50 ml (as an example) of fresh blood containing 5 ml sterile sodium citrate solution as anticoagulant.
4. Add 5 ml warmed ACD to the blood; transfer it into the polypropylene tubes and centrifuge them at $200\,g$ for 20 min at room temperature to obtain platelet-rich plasma (PRP).
5. Take the upper third of the PRP into a 50-ml falcon centrifuge tube, taking care not to take any red cells or leukocytes.
6. Add $10\,\mu g$ prostacyclin ($10\,\mu l$ stock solution at 1 mg/ml), mix gently by inversion, and immediately centrifuge at $1000\,g$ (*see* **Note 7**).
7. Decant the supernatant into a falcon tube, seal, and dispose as clinical waste. Resuspend the platelet pellet in 1 ml MTB containing $150\,\mu l$ ACD. Increase the volume by adding 25 ml MTB and 3 ml ACD. Use $5\,\mu l$ to count platelets.
8. Repeat step 6.
9. Decant the supernatant as above. Resuspend the pellet in MTB plus 1 mM EGTA to a final concentration of 10^9 platelets per milliliter (*see* **Note 8**). Incubate for 30 min at room temperature.
10. Spin down aliquots of 1 ml resuspended platelets at $10,000\,g$ for 2 min (room temperature) in the absence of prostacyclin. Rid of supernatants, and add $5\,\mu l$

protease inhibitor cocktail plus 5 μl phosphatase inhibitor cocktail to the pellets. Immediately freeze the platelet pellets in liquid nitrogen prior to storage at −80 °C (*see* **Note 9**).

3.2. Sample Preparation for 2-DE: Protein Precipitation and Delipidation

1. Homogenize pellets of frozen platelets by vortexing in a small volume (200 μl) of deionized water for 2 min.
2. Add 100 μl 60% TCA in acetone to each sample (final concentration of TCA: 20%), vortex, and incubate on ice for 45 min before centrifuging for 2 min at 10,000 g (*see* **Note 10**).
3. Wash the protein pellets twice in 500 μl cold acetone.
4. Resuspend the pellets in 300 μl IEF sample buffer by vortexing for 2 min and/or very mild sonication, if needed, avoiding heat (few seconds on an 80% duty cycle at 25% power; Status 70 MS73 with SH70G tip; Philip Harris Scientific; Ashby-de-la-Zouch; Leicestershire, UK). Do protein assay (*see* **Subheading 2.2., step 5**).
5. For micro-preparative IEF and *zoom* gels, load around 700 μg protein per-IPG strip (18 cm length). Proteins should be finally solubilized in 375 μl sample buffer for IPG 4–5, 5–6, and 4–7 and in 500 μl (including 10% isopropanol) sample buffer for IPG 6–11. In all cases, add 1.5% of the corresponding carrier ampholytes, vortex for a couple of minutes, and spin down before loading (*see* **Note 11**).
6. Insert samples into the grooves of the reswelling tray (Amersham Biosciences) placing the IPG strips on top (gel side down) without trapping air bubbles. Cover the strips with 2 ml silicone oil (DryStrip Cover Fluid). Allow in-gel rehydration for 12–16 h (*see* **Note 12**).

3.3. IEF and Equilibration

1. IEF (first dimension) is carried out for 70 kVh at 17 °C using the Multiphor (Amersham Biosciences). Place the strips (gel side up) in different lanes in the Multiphor. Soak IEF electrode paper strips (2 cm long) with deionized water, blot against filter paper to remove excess of liquid, and place them on top of the aligned IPG strips near the cathodic and anodic ends. When running basic IPG strips (pI 6–11), an extra paper strip soaked with 0.4% DTT can be applied onto the IPG gel surface near the cathodic electrode strip *(6)*.
2. The running conditions are as follows: 300 V for 2 h; IEF to the steady state at 3500 V. Focusing time at 3500 V varies depending on the IPG strip: 24 h for pI 4–5 and 5–6 and 20 h for pI 4–7 and 6–11 (*see* **Note 13**). Strips are covered with silicon oil during the electrophoresis.
3. Following IEF, rid of excess of silicon oil and immediately equilibrate the IPG strips in 2 ml equilibration buffer for 15 min (*see* **Note 14**).

3.4. SDS–PAGE (Second Dimension): Gel Casting and Electrophoresis

1. These instructions assume the use of the Ettan DALT II gel system (Amersham Biosciences), which allows casting and running 12 25.5 × 20.5 cm gels under identical conditions in a vertical system. The instructions will be for casting and running homogeneous gels (*see* **Note 15**). Please refer to the Ettan DALT II systems manual for more details.

2. Make sure the entire gel casting system is clean, dry, and free of any polymerized acrylamide. It is critical that the glass plates for the gels are scrubbed clean with a rinsable detergent after use (e.g., Neodisher A8 from Miele, Princeton, NJ, USA) and rinsed extensively with distilled water. Wipe the plate surfaces with ethanol using a halved TechClean wipe (Techspray International, Kempston, Bedford, UK) and allow them to dry before use.

3. Prepare a sufficient volume of gel overlay solution (water-saturated *n*-butanol); about 14 ml is needed for casting a full 14-gel set.

4. For a full 14-gel set, make up 0.9 l acrylamide gel stock solution without adding APS or TEMED. This amount will provide sufficient volume to cast gels using either a funnel or a peristaltic pump.

5. Assemble the gel caster as recommended by Amersham Biosciences. The caster should be placed on a level bench or on leveling table, so that gel tops are level.

6. The feed tube should be connected to either a funnel held in a ring-stand above the top of the gel caster (about 30 cm) or a peristaltic pump. Insert the other end of the feed tube into the grommet in the bottom of the balance chamber.

7. Fill the balance chamber with 100 ml displacing solution.

8. Add the appropriate volumes of APS and TEMED only when ready to pour the gels, not before. Once these two components are added, polymerization begins, and the gel solution should be completely poured within 10 min. The homogeneous gel solution necessary for 1 mm 10% gel set (10% T; 900 ml) is the following: 300 ml acrylamide stock; 225 ml 1.5 M Tris–HCl, pH 8.8; 356 ml milliQ water; 9 ml 10% SDS; 9 ml 10% APS; 1.54 ml 10% TEMED (*see* **Note 16**).

9. Pour the gel solution (by means of either a funnel or a peristaltic pump) taking care to avoid introducing any air bubbles into the feed tube.

10. Pump gel solution into the caster until it is about 1–2 cm below the final desired gel height. Stop the flow of acrylamide and remove the feed tube from the balance chamber grommet. Once the feed tube is removed, the dense displacing solution flows down the connecting tube forcing the remaining acrylamide solution into the cassettes to the final gel height.

11. Immediately pipette 1 ml water-saturated butanol onto each gel. If using a peristaltic pump to pour the gels, rinse the gel solution from the pump before it begins to polymerize.

12. Allow the gels to polymerize for at least 2 h before disassembling the caster (5 h for gradient gels). Rinse the top of the gels with milliQ water to rid of butanol. Either directly disassemble the caster or leave it overnight at 4 °C with the top of

the gels covered with 0.1 M Tris–HCl, pH 8.8 solution. After disassembling the caster, rinse the gels with milliQ water and cover the top with bagging buffer. They can be kept for 2 weeks at 4 °C.

13. To set up the second dimension, proceed as follows per gel: pour few milliliters 0.5% melted agarose (in 1× SDS–PAGE running buffer) on top of the corresponding 2-D gel and immediately immerse the IPG strip into the melted agarose till reaching contact with the top of the gel. Use a knife or spatula to help achieving a perfect contact between the IPG strip and the top of the 2-D gel before the agarose polymerizes.

14. Proceed with the SDS–PAGE at the following running conditions: 10 °C, 20 mA per gel for 1 h, followed by 40 mA per gel for approximately 4 h. Running conditions may vary depending on the electrophoresis equipment, so standardization may be needed.

3.5. Fixing, Staining, and Scanning

1. After finishing the electrophoresis, rinse gels with milliRo water and immerse them in fixing solution overnight (*see* **Subheading 2.4., step 2**).
2. Stain the gels overnight with SYPRO Ruby Protein gel stain (*see* **Note 5**).
3. Wash the gels for 30 min in washing solution (*see* **Subheading 2.4., step 4**).
4. Before imaging, rinse the gels in milliQ water twice for 5 min.
5. Visualize proteins using a 300-nm UV transilluminator, a blue-light transilluminator, FujiFilm LAS-3000 plus, or a laser scanner. As an example, the 2-DE proteome map of the human platelet is shown in **Fig. 3** *(14)*.
6. Proceed with the image analysis (*see* **Note 17**).

3.6. In-Gel Digestion and Protein Extraction

1. All the solvents used for protein digestion are HPLC grade, including water. All the steps are at room temperature unless specifically stated.
2. Protein features assigned to mass spectrometric analysis are excised from the gel either manually or by a software-driven robot cutter.
3. Dry the gel pieces in a speed-vac for 1 h.
4. Wash: add 50 μl 20 mM ammonium bicarbonate in water. Shake for 20 min. Remove supernatant. Repeat this step another two times.
5. Reduction with DTT: add 10 mM DTT in 20 mM ammonium bicarbonate in water (volume should cover the gel piece). Shake for 45 min. Remove supernatant.
6. Alkylation: add 50 mM iodoacetamide in 20 mM ammonium bicarbonate in water (volume should cover the gel piece). Shake for 20 min in the dark. Remove supernatant.
7. Wash: add 50 μl 20 mM ammonium bicarbonate in 50% acetonitrile. Shake for 20 min. Remove supernatant. Repeat this step another two times.
8. Wash: add 50 μl acetonitrile. Shake for 20 min. Remove supernatant.

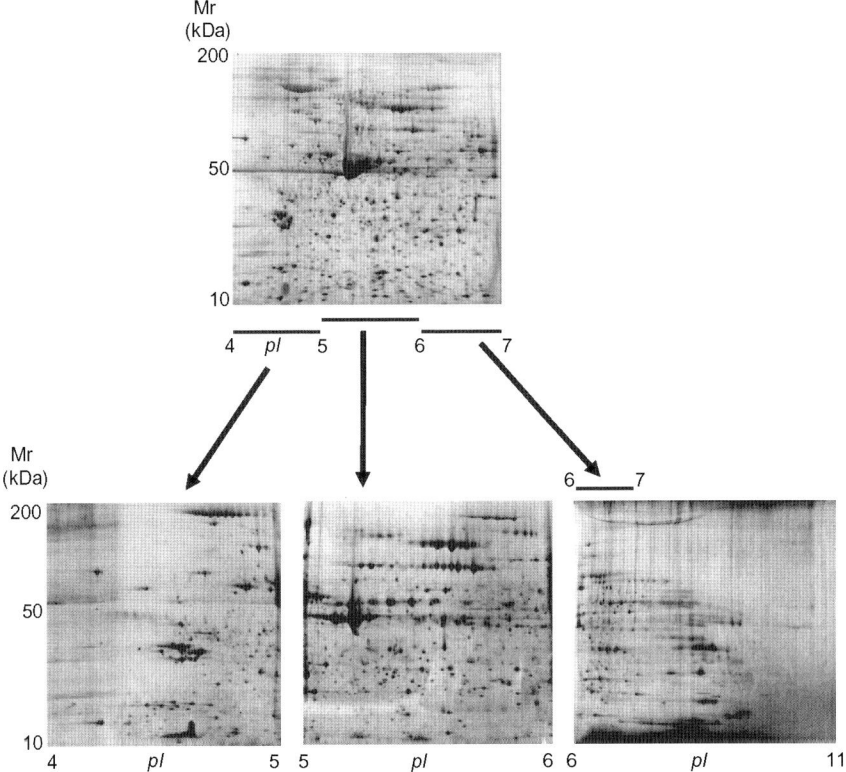

Fig. 3. Two-dimensional gel electrophoresis (2-DE) proteome map of the human platelet. The narrow pI range gels 4–5, 5–6, 4–7, and 6–11 are shown. Protein identifications are available at the following Web-based database: http: //www.bioch.ox.ac.uk/glycob/ogp. Modified with permission from *(14)*.

9. Dry gel pieces in the speed-vac for 1 h (gel pieces turn white).
10. Trypsin digestion: dissolve 100 μg trypsin-sequencing grade in 1 ml 10% acetonitrile in 1 mM HCl in water. Just before use, dilute this stock solution 10 times with 20 mM ammonium bicarbonate in water (final trypsin concentration: 10 ng/μl). Add 10 μl (volume may be increased based on size of gel piece) 10 ng/μl trypsin solution to each gel piece. After 5 min, remove excess of trypsin solution and add 10 μl 20 mM ammonium bicarbonate in water. Incubate overnight at 37 °C.
11. Extract peptides with 50 μl 5% formic acid in 50% acetonitrile in water. Shake for 20 min, and collect the supernatants in new vials. Repeat this step another two times, pooling the supernatants that come from the same gel piece.
12. Dry peptide extracts in speed-vac.

13. Reconstitution: dissolve the peptides in 6–10 µl 0.1% formic acid in water by vortex 3 s, shake for 20 min, and sonicate in a water bath for 2 min. Samples are now ready for LC-MS/MS analysis to get peptide sequencing (*see* **Fig. 2** and **Note 18**).

4. Notes

1. MilliQ water guaranties a resistivity of 18.2 MΩ cm and total organic content of less than five parts per billion. All the steps from protein precipitation to protein digestion should be carried out in a class 10,000 or ISO7 clean room to avoid dust and keratin contaminations.

2. The sample buffer should keep proteins perfectly solubilized and denatured but keeping their native charge intact. For that reason, samples cannot be boiled, and strong anionic detergents, such as SDS, cannot be used (those would mask the native charge of the proteins). A small amount of SDS (below 0.2%) can be used for very hydrophobic proteins, difficult to solubilize. Make sure that there is no urea crystals left in the sample buffer and that protein samples are without any salts or lipids, because those would interfere during the IEF. Insufficient solubility of particular proteins during IEF may cause horizontal streaks on the 2-D gel. Vertical streaks in the 2-D pattern can be caused by salt fronts, protein aggregates, and/or incomplete focusing in the first dimension.

3. Add the SDS at the end and spin down to avoid the presence of bubbles.

4. TEMED is best stored at room temperature in a desiccator. Buy small bottles as it may decline in quality (gels will take longer to polymerize) after opening.

5. Fluorescent dyes, such as SYPRO Ruby and SYPRO Orange (both from Invitrogen), are recommended because of their high sensitivity (detection levels of 1 ng protein in a gel piece with SYPRO Ruby), linearity, reproducibility, and MS compatibility. Staining with these dyes is noncovalent and can be accomplished in a single one-step procedure. When doing a differential analysis (e.g., comparing basal versus activated platelets), 2-D Fluorescence Difference Gel Electrophoresis (2-D DIGE) – where protein samples are labelled with cyanine dyes cy3, cy5, and cy2 ahead of the electrophoresis – could be used instead single staining. In that case, the protocol here presented would change according to the manufacter's descriptions for CyDye DIGE Fluor minimal dyes (GE Healthcare). Silver staining, although highly sensitive, has several drawbacks, such as (i) the poor reproducibility of several stains, (ii) limited dynamic range, and (iii) the fact that certain proteins stain poorly, negatively, or not at all. Moreover, silver staining procedures are quite labor-intensive.

6. MTB is a buffer appropriated for platelet isolation and has been successfully used in platelet proteomics research *(13–17)*. ACD is used to lower the pH to around 6.6, under which conditions platelet aggregation does not occur, allowing further biochemical studies.

7. Prostacyclin avoids unwanted platelet activation during centrifugation by raising the intracellular concentration of cAMP. That allows achieving the concentration of platelets needed for proteomics studies.

8. EGTA is used to prevent platelet aggregation at the end of the preparation, which would interfere with many biochemical studies. EGTA reduces the concentration of Ca^{2+} to below 1 nM in the absence of added calcium. At this concentration, the integrin GPIIbIIIa is unstable and so cannot support aggregation.

9. Purity of the isolated platelets can be controlled by microscopic inspection, guaranteeing only cell preparations of at least 99.9% to be subjected to further analysis. The effect of contaminating cells can be investigated by purification of platelets by centrifugation through leukocyte removal filters.

10. Protein precipitation and delipidation are used to rid of any unwanted detergents, salts, or lipids that would interfere with the IEF, especially when using alkaline IPG strips, such as 6–11. Just after adding the TCA stock, a mild sonication can be used instead vortexing (keeping the sample on ice) (few seconds on an 80% duty cycle at 30% power; Status 70 MS73 with SH70G tip; Philip Harris Scientific; Ashby-de-la-Zouch). Standardization is needed in each particular case to find out the best conditions for sample preparation. For example, in some cases, especially with pI 4–7 and 3–10 gels, it works well to dissolve the cell pellet directly in sample buffer, skipping the protein precipitation step.

11. For micro-preparative purposes, 18 or 24 cm IPG strips are recommended. For pI 3–10 gels, 500 μg protein should be enough for the analysis. Carrier ampholites are used to help proteins to keep solubilized during IEF: some proteins that require salt for solubility tend to precipitate at their pI during the IEF (where salt should be avoided) and carrier ampholytes help to counteract this insufficiency of salt in the sample.

12. In-gel rehydration is appropriate for micro-preparative IEF. By applying low voltage (30 V) during reswelling using the IPGphor (Amersham Biosciences), protein entry, especially of high M_r proteins, could be improved. For very alkaline IPG strips, such as 10–12, cup-loading at the anode is recommended *(6)*.

13. Depending on the length of the IPG strip, and if the IEF is analytical or micro-preparative, the electrophoresis running conditions will vary *(6)*. Insufficient solubility of particular proteins during IEF may cause horizontal streaks on the 2-D gel. Vertical streaks in the 2-D pattern can be caused by salt fronts, protein aggregates, and/or incomplete focusing in the first dimension.

14. Prolonged equilibration time is important to completely load all proteins with SDS for improved protein transfer from the first to the second dimension. Conditions should be standardized for a given sample *(6)*.

15. Homogenous gels are easier to prepare and more reproducible, whereas gradient gels offer a higher resolution and are recommended despite being necessary more time to standardize the right conditions for the whole 2-D platform to work in a reproducible way. Gels can be attached to the back glass plate by using bind silane

(Amersham Biosciences). In that case, step (front) plates should be treated with repel-silane (Amersham Biosciences); so, after running the second dimension, it would be possible to proceed further with the gels just attached to the back plates. That would improve the reproducibility of the study and help with the image analysis (assuming the glass plates are compatible with the dye).

16. Gel solution composition will vary depending on the %T (see Amersham Biosciences manual for Ettan DALT II gel system). Standardization to get the best conditions is required. Perform a small-scale test before using a new composition to check that the solution polymerizes in about 10 min.

17. Image analysis is a key step in 2-DE-based proteomics, allowing the comparison between different sets of gels. It is especially important when doing a differential image analysis (e.g., comparing "healthy" versus "diseased" gels). There are different software packages in the market, such as MELANIE III (Bio-Rad), Progenesis Workstation (nonlinear dynamics), and DeCyder™ (Amersham Biosciences). Some of the companies that offer these software packages also offer software-driven robot cutters for gel excision.

18. Peptide sequencing by MS/MS is the recommended option in proteomics and can be used in conjunction with peptide mass fingerprinting *(18)*. Ion trap and quadrupole-time-of-flight (Q-TOF) are the recommended mass spectrometers for peptide sequencing, in combination with ESI. For peptide mass fingerprinting, MALDI in combination with TOF MS would be the preferred option. For MALDI, peptides tend to be reconstituted in 0.1% trifluoroacetic acid (TFA).

Acknowledgments

The author thanks Dr. Nicole Zitzmann, Prof. Raymond A. Dwek, and Prof. Steve P. Watson for their continuous support as well as the Oxford Glyco-biology Institute Endowment for the funding. The author is a Parga Pondal Research Fellow (Xunta de Galicia, Spain).

References

1. Tyers, M., and Mann, M. (2003) From genomics to proteomics. *Nature* **422**, 193–197.

2. Fenn, J.B., Mann, M., Meng, C.K., Wong, S.F., and Whitehouse, C.M. (1989) Electrospray ionization for the mass spectrometry of large biomolecules. *Science* **246**, 64–71.

3. Karas, M., and Hillenkamp, F. (1988) Laser desorption ionization of proteins with molecular mass exceeding 10000 daltons. *Anal. Chem.* **60**, 2299–2301.

4. Aebersold, R., and Goodlett, D.R. (2001) Mass spectrometry in proteomics. *Chem. Rev.* **101**, 269–295.

5. Aebersold, R., and Mann, M. (2003) Mass spectrometry-based proteomics. *Nature* **422**, 198–207.

6. Görg, A., Obermaier, C., Boguth, G., Harder, A., Scheibe, B., Wildgruber, R., and Weiss, W. (2000) The current state of two-dimensional electrophoresis with immobilized pH gradients. *Electrophoresis* **21**, 1037–1053.
7. Görg, A., Weiss, W., and Dunn, M.J. (2004) Current two-dimensional electrophoresis technology for proteomics. *Proteomics* **4**, 3665–3685.
8. Herbert, B. (1999) Advances in protein solubilisation for two-dimensional electrophoresis. *Electrophoresis* **20**, 660–663.
9. Hoving, S., Voshol, H., and van Oostrum, J. (2000) Towards high performance two-dimensional gel electrophoresis using ultrazoom gels. *Electrophoresis* **21**, 2617–2621.
10. Lopez, M.F., Kristal, B.S., Chernokalskaya, E., Lazarev, A., Shestopalov, A.I., Bogdanova, A., and Robinson, M. (2000) High-throughput profiling of the mitochondrial proteome using affinity fractionation and automation. *Electrophoresis* **21**, 3427–3440.
11. Link, A.J., Eng, J., Schieltz, D.M., Carmack, E., Mize, G.J., Morris, D.R., Garvik, B.M., and Yates, J.R., III. (1999) Direct analysis of protein complexes using mass spectrometry. *Nat. Biotechnol.* **17**, 676–682.
12. Washburn, M.P., Wolters, D., and Yates, J.R., III. (2001) Large-scale analysis of the yeast proteome by multidimensional protein identification technology. *Nat. Biotechnol.* **19**, 242–247.
13. García, A., Zitzmann, N., and Watson, S.P. (2004) Analyzing the platelet proteome. *Semin. Thromb. Hemost.* **30**, 485–489.
14. García, A., Prabhakar, S., Brock, C.J., Pearce, A.C., Dwek, R.A., Watson, S.P., Hebestreit, H.F., and Zitzmann, N. (2004) Extensive analysis of the human platelet proteome by two-dimensional gel electrophoresis and mass spectrometry. *Proteomics* **4**, 656–668.
15. García, A., Prabhakar, S., Hughan, S., Anderson, T.W., Brock, C.J., Pearce, A.C., Dwek, R.A., Watson, S.P., Hebestreit, H.F., and Zitzmann, N. (2004) Differential proteome analysis of TRAP-activated platelets: involvement of DOK-2 and phosphorylation of RGS proteins. *Blood* **103**, 2088–2095.
16. García, A., Watson, S.P., Dwek, R.A., and Zitzmann, N. (2005) Applying proteomics technology to platelet research. *Mass Spectrom. Rev.* **24**, 918–930.
17. García, A., Senis, Y.A., Antrobus, R., Hughes, C.E., Dwek, R.A., Watson, S.P., and Zitzmann, N. (2006) A global proteomics approach identifies novel phosphorylated signalling proteins in GPVI-activated platelets: involvement of G6f, a novel platelet Grb2-binding membrane adapter. *Proteomics* **6**, 5332–5343.
18. Mann, M., Hendrickson, R.C., and Pandey, A. (2001) Analysis of proteins and proteomes by mass spectrometry. *Annu. Rev. Biochem.* **70**, 437–473.

24

Stem Cell Therapy in the Heart and Vasculature

Loren E. Wold, Wangde Dai, Joan S. Dow, and Robert A. Kloner

Summary

Stem cell therapy is a progressive approach to a pervasive clinical problem; cardiovascular disease is the number 1 killer in the USA and other developed countries, and aspects of it are amenable to stem cell therapy. Many types of stem cells have been used in treating the heart during myocardial infarction, and here, we describe our approach of direct myocardial injection of bone marrow-derived mesenchymal stem cells into the infarct of rats. We will also briefly introduce the methods we have used to inject neonatal cardiomyocytes into the aorta as a first step in attempting to produce an external cardiac pump. Proper surgical technique and postoperative care are as important as adequate injection of the cells and will greatly improve the survival of the animal after surgery. By carefully following the methods presented in this chapter, the reader will be able to perform direct myocardial and vascular injection of stem cells into rats.

Key Words: Stem cell; myocardial infarction; vascular disease.

1. Introduction

Stem cells are a group of cells that are undifferentiated and capable of self-renewal. There are many different types of stem cells that have been characterized, including those derived from adult bone marrow (mesenchymal, hematopoietic, and endothelial precursor), adipose tissue, peripheral and umbilical cord blood, embryonic stem cells derived from blastocysts as well as the lineage-committed cells of skeletal muscle (although not stem cells by definition, but included for comparative purposes). The heart itself has been shown to contain resident cardiac stem cells (*1,2*) that have been postulated to be an important component of cardiac homeostasis. Stem cells have been

From: *Methods in Molecular Medicine, Vol. 139: Vascular Biology Protocols*
Edited by: N. Sreejayan and J. Ren © Humana Press Inc., Totowa, NJ

used to regenerate many different tissue types including skeletal muscle, bone, neural tissue *(3)*, and most recently the myocardium *(4,5)*.

Cellular cardiomyoplasty has been investigated using several different types of host cells. A study by Orlic and colleagues *(6)* showed that transplanted Lin⁻- c-kit⁺ cells regenerated myocardial tissue and improved cardiac performance in a mouse model of coronary artery occlusion. Murry and colleagues *(7)*, however, showed through genetic tracking techniques that hematopoietic stem cells injected into the infarcted mouse heart were not able to differentiate into cardiomyocytes.

Embryonic stem cells have also been shown to improve ventricular function and contractility six months after transplantation, in a rat model of myocardial infarction *(8)*, with the ability to exhibit structural and functional character- istics analogous to cardiac tissue *(9)*. The use of embryonic stem cells in the clinic is hampered by the requirement of immunosuppression to avoid rejection of the transplanted cells *(10)* and the development of teratomas *(11)*. Skeletal muscle-derived satellite cells, although not truly stem cells as their lineage is committed, have been shown to decrease ventricular remodeling and increase post-exercise cardiac function in a rat model of myocardial infarction *(12)*. Rubart and colleagues *(13)* showed that transplanted skeletal myoblasts were able to fuse with host cardiomyocytes and express connexin 43, suggesting the ability to electrically couple. However, only a small percentage of donor skeletal myoblasts were able to fuse with the host cardiomyocytes, which is necessary for the proper electrical functioning of the donor cells within the heart.

One of the pivotal questions that remains is the most optimal mode of delivery of stem cells to the site of injury (such as into the scar of a myocardial infarction). Numerous methods have been used, including direct cardiac intra- muscular injection, intravenous or intracoronary administration as well as transendocardial or transepicardial delivery. The easiest mode of delivery is intravenous administration; however, the length of time it takes for the cells to migrate to the injured tissue and the percentage of cells injected that actually "home" to the area of injury could be limitations to this approach. The method that we describe in this chapter is direct intramuscular injection of stem cells, which allows direct visualization of the injured area and control over the site of injection. This technique has been used with beneficial effects on the heart in both the clinical *(14)* and the basic biomedical *(15)* settings.

Stem cells are being studied with increased frequency in both the laboratory and the clinic. Numerous clinical studies utilizing stem cells are either underway or recruiting patients, with sponsorship by both the private sector and the US government *(16)*. Therefore, it is important to understand the most efficacious type of cell and the best mode of delivery of these cells to optimize the benefits of stem cell therapy.

2. Materials

2.1. Equipment

1. Fischer CDF rats or equivalent (approximately 120–200 g).
2. Operating table.
3. Ventilator (Harvard rodent ventilator, Model 683, Harvard Apparatus, South Natick, MA, USA) and intubation tube (PE200 or smaller).
4. Small flat spatula for intubation.
5. Sterile surgical instruments.
6. Sterile non-absorbable suture (4-0 or 6-0; Ethicon, Somerville, NJ, USA).
7. Tuberculin syringe (28 gauge or smaller).
8. Sterile absorbable suture (2-0 PDSII; Ethicon).
9. Sterile skin staples (Michel clips 7.5 mm; Miller Instruments Company, Bethpage, NY, USA).
10. Heating pad (one for surgery and one for recovery).
11. Sterile hood for the preparation of stem cells.
12. Cell counter.
13. Standard tabletop centrifuge.
14. Trypan blue dye (to stain cells; Gibco BRL, Grand Island, NY, USA).
15. Sterile Eppendorf tubes.

2.2. Reagents

1. Ketamine (75 mg/kg; Abbott Laboratories, North Chicago, IL, USA).
2. Xylazine (5 mg/kg; Ben Venue Laboratories, Bedford, OH, USA).
3. Betadine/Chlorhexidine scrub.
4. Alcohol swabs.
5. bupivacaine (0.1 mg/kg body weight intercostal muscle; AstraZeneca LP, Wilmington, DE, USA).
6. Stem cells for injection.
7. Normal saline (post-op administration of 1 ml/100 g body weight).
8. Buprenorphine (0.001 mg/100 g body weight; Reckitt Benckiser Pharmaceuticals, Richmond, VA, USA).

3. Methods

3.1. Bone-Marrow Derived Mesenchymal Stem Cells

We obtain our adult bone marrow-derived mesenchymal stem cells from Osiris Therapeutics of Baltimore, MD. These cells are obtained from the femoral and tibial bones of ACI rats (250–300 g). They are expanded to passage 3 before infusion to recipient animals. At the time of this passage, cultures are approximately 95% homogenous for rat mesenchymal stem cells *(17,18)*. They are also labeled with DiI to track the cells after injection. In our hands, these cells do not induce an immunologic reaction when injected into Fischer CDF rats.

These cells arrive at the Good Samaritan Hospital Heart Institute on dry ice at $-80\,°C$ where they are dissociated and resuspended in sterile saline immediately before injection. It is imperative to quickly thaw and warm the frozen cells to $37\,°C$ (we use a water bath and place the small Eppendorf tubes with the cells into the bath for 90 s).

3.2. Rat Myocardial Infarction Model

Our group has used both Sprague-Dawley as well as Fischer CDF rats with equal success; however, when injecting stem cells, the investigator must be cognizant of immunorejection. The details of the procedure are as follows:

1. Anesthetize the rat with ketamine (75 mg/kg, i.p.) and xylazine (5 mg/kg, i.p.) and shave. Make sure to inject into the peritoneum because intramuscular injection may cause an abscess to develop.
2. Intubate the animal and ventilate using a mechanical ventilator with 1 ml room air/100 g body weight (60 cycles/min).
3. Perform left thoracotomy and make an incision from the base of the sternum toward the armpit.
4. Using blunt dissection, navigate around the edges of the incision and free the skin from the underlying tissue. Dissect between the muscles of the chest, moving at the same angle as the skin incision. Using a suture, retract the chest muscle back toward the sternum.
5. The next layer of muscle runs lengthwise over the ribs. Grasp the upper edge of this muscle, dissect beneath it and down, exposing the ribs. Inject bupivacaine (0.1 mg/kg body weight intercostal muscle) into the fourth, fifth, and sixth intercostal muscles.
6. Insert rib spreaders into the fourth intercostal space with the teeth facing up and upon entering the chest cavity, rotate the spreaders toward the head of the animal. This technique will avoid trapping the lungs in the teeth of the rib spreader. The heart should now be visible (*see* **Fig. 1A**).
7. Remove the pericardium and locate the coronary artery. The pericardium can often be cut and rotated away from the heart by using a sterile Q-tip, exposing the heart.
8. Place a non-absorbable suture around the artery, using extreme caution to not puncture the atrial appendage (*see* **Note 1** and **Fig. 2**). Tie this suture in a double surgeons knot. Cut the suture; however, leave approximately 0.5 cm for use during cell injection.
9. Remove the rib spreaders and suture the chest closed with absorbable suture. Air removal from the chest is accomplished by gently squeezing the chest while tying the absorbable suture. Close the chest with skin staples, administer normal saline

A

B

C

D

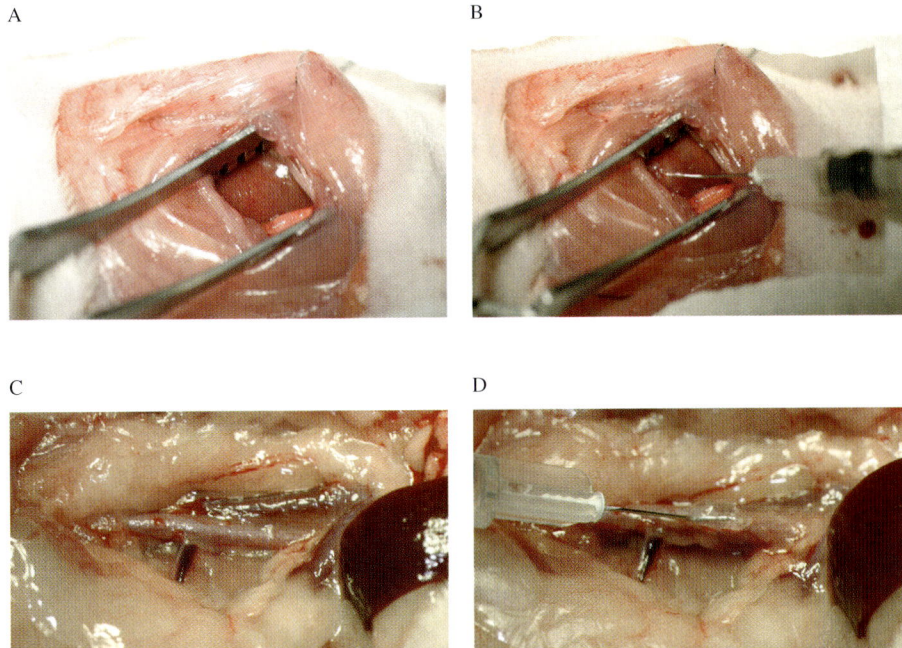

Fig. 1. (**A**) Representative picture of the open chest showing the heart exposed through the fourth intercostal space. (**B**) Injection of stem cells into the heart forms a noticeable "bleb" on the surface of the heart. Notice that the tip of the needle is still visible within the anterior free wall of the heart. (**C**) Exposure of the abdominal aorta for injection of stem cells. (**D**) Injection of stem cells into the wall of aorta. Visible bleb or blister of cells is seen at distal needle site. Note the 45° angle of the needle facilitates maneuverability of the needle while in the abdominal cavity of the rat.

subcutaneously (1 ml/100 g body weight, s.q.), and place the animal on a heating pad until it fully recovers. Buprenorphine (0.001 mg/100 g body weight, s.q.) is given for postoperative pain management.

10. For cell injection, follow the same procedures as outlined in **Subheadings 3.2.1.–3.2.9.**, however enter the chest cavity at the fifth intercostal space. The first surgery may have caused the heart to adhere to the chest wall. Adhesions are gently removed with the tip of a sterile Q-tip.

11. Once the heart is exposed, grasp the 0.5-cm suture used to make the occlusion and hold for stabilizing the heart during cell injection.

12. The stem cells should be injected using a 28 gauge tuberculin syringe or smaller to minimize the size of the injection site (*see* **Notes 2–4**; **Figs 1B** and **3A and B**). It may be useful to bend the distal end of the needle at a 45° angle. A 150 g rat can be injected with approximately 70 µl of cells with minimal fluid loss.

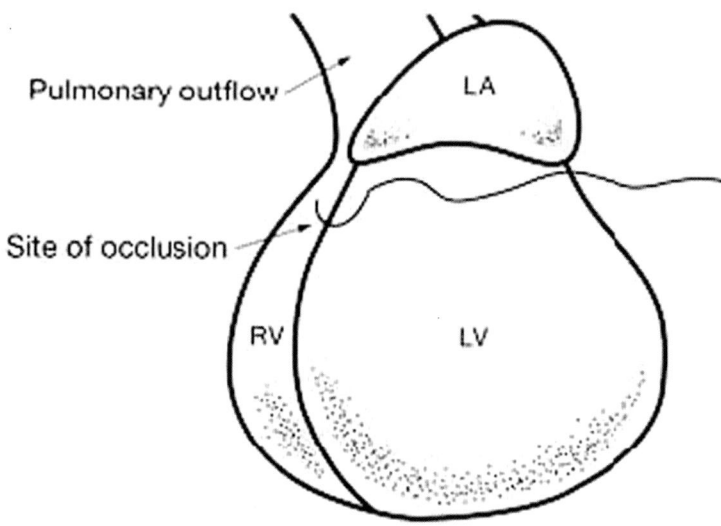

Fig. 2. Schematic diagram of the placement of the sterile non-absorbable suture under the left coronary artery at the interventricular groove, immediately below the left atrial appendage.

13. Carefully enter the left ventricular free wall with the tip of the needle parallel to the epicardial plane a short distance from the infarct and slowly advance the needle into the infarcted region. If the tissue has been successfully infarcted, it will have a pale, gray color and not be contracting. If there is a 45° angle in the needle, the needle can be bent slightly backward at this point, and the tip can be visualized under the epicardial layer, assuring that the tip has not penetrated into the cavity. Inject approximately 75% of the fluid in the needle, watching for the formation of a pale bleb (*see* **Figs 1B and 3B**); Note 5 under the epicardium. Slowly remove the syringe while injecting the remaining 25% of the cells. This will aid in even distribution of the injected fluid. If desired, numerous injection sites can be used to deliver the cells. However, we have found that injecting the cells with one injection reduces the chance of leakage (*see* **Note 2**).
14. After cell injection, follow the same guidelines as the first surgery for closing the chest and animal recovery.

3.3. Isolation and Injection into the Aorta

1. Follow the same procedures listed above for anesthesia. The left lumbar area is shaved and prepped for sterile surgery.
2. Place the rat in the right lateral decubitus position and a vertical dorsal lumbotomy incision is made along the lateral margin of the erector spinae muscle from the

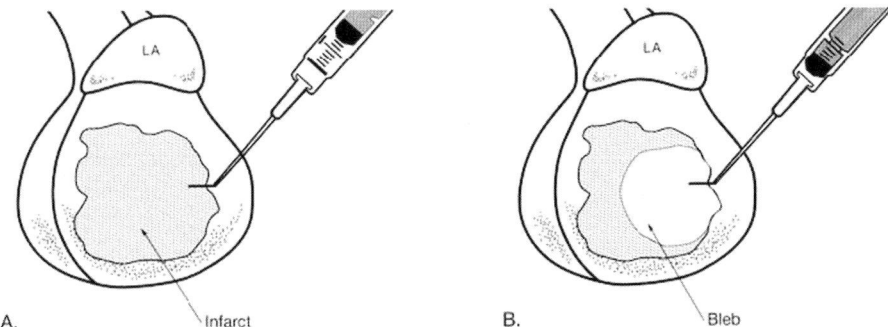

A.   Infarct   B.   Bleb

Fig. 3. (**A**) Schematic drawing of the placement of the needle at the tip of the infarct. (*see* **note 5**) (**B**) Formation of a bleb on the surface of the heart. Note the even spread of the injected fluid within the infarct zone.

margin of the 12th rib to the iliac crest. Incise the skin and underlying muscle and expose the peritoneum. Dissect the retroperitoneal plane behind the kidney to expose the left renal artery and aorta. The site of injection of immature cardiac cells is approximately 3 mm above the level of the start of the renal arteries. Carefully dissect the fascia surrounding the aorta (*see* **Fig. 1C**).

3. Inject the cells into the wall of aorta in the same manner as that used to inject into the heart. Successful injection is shown by a raised pale bleb that extends 2–3 mm along the length of the vessel and protrudes approximately 2 mm above the surface of the vessel (*see* **Fig. 1D**). Repeat the injection four times around the surface of the aorta, to insure an even and homogenous cuff of cells around the vessel.

4. Suture and close the abdominal wall of the rat as described above and staple the skin incision. The rat is maintained on a heating pad with adequate pain management achieved through buprenorphine (0.001 mg/100 g body weight).

4. Notes

1. Ligation of the coronary artery: The coronary artery is intramural and therefore not readily visible on the surface of the heart. To properly ligate the coronary artery to produce a myocardial infarction, advance the suture into the muscle at the interventricular groove (between the left and right ventricles) just below the left atrial appendage. It is often helpful to bend the needle on the suture to facilitate advancing the suture through the surface of the muscle or to use a curved needle (*see* **Fig. 2**).

2. Loss of injected fluid: When injecting stem cells into either the myocardium or the vasculature, it is important to inject an adequate number of cells (we use ~2 million bone marrow-derived mesenchymal stem cells per injection). Unfortunately, cells may be lost to the vasculature after injection, and therefore, this should be taken

into account when deciding on the adequate number of cells to inject. Removing the needle in a gentle fashion following injection—without tearing the tissue—will also minimize loss of cells because of leakage from the injection site. We have found that practicing the injection technique on grapes is very easy to do and helpful. Simply advance the needle under the skin of the grape and inject; the bleb

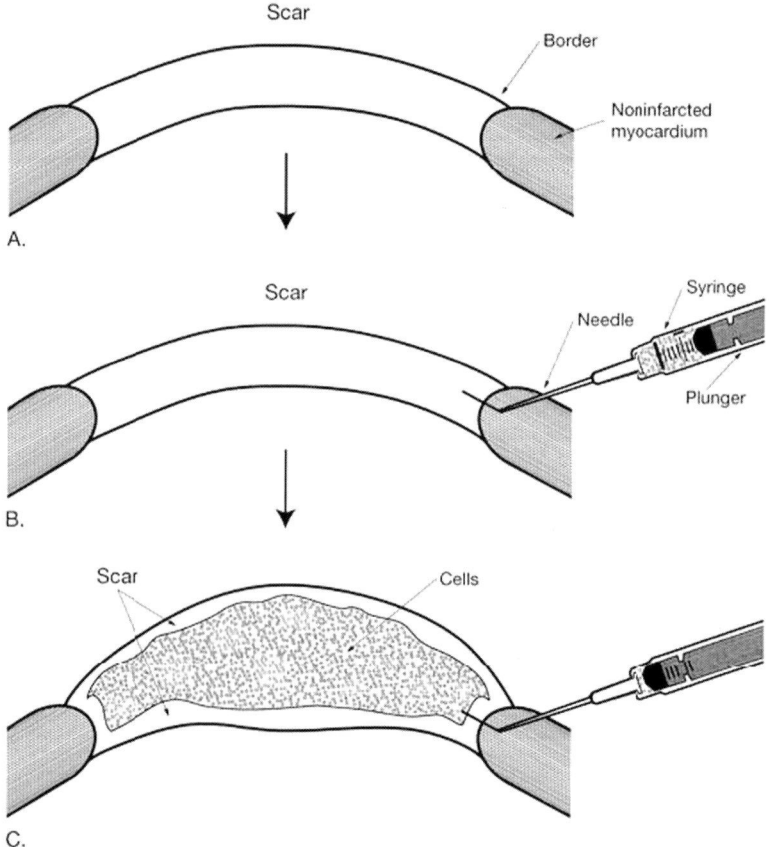

Fig. 4. (**A**) Cross-sectional schematic of the heart with a thin anterior free wall (scar) representing a myocardial infarction. (**B**) Schematic of a heart slice showing the tuberculin syringe located within the anterior free wall (scar) of the infarction. Bend the tip of the needle up slightly to ensure proper placement of the needle within the ventricular wall (*see* **Note 3**). (**C**) Schematic of a heart slice representing the end of injection. Notice that the fluid injected has formed a large bleb completely confined within the wall of the infarct. The cells are sandwiched between an endocardial and epicardial layer of infarct scar. (Syringe and needle not drawn to scale)

will form if injected properly. This is an easy and cost-efficient way to practice the technique of injection of cells into the wall of the heart.

3. Problems with the syringe: A tuberculin syringe is the best size syringe to use as it has very little dead space, and the injection site is very small. Make sure to bend the needle approximately 45° to allow ease of manipulation during injection. Once you have entered the myocardium, tip or tilt the syringe back slightly and try to visualize the tip of the needle under the epicardium (*see* **Fig. 4A–C**). This will help ensure that you have not penetrated into the cavity. This maneuver will cause the epicardium to "tent" over the needle tip. When the needle is in place in either the heart or the vasculature, draw back slightly. If blood is present during drawback, it is likely that the needle has passed into either the ventricle or the cavity of the vessel. If the needle damages the tissue (evident by excessive bleeding from the myocardium), remove it and start anew. The needle can possibly pass through the wall of the heart or vasculature and come back out. Usually any bleeding that occurs can be controlled with gentle pressure from a sterile Q-tip.

4. Problems with the fluid injected: Cells dispersed in normal medium (we use Iscove's Modified Dulbecco's Medium) will tend to leak from the injection site; so, minimizing the size of the injection site is pivotal. Increasing the viscosity of the injected fluid will enhance the amount of fluid that will remain in the tissue. Be careful not to inject too fast or the injected fluid will tend to leak. Watch for the formation of a bleb over the injection site (*see* **Fig. 1B**).

5. What to expect if the coronary occlusion has been successful? If ligation of the coronary artery was successful, the anterior free wall will be necrotic, evidenced by a pale, thin, and fibrotic scar posterior to the suture. Necrosis is fully developed within 24–48 h after a permanent coronary occlusion. For post-mortem analysis, triphenyltetrazolium chloride (TTC) can be used to better define the infarct region. If the ligation of the coronary artery was not successful, no necrosis will be evident. This unfortunately happens in about 10% of all attempts. Also, proper coronary artery occlusion can sometimes result in death of the animal because of either ventricular arrhythmias or congestive heart failure.

Acknowledgments

The authors thank Dr. Bradley J. Martin of Osiris Therapeutics, Baltimore, MD, for supplying the adult bone marrow-derived mesenchymal stem cells as well as Sharon L. Hale for expert technical assistance in preparation of the cells for injection and Kerry Grindle, Art Director for assistance with figure illustrations. We also acknowledge grant support from the National Institute of Health to R.A.K. (NIH-1RO1-HL073709-01, NIH-1RO1-HL61488, and NIH-1RO1-HL071965).

References

1. Anversa, P., Kajstura, J., and Leri, A. (2004) Circulating progenitor cells: search for an identity. *Circulation* **110**, 3158–3160.
2. Beltrami, A. P., Barlucchi, L., Torella, D., Baker, M., Limana, F., Chimenti, S., et al. (2003) Adult cardiac stem cells are multipotent and support myocardial regeneration. *Cell* **114**, 763–776.
3. Donovan, P. J., and Gearhart, J. (2001) The end of the beginning for pluripotent stem cells. *Nature* **414**, 92–97.
4. Kudo, M., Wang, Y., Wani, M. A., Xu, M., Ayub, A., and Ashraf, M. (2003) Implantation of bone marrow stem cells reduces the infarction and fibrosis in ischemic mouse heart. *J. Mol. Cell Cardiol.* **35**, 1113–1119.
5. Pittenger, M. F., and Martin, B. J. (2004) Mesenchymal stem cells and their potential as cardiac therapeutics. *Circ. Res.* **95**, 9–20.
6. Orlic, D., Kajstura, J., Chimenti, S., Jakoniuk, I., Anderson, S. M., Li, B., et al. (2001) Bone marrow cells regenerate infarcted myocardium. *Nature* **410**, 701–705.
7. Murry, C. E., Soonpaa, M. H., Reinecke, H., Nakajima, H., Nakajima, H. O., Rubart, M., et al. (2004) Haematopoietic stem cells do not transdifferentiate into cardiac myocytes in myocardial infarcts. *Nature* **428**, 664–668.
8. Min, J. Y., Yang, Y., Converso, K. L., Liu, L., Huang, Q., Morgan, J. P., et al. (2002) Transplantation of embryonic stem cells improves cardiac function in postinfarcted rats. *J. Appl. Physiol.* **92**, 288–296.
9. Kehat, I., Kenyagin-Karsenti, D., Snir, M., Segev, H., Amit, M., Gepstein, A., et al. (2001) Human embryonic stem cells can differentiate into myocytes with structural and functional properties of cardiomyocytes. *J. Clin. Invest.* **108**, 407–414.
10. Dengler, T. J., and Katus, H. A. (2002) Stem cell therapy for the infarcted heart ("cellular cardiomyoplasty"). *Herz* **27**, 598–610.
11. Lee, M. S., Lill, M., and Makkar, R. R. (2004) Stem cell transplantation in myocardial infarction. *Rev. Cardiovasc. Med.* **5**, 82–98.
12. Jain, M., DerSimonian, H., Brenner, D. A., Ngoy, S., Teller, P., Edge, A. S., et al. (2001) Cell therapy attenuates deleterious ventricular remodeling and improves cardiac performance after myocardial infarction. *Circulation* **103**, 1920–1927.
13. Rubart, M., Soonpaa, M. H., Nakajima, H., and Field, L. J. (2004) Spontaneous and evoked intracellular calcium transients in donor-derived myocytes following intracardiac myoblast transplantation. *J. Clin. Invest.* **114**, 775–783.
14. Kinnaird, T., Stabile, E., Burnett, M. S., and Epstein, S. E. (2004) Bone marrow-derived cells for enhancing collateral development: mechanisms, animal data, and initial clinical experiences. *Circulation* **95**, 354–363.
15. Bel, A., Messas, E., Agbulut, O., Richard, P., Samuel, J. L., Bruneval, P., et al. (2003) Transplantation of autologous fresh bone marrow into infarcted myocardium: a word of caution. *Circulation* **108**, II247–II252.
16. Wold, L. E., Dai, W., Sesti, C., Hale, S. L., Dow, J. S., Martin, B. J., et al. (2004) Stem cell therapy for the heart. *Cong. Heart Fail.* **10**, 293–301.

17. Haynesworth, S. E., Goshima, J., Goldberg, V. M., and Caplan, A. I. (1992) Characterization of cells with osteogenic potential from human marrow. *Bone* **13**, 81–88.
18. Pittenger, M. F., Mackay, A. M., Beck, S. C., Jaiswal, R. K., Douglas, R., Mosca, J. D., et al. (1999) Multilineage potential of adult human mesenchymal stem cells. *Science* **284**, 143–147.

Index

Printed in the United States of America